OFFSHORE GAS HYDRATES

OFFSHORE GAS HYDRATES

ORIGINS, DEVELOPMENT, AND PRODUCTION

RUDY ROGERS
Professor Emeritus
Mississippi State University

ELSEVIER

Amsterdam • Boston • Heidelberg • London • New York • Oxford
Paris • San Diego • San Francisco • Singapore • Sydney • Tokyo
Gulf Professional Publishing is an imprint of Elsevier

GP
PP

Gulf Professional Publishing is an imprint of Elsevier
225 Wyman Street, Waltham, MA 02451, USA
The Boulevard, Langford Lane, Kidlington, Oxford, OX5 1GB, UK

Library of Congress Cataloging-in-Publication Data
A catalog record for this book is available from the Library of Congress

British Library Cataloguing-in-Publication Data
A catalogue record for this book is available from the British Library

ISBN: 978-0-12-802319-8

For information on all Gulf Professional Publishing publications
visit our website at http://store.elsevier.com/

DEDICATION

Offshore Gas Hydrates is dedicated to the late Bob Woolsey, geologist extraordinaire who was a leader in extending gas hydrate knowledge from a laboratory curiosity to practical offshore applications. He was a pioneer in forging this giant step for the advancement of gas hydrate knowledge – a step toward alleviating worldwide energy deficiencies while introducing intriguing insights into seafloor ecologies and continental slope instabilities. Dr Woolsey envisioned an offshore Gas Hydrate Observatory for hydrate studies and single-handedly procured funding, established site location at Mississippi Canyon lease block 118, and assembled a Gulf Coast consortium of government, industry, and academe to fulfill research goals. The book is a salute to this outstanding individual.

CONTENTS

Bonus videos available in the E-book only and on ScienceDirect: (1) hydrates, carbonates, chemosynthetic communities, and natural hydrocarbon leakages on the seafloor at the Mississippi Canyon hydrate observatory site (2) hydrate nucleation, migration and self-packing in a laboratory test cell under the influence of anionic surfactants.

PREFACE

A new, clean energy source is discovered in contemporary times, so abundant that its resident carbon may exceed carbon in all known fossil energy sources combined. Impossible you say? Enter, gas hydrates. Forming worldwide in the frontier of deep ocean floors, the ice-like hydrate scavenges and stores both thermogenic and biogenic methane. Perhaps its existence in extreme environments devoid of free oxygen, or perhaps its decomposition during sediment retrieval with traditional unpressurized coring tools, is responsible for the delayed discovery. In any case, there are fascinating consequences of the offshore hydrate discoveries now presented as a unit for the first time in this textbook.

Chapter 1 presents historic timelines of gas hydrate awareness and presents estimates of in-place hydrate gas regionally and universally. It assumes readers approach the book already well versed in gas hydrate structural fundamentals.

The primary excitement of offshore hydrate discoveries derive from prospects of producing hydrate's methane as a fuel, for the ubiquitous hydrate accumulates near many nations whose fuel shortages hinder national progress. Solutions are rapidly being made to technical, economic, and environmental problems involved with producing hydrate gas. The text addresses these production issues and, while doing so, discusses a suite of associated but absorbing discoveries concerning science of extreme environments.

Chapter 2 initiates the production quest by discussing sediment–hydrate relationships in seafloor hydrate zones critical to understanding search and evaluation of future quality reservoir sites for hydrate-gas production. Because of its extensive infrastructure for conventional gas production, the Gulf of Mexico is highlighted in Chapter 3 as a gas hydrate province having prolific hydrate accumulations, occluding both biogenic and thermogenic gases. It affords practical study for prospective hydrate-gas production. Discussed are Joint Industry Program (JIP) evaluations of seismic, logging, and drilling at selected sites to establish protocol for hydrate reservoir development. Then, Chapter 4 covers process and technique combinations now being devised to produce hydrate gas: computer simulations, limited Arctic production, conventional process modifications, and laboratory evaluations of processes are made. A useful discussion ensues on a process to store and transport gas produced from remote offshore locations. Chapter 5 reviews progressively increasing hydrate inhibitions required for hydrocarbon production in ultra-deepwater operations – both conventional and hydrate-gas productions. Besides state-of-the-art inhibitors, advanced inhibitors such as antifreeze proteins and microbial cell wall materials are discussed.

Contemporary offshore gas hydrate discoveries, and their accompanying focused seafloor studies, give insight into hydrate-associated seafloor instabilities. Gas hydrates collect thermogenic and biogenic gases generated in deep ocean sediments, and have been doing this over geologic time, to form vast repositories primarily of methane and secondarily of carbon dioxide. The natural storage process of seafloor hydrates beneficially modulates the ejection of these gases into the atmosphere. But there is a caveat. The stored gases, as a solid clathrate solution, generate extraordinarily high pressures if the hydrates decompose in confined spaces from inadvertent heat injection, for example, when hydrates surrounded by impermeable clayey sediments are heated by a penetrating drill bit. Or on a larger scale, the instability phenomenon is evidenced by prehistoric submarine landslides having occurred on continental slopes when warmer or shallower waters triggered hydrate dissociations that weakened shear strengths of inclined sediments. In these and other diverse cases, then, hydrates initiate seafloor instabilities. In Chapter 6 variations in instability manifestations are discussed.

Gas hydrate formation is a crystallization process, where nucleation sites classically initiate crystal growth. But additional site criteria on a microscopic scale exist for hydrate nucleation. Sufficient gas molecules must propitiously converge with oriented water molecules properly arranged by hydrogen bonding – a stochastic occurrence often involving lengthy times. Offshore hydrate studies reveal microbe–mineral–hydrate synergies that create unique nucleation sites to accelerate hydrate initiation. This fascinating phenomenon affects hydrate formation and stability in extreme environments and is developed in the text. Chapter 7 introduces the biogenic hydrate provinces. Blake Ridge, Cascadia margin subduction zone, Nankai Trough, and other regions where offshore hydrates primarily collect biogenic gases are characterized. Chapter 8 discusses the intriguing role of microbes in catalyzing gas hydrate formations in seafloors, followed by mechanisms whereby the microorganisms enter and thrive in interstitial spaces of hydrate masses.

Explorations for hydrate gas conducive to production have advanced deep-sea knowledge. Amid gas hydrate accumulations, which often protrude above the seafloor, unique ecologies develop with complex interdependencies. Why? Carbon, essential to microbial metabolism but generally scarce in these extreme environments, is abundant in the hydrocarbons of hydrate-zone cold seeps and methane collected in gas hydrates. Consequently, extensive microbial species proliferate around these oases of carbon. For example, Archaea and sulfate-reducing bacteria team in clusters to anaerobically oxidize methane (AOM), utilizing cold-seep gases and seawater sulfates/nitrates. As a result, massive carbonates litter the seafloor, marking current and previous seep locations. Although AOM produces hydrogen sulfide, the toxic compound promotes chemosynthetic communities and on occasion gas hydrates. *Beggiatoa* mats spread over hydrate outcrops and seafloor, extracting and converting hydrogen sulfide to help replenish sediment sulfates in a sustaining cycle. Extensive root systems of tube worms scavenge hydrogen sulfide from sediments to support their internal symbiotic microbes. The text describes gas hydrates fitting into a complex ecological scenario. Chapter 9 is devoted to the ecology of the hydrate zone.

Bolstered by contemporary feedback from planetary scientific probes, interest intensifies in the possibility of extensive gas hydrates within Martian subsurfaces. By analogies based on hydrate information acquired from Earth's extreme ocean and polar environments, the final Chapter 10 reviews gas hydrate feasibility under Martian conditions. If hydrates form in the Martian subsurface environment, as the data review indicates probable, our analogy with seafloor hydrates introduces an intriguing circumstance where any microbes on Mars would have the best protection and sustenance for longevity provided within the subsurface hydrates.

ACKNOWLEDGMENTS

The author gratefully acknowledges support of individuals and groups that made the work possible. The Gulf Coast Gas Hydrate Consortium, administered by MMRI of the University of Mississippi, provided a valuable forum over the years to present and discuss university, industry, and government hydrate research. The interest, support, and funding of NETL, NOAA, and Department of the Interior for hydrate research were irreplaceable. The author is appreciative of exceptionally good graduate students who collaborated with him in gas hydrate research at Mississippi State University. Finally, the work was made achievable by the kind support of the author's wife, Beth.

CHAPTER ONE

Introduction

1.1 THE MYSTIQUE OF GAS HYDRATES

For more than a century after Sir Humphrey Davy reported the first gas hydrate (chlorine) to the Royal Society (Davy, 1811), gas hydrates remained curiosities in informal, disparate laboratory experiments. But during October 1931, on activation of the first natural gas pipeline, an expensive 24 in. diameter line operating under 600 psi and traversing 1000 miles from the newly discovered energy riches of the Texas Gulf Coast to Chicago (Ziegenhain, 1931), inaugural gas failed to flow. A massive, solid plug was found blocking the line. The plug resembled ice, yet at no time or place had line temperatures dropped below 40°F. The final verdict: a massive Davy-type gas hydrate was the culprit (Hammerschmidt, 1934).

During the next 50 years as oil and gas drilling expanded to offshore and proceeded into deeper waters, hydrate research focused on economically preventing clathrate blockages from disrupting offshore operations – an elusive goal still being pursued. But the insight generated by this preoccupation perhaps contributed to the next in a series of gas hydrate discoveries.

The Ocean Drilling Program (*ODP*), funded to advance scientific knowledge of deep ocean environments, discovered ice-like gas hydrates in ocean sediments – found them to exist superabundantly. Consider the massive hydrate protuberances photographed from manned submersibles offshore Vancouver as well as the Gulf of Mexico (GOM). For example, a hydrate mound discovered and photographed by Professor Chapman and associates is presented in Figure 1.1 (Chapman, R.N., 2015. Professor, School of Earth and Ocean Sciences, University of Victoria, Victoria, BC, personal communication).

Progressively improving analyses aided by advanced coring, drilling, seismic, and logging techniques gradually revealed seafloor hydrate deposits spanning all latitudes. Calculations accompanying discoveries indicated hydrate-occluded methane perhaps on a scale of the combined energy found in all fossil fuels. In seacoast countries long dependent on imported hydrocarbons, the discoveries introduced national hopes of fuel independence.

Offshore Gas Hydrates. http://dx.doi.org/10.1016/B978-0-12-802319-8.00001-2

Figure 1.1 ***Massive gas hydrate mound discovered on seafloor, Cascadia margin.*** *Beggiatoa* mats cover white hydrates protruding into ocean. *(Source: Chapman, R.N., 2015. Professor, School of Earth and Ocean Sciences, University of Victoria, Victoria, BC, personal communication.)*

How a new energy source, an elephant in the room, remained so long unknown adds to an evolving gas hydrate enigma. One physical property helped disguise the presence. Because hydrates decompose as pressures are lowered or temperatures are raised beyond an instability threshold, early nonpressurized coring tools allowed hydrate decomposition during extraction–retrieval of cores from deepwater ocean sediments.

Early discoveries initiated quests for seafloor hydrate fundamentals, especially mechanisms of formation, dissociation, and kinetics. But the discoveries initiated additional interests that will be outlined in the remaining section.

One way the hydrate mystique reemerges is a microbial association with seafloor hydrate accumulations. In seafloor sediments, microbes proliferate around near-surface gas hydrates. Why? Microbes require carbon for metabolism and an inordinate carbon supply is stored in hydrates. It is not surprising, therefore, that the microorganisms promote hydrate accumulations by providing catalysis and hydrate crystal nucleation sites. In fact, the activity creates an intriguing phenomenon of seafloor synergy between microbes, minerals, and hydrates. Throughout the text, threads of microbe–hydrate–mineral synergistic interactions are discussed.

Associated with hydrates on the ocean floor are communities of organisms sustaining life by chemosynthesis in the absence of light or oxygen – tubeworms, ice worms, clams, and *Beggiatoa* mats. Aided by symbiotic bacteria, an extravagant and fascinating ecosystem exists around seafloor gas hydrates.

Gas hydrates have caused seafloor instabilities of catastrophic proportions over geologic time. Such episodes are even woven into the story fabric of works of fiction. On continental slopes, gas hydrates cement unconsolidated sediments to impart additional shear strength and stability. Warmer or shallower ocean waters eventually breach the temperature–pressure threshold of hydrate stability to lower shear strengths along the slope's bottom plane. Ultimately, mudslides propel downslope the hydrate-sediment overburden, releasing from ensuing rubble large quantities of methane gas. Evidence indicates this happened in the ancient Storegga Slide offshore Norway. Possible consequences of submarine mudslides include tsunamis and sudden large imposition of greenhouse gases into the atmosphere.

Consider some dangers gas hydrates pose to deepwater drilling: (1) 1 m^3 of solid hydrate may contain 150–180 m^3 of hydrocarbon gases at standard temperature and pressure (STP); (2) gas hydrates in sediments decrease permeability to the flow of gas; (3) energy input decomposes hydrates, perhaps from drilling-bit friction or exothermic well-cementing reactions. Consequently, hydrates decomposed within surrounding porous media made impermeable by hydrates and shales build extraordinary pressures.

Voluminous gas discharges into deep ocean waters from naturally occurring submarine mud volcanoes or from well blowouts create plumes of gas, bounded by the upper limit of hydrate stability, drifting long distances in the water subsurface. Hydrate stability may extend upward in the water column to within 400–500 m of the surface. On initial emission into the high pressures and low temperatures of deep waters, each bubble of hydrocarbon gas is rapidly enveloped by hydrate, delaying gas dissipation. Efforts to control the Deepwater Horizon blowout in 1544 m deep waters of the GOM experienced debilitating interference by extensive hydrate formation at the point of seafloor eruption.

Finally, many scientists now believe gas hydrates exist in the Martian subsurface, heightening the hydrate enigma while exploration proceeds on the planet. Scientific knowledge gained from studying Earth's seafloor hydrates points favorably toward hydrates forming under extreme Martian conditions. For the first time, a hypothesis connects the significance of hydrates to an enhanced possibility of microbes on Mars. Sequestered within Earth's seafloor hydrate masses and accessible to occluded carbon as well as a lifeline of diffusing nutrients, microbes thrive while sheltered from the vagaries of changing and hostile exterior environments. Does an analogy hold for Mars? The last chapter discusses the possibility.

Figure 1.2 *Geometrical similarity of truncated icosahedron with soccer ball (Rotenberg, 2014).*

1.2 GAS HYDRATES, A BASIC MOLECULAR STRUCTURE FOUND IN NATURE

In nature the geometry of a forming object that establishes minimum surface area for a given volume requires lowest energy for formation and continued stability. That is, water drops preferentially form as spheres to minimize surface energy.

Of the five basic regular solids (tetrahedron, cube, octahedron, dodecahedron, icosahedron), the truncated icosahedron most closely approaches a sphere and the dodecahedron approximates it. Consequently, these geometries tend to occur naturally on a molecular basis (Stewart, 2011).

In Figure 1.2 is depicted a truncated icosahedron on a scale to show its geometry similar to a soccer ball (Rotenberg, 2014).

Gas hydrates are akin to the polyhedron structures. Quite interesting is the occurrence in nature of other natural entities having the structure. In the following sections, hydrates are compared with two diverse natural products having polyhedron structures – fullerenes and virus coatings. All of them exist in nature because their common configuration is thermodynamically preferred.

1.2.1 Buckminster Fuller's Geodesic Dome

Propounded by Buckminster Fuller, geodesic domes, the fractional part of a geodesic sphere, are especially strong and surface-minimizing structures. With connecting polyhedrons, Fuller's approximately spheroidal structures distribute support equally of an applied load. For construction purposes, the geodesic domes are stronger than other designs and achieve that strength with fewer materials. As spheroids, they provide minimum surface areas per unit volume – an attribute for energy efficiency. Fuller in his original patent claims a geodesic structure requiring only 0.78 wt/unit area compared with 50 wt/unit area for common construction materials; in addition, his geodesic dome of 14.9 m diameter was touted to withstand wind velocities of 93.2 km/h (Fuller, 1954).

Three properties of the geodesic dome excel: (1) exceptional strength of the structure, (2) energy efficiency of the structure, and (3) minimization of construction materials. Likewise, the three properties are the essence of geodesic domes occurring in nature – the first discussed being the fullerene, or buckyball, named in honor of the scientist.

1.2.2 Fullerenes (Buckyballs)

In addition to graphite and diamond, fullerenes are the third known carbon form. Approximately spheroidal and hollow, these truncated icosahedra are formed of a shell of covalently bonded carbon in geodesic sphere patterns. Receiving a 1996 Nobel Prize in chemistry for the discovery, Professor Smalley of Rice University first synthesized and identified fullerenes.

The most common fullerene structure C_{60} has 60 carbon atoms arranged as 20 regular hexagons and 12 regular pentagons, forming a geodesic sphere of outer diameter 7.1 Å and an interior cavity diameter of 3.7 Å (Kurotobi and Murata, 2011). See Figure 1.3.

Buckyballs are exceptionally strong and lightweight. Configurations other than C_{60} have emerged. For example, carbon nanotube fibers are long hollow tubes, 1 nm diameter by thousands of nanometers length, terminated with C_{60} hemispheres. At a 1.3–1.4 specific gravity, these single fibers are lighter than aluminum. Single carbon nanotube strands have specific tensile strengths exceeding that of high-carbon steel by a factor of 312, ranking among the strongest of materials (Collins and Avouris, 2000; Callister, 2007). With unique physical and electrical properties, nanotubes portend myriads of improved commercial products.

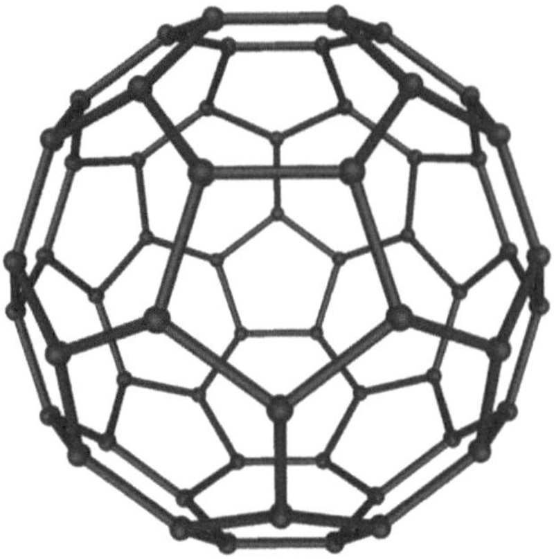

Figure 1.3 *Fullerene C_{60} (Ströeck, 2006).*

1.2.3 Virus Coatings

Many viruses are shaped as truncated icosahedral (Stewart, 2011). Their DNA and RNA are coated with roughly spherical proteins and arranged to give an approximately spherical coating; their truncated icosahedron becomes the closest possible arrangement to a sphere.

These virus coatings of geodesic dome structures are made from proteins indistinguishable from one another. Because the protein associations approximate spheres, the virus coatings fulfill thermodynamic tendencies of natural processes to assume geometries of minimum energy (Stewart, 2011).

1.3 SEAFLOOR HYDRATE STRUCTURES

1.3.1 Gas Hydrate Structures

Gas hydrates are geodesic spheres that resemble fullerenes, as well as those virus coatings with characteristic soccer ball appearances. Instead of the strong carbon–carbon bonds of fullerenes, much weaker hydrogen-bonded water molecules form the superstructures of gas hydrates. The hydrate superstructures rely for stabilization on properly sized gas molecules occupying inner cavities. Again, groupings of hydrogen-bonded water molecules as geodesic dome structures are thermodynamically preferred.

Occluded gas molecules are effectively size-screened by the dimensions of individual hydrate cages. For example, the smallest unit cage of dodecahedron admits a single molecule having a size in which methane or argon easily fits but excludes larger molecular species. Therefore, under a water column affording sufficiently high hydrostatic pressures and low temperatures, hydrates form in seafloor sediments when typically stabilized by low-molecular-weight alkanes, carbon dioxide, or hydrogen sulfide.

Because sII hydrates form at lower pressure–higher temperature combinations than sI hydrates, ocean water depth helps dictate hydrate structure in seafloor sediments, especially in the GOM where both thermogenic and biogenic hydrocarbon gases abound. As a rule of thumb, minimum water depth for methane sI hydrate stability is 600 m, while sII hydrates from low-molecular-weight, thermogenic hydrocarbons may stabilize in water depths as low as 300 m (Collett, 1995; Milkov and Sassen, 2000). However, these values vary significantly with geothermal gradient, water salinity, and gas compositions.

Whereas fullerene structures spanning numerous molecular weights have been identified, only three gas hydrate structures are known: sI, sII, and sH (Ripmeester et al., 1987). Seafloor hydrates form as sI where biogenic gases heavily weighted with methane are prevalent; sII structures commonly form when guest gases of molecular size such as propane or *i*-butane accompany the methane in concentrations on the order of greater than 1%. The hydrates of sH are omitted from discussions in this text because of their rarity in seafloors.

1.3.2 Seafloor Hydrates sI

A combination of pentagonal dodecahedron and truncated icosahedron (tetrakaidecahedron) forms the unit cell of sI gas hydrates. Each geodesic sphere forms a specific cavity size stabilized by a single guest molecule. Twelve pentagonal faces of pentagonal dodecahedron 5^{12} create a small cavity, whose size accommodates a molecule of CH_4 or H_2S. A slightly larger cavity develops within tetrakaidecahedron $5^{12}6^2$ (12 pentagonal faces, 2 hexagonal faces), whose size fits C_2H_6 or CO_2 or smaller molecules. Hydrogen bonding connects 5^{12} and $5^{12}6^2$ adjacent polyhedra into a close-packed, unit-cell arrangement comprising 2 cavities of 5^{12} and 6 cavities of $5^{12}6^2$ (Ripmeester et al., 1987; Makogon, 1981; Sloan, 1998). The sI type of hydrate represents a body-centered cubic crystal lattice (Callister, 2007). Those seafloor sI hydrates typically found at Blake Ridge, Cascadia margin, or Nankai Trough are stabilized predominantly by methane from microbial activity.

Although gas hydrates containing biogenic gas origin are composed primarily of methane in a sI structure, other products of microbial activity, such as carbon dioxide and hydrogen sulfide, may be present. The mixing of either gas with methane substantially reduces stringencies of *P–T* combinations required for hydrate formation. However, whenever anaerobic oxidation of methane (AOM) occurs within a seafloor hydrate zone, sulfide may be reconverted to sulfate by chemosynthetic communities rather than going directly to stabilize hydrates.

Nevertheless, there are many examples where H_2S appears in hydrate recoveries. Hydrogen sulfide is found in concentrations of 10–15% in gas hydrates on the continental margin of South Australia (Swart et al., 2000). Significant amounts (5–6%) of H_2S are contained in gas hydrates of the Black Sea (Popescu et al., 2006). Up to 10% of H_2S in methane hydrates has been reported in isolated cores from the Cascadia subduction zone (Kastner et al., 1998). Minor sulfide amounts are occluded in methane hydrates extracted from the eastern margin of the Sea of Japan (Lu et al., 2011); because hydrate samples were retrieved from below the sulfate–methane interface (SMI), the H_2S possibly comes from AOM at a time when the SMI was deeper, or the sulfides may have come from volcanic activity.

In a like manner representative CO_2 contents of methane hydrates are reported: (1) 0–9% (v/v) of carbon dioxide was within GOM hydrate sediments of pressurized cores retrieved as deep as 300 mbsf in Atwater Valley (Lorenson et al., 2005); (2) gas hydrate samples from Hydrate Ridge have been reported with small but insignificant CO_2 volumes (Suess et al., 1999); (3) some Nankai Trough samples indicate 0–0.63% of CO_2 (v/v) (Waseda and Uchida, 2002).

Instead of incorporation into gas hydrates, carbon dioxide product from the AOM reaction is deposited as extensive carbonate precipitates at hydrate sites, while hydrogen sulfide is converted by *Beggiatoa* to replenish sulfates in near-surface sediments.

1.3.3 Seafloor Hydrates sII

Type II gas hydrates form a diamond cubic crystal lattice in which 1 unit cell has 16 pentagonal dodecahedron 5^{12} plus 8 hexakaidecahedron $5^{12}6^4$. The 4.683 Å radius of the $5^{12}6^4$ cavity accepts molecular sizes of propane or *i*-butane for stabilization (Ripmeester et al., 1987; Makogon, 1981; Sloan, 1998). In hydrate provinces such as the GOM where natural gas seeps as well as prolific microbial activity are common, both sI and sII hydrates are found. At the seafloor surface, sII hydrates stabilize in significantly shallower water than the sI type.

1.4 TIMELINE OF GAS HYDRATE RESEARCH AND DEVELOPMENT PROJECTS

A timeline of gas hydrate development is informative. Hydrate gas estimates viewed chronologically show estimate improvements as hydrate information assimilation increased rapidly over a relatively short period during accelerated research efforts. Early estimates cover in-place gas; recent estimates concentrate more on recoverable hydrate gas.

The Deep Sea Drilling Project and the International Ocean Drilling Program first retrieved cores pressurized to preserve retrieved gas hydrates. By the year 2000, 53 sites in the GOM having gas hydrates had been identified (Milkov and Sassen, 2000). Seismic data facilitated hydrate discovery off the South Carolina coast at Blake Ridge during the Deep Sea Drilling Project (Max and Lowrie, 1996). An early estimate placed 4000 trillion cubic feet (tcf) of in-place hydrate gas at Blake Ridge in an area about the size of Maryland (Dillon and Paull, 1983).

In 1998 the Japanese National Oil Company in partnerships with the United States and Canada drilled an experimental Mallik well through permafrost of the Mackenzie Delta. Logging and coring revealed gas hydrates in depth intervals of 897–1110 m. In fact, gas hydrates have been located on continental slopes throughout most of the world (Trehu et al., 1999; Taylor et al., 2000; Shelander et al., 2011). Occurrences have been identified in arctic regions, offshore equatorial locations, Southern Hemisphere, and Northern Hemisphere.

A historical gas hydrate timeline is presented in Table 1.1.

1.5 EARLY ESTIMATES OF WORLDWIDE HYDRATE GAS

1.5.1 Calculating In-Place Hydrate Gas

The first attempts to estimate worldwide hydrate gas accumulations came from the USSR, beginning in 1973 (Trofimuk et al., 1973) before widespread hydrate-related drilling, coring, and seismic provided substantial data for the estimates to have a reliable basis.

Milkov (2004) summarized the predictive volumetric equations that Makogon (1981) applied to gas hydrate reservoirs, adapted from conventional oil and gas volumetric calculations (Craft and Hawkins, 1991).

In Equation 1.1 the global in-place hydrate gas is estimated from total gas hydrate zone (GHZ) volume V_{ghz} in seafloor sediments and the yield of hydrates from those sediments, D:

Table 1.1 Historical timeline of gas hydrate events

1811	Sir Humphrey Davy reports discovery of gas hydrates. Yellow, chlorine gas crystals
1810–1931	Gas hydrates remain a laboratory curiosity
1931	Hydrate blocks inaugural Texas–Chicago natural gas pipeline
1950 to present	Extensive research to prevent hydrates forming in offshore oilfield operations
1972	Discovery of natural gas hydrates, North Slope of Alaska
1966–1973	Makogon and Trofimuk of the USSR developed technique of estimating hydrate gas
1984	First reported discovery of Gulf of Mexico gas hydrates at Bush Hill, GC-185
1984	First reported discovery of cold seep chemosynthetic communities in GOM
1983–1990	Deep Sea Drilling Project and Ocean Drilling Program indicate extensive gas hydrates in ocean sediments
2000	Microbial influence on hydrates found. Microbe–hydrate–mineral synergy
2002	First reported deep, gas hydrates in the GOM; aligned bright spots interpreted as Bottom-Simulating Reflector (BSR)
2008	First hydrate gas production; onshore Mallik
2009	JIP drills seven deep GOM hydrate wells; first synchronization of logs, seismic, cores
2009	First thick, coarse-sand reservoirs with high S_{gh} discovered lucrative for hydrate gas production. JIP in GOM
March 2013	First offshore hydrate production test (depressurization), eastern Nankai Trough
2015–2020	Process refined in offshore field tests (combinations: dissociation, CO_2 inject, heat inject)

$$V_{\text{global}} = V_{\text{ghz}} D \tag{1.1}$$

The global volume of in-place gas can be further delineated by Equation 1.2:

$$V_{\text{global}} = [A \times \Delta z]\lfloor \phi \times S_{\text{gh}} \times G \rfloor \tag{1.2}$$

where V_{global}, global in-place hydrate gas (gas volume is at STP); V_{ghz}, global sediment volume containing gas hydrates; D, (gas volume at STP)/(global sediment volume containing hydrates); A, areal extent of gas hydrate–bearing sediments in oceans; Δz, thickness of GHZ; ϕ, sediment porosity; S_{gh}, fractional hydrate pore saturation; and G, gas occluded in hydrates (volume of gas at STP)/(volume of hydrates).

The global in-place hydrate gas was considered the sum of continental shelf, continental slope, and abyssal plain evaluations as given in Equation 1.3:

$$V_{\text{global}} = V_{\text{shelf}} + V_{\text{slope}} + V_{\text{abyssal}} \tag{1.3}$$

Ingenious means to estimate unknown parameters in the aforementioned equations were often made prior to the availability of reliable drilling, seismic, and coring data. Despite early approximations that might incur large errors, the immensity of all estimates suggested a great hydrate gas potential as a new energy source.

1.5.2 Early Estimates, Prior to Drilling, 1970s to Late 1980s

Trofimuk et al. (1973) published the first in-place estimate of hydrate gas on a worldwide basis as 3×10^6 tcm. Here, the most egregious assumptions made in Equation 1.1 were that areal extent involved 93% of all world oceans with greater than 500 m water depth and S_{gh} was an unrealistic 100% (Milkov, 2004).

Some early estimates representative of the 1970s and 1980s are presented in Table 1.2.

It should be reiterated that drilling, seismic, logging, and coring data related to hydrates were notably lacking during this time period.

1.5.3 Estimates, Late 1980s to Mid-1990s

During the time period of late 1980s to mid-1990s, several independent estimates derived similar values, even though the estimating techniques varied widely. A consensus value was considered to be 21,000 tcm of worldwide,

Table 1.2 Representative estimates of worldwide, in-place hydrate gas, 1970s to 1980s

References	In-place gas (tcm)	Significant assumptions (Milkov, 2004)
Trofimuk et al. (1973)	3×10^6	80/20 CH_4/CO_2; >500 m water depth; 93% of world ocean area
McIver (1981)	3,100	CH_4 origin: microbes convert 1% of 0.5% organic C content
Kvenvolden and Claypool (1988)	40,000	CH_4 origin: microbes convert 1–2% TOC; 10% S_{gh}; $\phi = 50\%$

Table 1.3 Representative estimates, late 1980s to mid-1990s

References	tcm	Significant assumptions (Milkov, 2004)
Kvenvolden (1988)	20,000	Extrapolated from Arctic Ocean hydrate gas estimates to apply to world continental slopes; $\phi = 0.3$; $S_{gh} = 100\%$; $D = 42\%$
MacDonald (1990)	21,000	Water depths = 200–300 m; $\Delta z = 500$ m; $\phi = 0.4$; $S_{gh} = 10\%$; $D = 6.2\%$
Gornitz and Fung (1994)	26,400	Hydrate forms in sediments >0.5% TOC; ϕ varies from 50% at BGHS to 0% at top Gas Hydrate Stability Zone (GHSZ)

in-place hydrate gas. Representative estimates and major assumptions are presented in Table 1.3.

To help place in perspective these hydrate gas quantities, Masters et al. (1991) estimate 250 tcm (8829 tcf) of known conventional natural gas worldwide that could be recovered, indicating perhaps 80 times more gas in hydrates than in conventional gas reserves.

1.5.4 Later Estimates of Worldwide In-Place Hydrate Gas

Klauda and Sandler (2005) estimate 44,000–120,000 tcm of in-place hydrate gas throughout the world. Their assumptions include the following:

1. All continental margins included
2. Thermodynamic model used to better define Bottom Gas Hydrate Stability (BGHS)
3. Capillary effect on hydrate-forming pressure included
4. Salinity effect on equilibrium P–T included
5. Omitted high-gas-flux areas
6. Only greater than 0.4% of total organic carbon (TOC) areas included
7. Used varying ocean temperatures
8. S_{gh} averaged 3.4%, but accounted for local variations
9. Used a 1° latitude × 1° longitude global grid

This estimate moderately counters an otherwise downward trend with time of estimated in-place gas from all other evaluations of previous years.

1.6 PETROLEUM SYSTEMS APPROACH TO ESTIMATING HYDRATE GAS

The practicality of hydrate gas estimates improves with time. Well logs, advanced seismic, pressure-stabilized deep cores, manned submersible expeditions, simulations, and small-scale production wells increasingly add a wealth of worldwide gas hydrate data. Later emphasis shifted to a petroleum systems approach, that is, estimating the hydrate gas that could potentially be produced from a technical standpoint but also having an end view of economical production (Boswell and Collett, 2006; Johnson, 2011; Max et al., 2006).

In the petroleum systems, basic parameters sought in conventional oil and gas reservoirs were superposed on gas hydrate accumulations. Primarily, hydrate reservoirs must have an accessible gas source feeding a trapping mechanism within which is a porous media of good permeability and porosity. This means the parameters D and V_{ghz} in Equation 1.1 are modified to incorporate only sediments with appreciable sands bearing gas hydrates; fine-grained muds and shales are excluded, although containing dispersed hydrates and fracture-filled hydrates.

Wood and Jung (2008) of the US Naval Research Laboratory developed a model by which to estimate the volume of hydrate-bearing sands. Johnson (2011) used the Wood and Jung model, employing a 2° latitude × 2° longitude grid, to estimate the gas hydrate content in quality sand reservoirs. Johnson estimates a median hydrate gas accumulation of 122.6 tcm, ranging up to a maximum of 889.1 tcm, which represents hydrate gas contained in sand-based reservoirs of continental margins. As a comparison, consider that worldwide, conventional natural gas reserves are estimated as 187 tcm in the same time frame (Shakirov and Obzhirov, 2011).

Although there is considerably more hydrate gas dispersed in shales and filling veins/fractures of fine-grained muds, it becomes apparent from the petroleum systems approach (Boswell and Collett, 2006; Johnson, 2011; Max et al., 2006) that these hydrates are least likely to be produced – at least with present technology. The large in-place gas estimated in early years was unduly influenced by large amounts of hydrate gas held in the fine-grained sediments that lack economical production means for the foreseeable future.

1.7 ESTIMATED REGIONAL IN-PLACE HYDRATE GAS

1.7.1 Western Rim, Pacific Ocean

Although a paucity of hydrate wells exist in the western rim of the Pacific Ocean, seismic and coring indicate prolific gas hydrates in the region extending from the Bering Straits to New Zealand. Except for Australia and Antarctica, the region lies within subduction zones, where continental plate edges override and are covered with sediments as deep as 10 km. Frequent seismic activity generates faults and fractures (Shakirov and Obzhirov, 2011). As a consequence of these geologic factors, hydrates are most often found in fractures and veins, between strata, and dispersed between sand grains.

Table 1.4 Estimated in-place hydrate gas in Western Pacific

Location	In-place hydrate gas (tcm)	References
Western Pacific	50	Shakirov and Obzhirov (2011)
Okhotsk	2	Shakirov and Obzhirov (2011)
Surrounding Japan	1–10	Tanahashi (2011)
Eastern Nankai Trough*	1.13	Fujii et al. (2008)

*For a 4678 km^2 section of the Eastern Nankai Trough based on well logs, seismic data, and core samples from ~30 boreholes in 16 locations.

Rough estimates of in-place hydrate gas in the region are presented in Table 1.4.

1.7.2 Gulf of Mexico

GOM is another region of abundant gas hydrates. Here, 1753 conventional oil and gas, deepwater (>457 m) wells had been drilled through the hydrate zone by the end of 2010 (Boswell et al., 2012), whose seismic, logging, and core data add resourcefulness to GOM hydrate gas estimates. Representative estimates of in-place hydrate gas made during the 23 years after initial hydrate discovery range from 67 to 680 tcm. Some of the estimates with accompanying assumptions are listed in Table 1.5.

The extensive supporting data used by the Bureau of Ocean Energy Management (BOEM) lends credence to their estimate of 607 tcm of hydrate gas within the northern GOM. It is especially noteworthy that Frye (2008) estimates 190 tcm of in-place hydrate gas within quality sand reservoirs that would be considered producible by conventional gas standards. That is, 31% of the in-place gas exists in sand reservoirs with high permeability, porosity, and S_{gh} values – quite similar to an evaluation of the Nankai Trough (Boswell et al., 2012; Fujii et al., 2008).

1.8 ESTIMATES OF RECOVERABLE HYDRATE GAS

Johnson (2011) used conventional oil and gas exploration criteria to delineate hydrate gas reserves in an extensive study that included estimates of potentially recoverable gas from at least two provinces of the United States – GOM and Alaskan North Slope. An estimate of potential gas recovery from permeable sands having attractive permeabilities, porosities, trapping mechanisms, and gas sources of the GOM amounts to 6717 tcf (Frye, 2008). Also, an estimate of gas that could be produced economically

Table 1.5 Estimates of in-place hydrate gas, northern Gulf of Mexico

References	In-place gas (tcm)	Assumptions of estimates
Krason et al. (1985)	67	S_{gh} = 5% average over entire height of hydrate stability zone
Collett (1995)	680	Northern GOM. Incorporated regional heterogeneities: GHSZ height, temperature and pressure variations, and reservoir quality variations
Milkov and Sassen (2001)	12	Restricted hydrate gas producibility to shallow hydrates, mounds, hydrates around vents, and within fine mud fractures
Klauda and Sandler (2003)	120	From numerical simulations
Frye (2008)	607	Mean of estimates made by BOEM for 450,000 km^2 northern GOM. Extensive drilling, logging, seismic data used from conventional wells
Frye (2008)	190	Sand reservoirs; producible by conventional means

from hydrates on the North Slope of Alaska amounts to 85.4 tcf (Collett et al., 2008).

1.9 ENERGY NEEDS COMPARED WITH GAS HYDRATE SUPPLY

1.9.1 The United States

The Energy Information Administration (EIA) reports natural gas usage, as trillion cubic feet at STP, in the United States for the period since the mid-1970s. See Figure 1.4.

Since 1986 a general trend of increasing natural gas consumption is evident in Figure 1.4.

A representative 19.9 tcm (mean) in-place hydrate gas is estimated to be in potentially producible sands of the United States (Johnson, 2011). Consider that the United States consumed 0.739 tcm (26.1 tcf) of conventional natural gas per year in 2013 (EIA, 2015). At the stated usage rate, the in-place hydrate gas would ideally cover 27 years of the country's natural gas consumption.

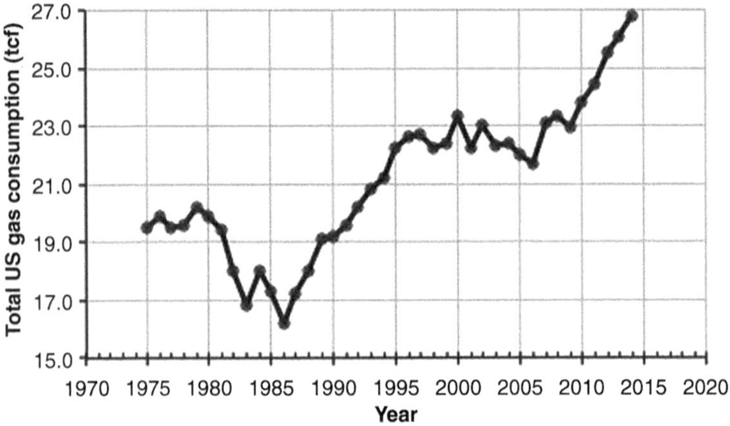

Figure 1.4 ***Natural gas total consumption, the United States.*** *(Source: US Energy Information Administration, 2014.)*

Since 2000, natural gas for electrical power generation has increased 64% in the United States (Liss, 2012), indicating it to be the contemporary fuel of choice for new electrical power plants. Main advantages for the natural gas choice include: (1) lower capital investments, (2) less time to construct plant, (3) stop and start capability on demand, and (4) low emissions.

The impacts of low emissions of natural gas are seen in Table 1.6.

It is evident from Table 1.6 that natural gas performs well as an environmentally competitive fuel for power plants: lower greenhouse gases (carbon dioxide, nitrogen oxides), very large decrease in acid-rain components (sulfur dioxide, nitrogen oxides), smog reduction (nitrogen oxides, carbon monoxide), large improvement in air quality (particulates), and elimination of mercury.

The annual amount of gas used in US power plants since the year 2000 is graphed in Figure 1.5.

Table 1.6 Percentage excess air pollutant emissions over natural gas

Pollutant	Oil (% increase)	Coal (% increase)
Carbon dioxide	40	78
Carbon monoxide	−18	420
Nitrogen oxides	387	397
Sulfur dioxide	112,100	259,000
Particulates	1,100	39,100

Source: Energy Information Administration.

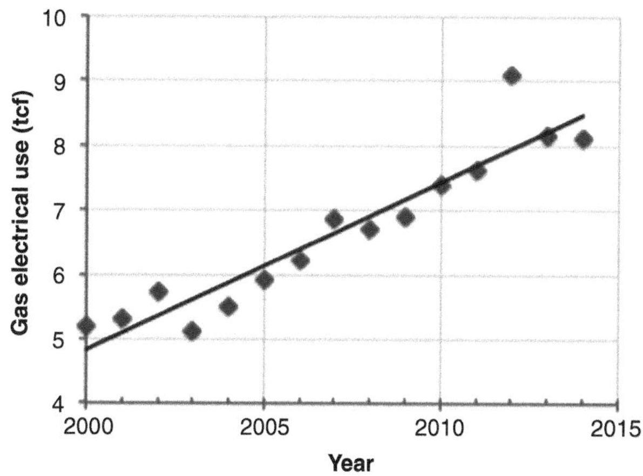

Figure 1.5 ***Electric power generation from natural gas, the United States.*** *(Source: Energy Information Agency, 2014.)*

1.9.2 Japan

To put in perspective the importance to a country of hydrate gas reserves, consider Japan.

As an importer of energy, Japan ranks first in liquefied natural gas (LNG), second in coal, and third in net oil (EIA, 2014). Their natural gas consumption compared with production is presented in Figure 1.6.

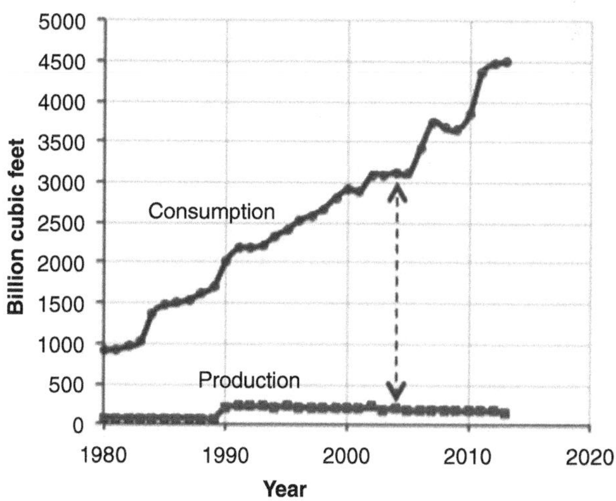

Figure 1.6 ***Japan's conventional natural gas.*** *(Source: US Energy Information Administration, 2013.)*

Although Japan's proven conventional natural gas reserves declined from 1.4 tcf (0.040 tcm) in 2007 to 0.738 tcf (0.021 tcm) in 2010, confirmation has been given of 40 tcf (1.14 tcm) hydrate-methane reserves in Nankai Trough off their southeast coast (EIA). This represents a significant potential impact for them on critical national energy needs.

REFERENCES

Boswell, R., Collett, T.S., 2006. The gas hydrates resource pyramid. In: Fire in the Ice, Methane Hydrate Newsletter. Fall Issue. U.S. Department of Energy, Office of Fossil Energy, National Energy Technology Laboratory, Morgantown, WV, pp. 5–7.

Boswell, R., Collett, T.S., Frye, M., Shedd, W., McConnell, D.R., Shelander, D., 2012. Subsurface gas hydrates in the northern Gulf of Mexico. Mar. Pet. Geol. 34, 4–30.

Callister, Jr., W.D., 2007. Materials Science and Engineering, vol.7. John Wiley & Sons, Inc., New York, 721 pp.

Collett, T., 1995. Gas hydrate resources of the United States. In: Gautier, D., Dolton, G. (Eds.), National Assessment of US Oil & Gas Resources. USGS Ser. 30, p. 78.

Collett, T.S., 2008. Geologic and engineering controls on the production of permafrost-associated gas hydrate accumulations. In: Proceedings of the 6th International Conference on Gas Hydrates (ICGH 2008), Vancouver, BC, Canada, July 6–10. Paper 5365.

Collins, P.G., Avouris, P., 2000. Nanotubes for electronics. Sci. Am. 283 (6), 62–69.

Craft, B.C., Hawkins, M.F., 1991. Applied Petroleum Reservoir Engineering, second ed. Prentice Hall, Englewood Cliffs, NJ, p. 431 (revised by R.E. Terry).

Davy, H., 1811. On a combination of oxymuriatic gas and oxygen gas. Trans. R. Soc. Lond. 101, 1–35.

Dillon, W.P., Paull, C.K., 1983. Marine gas hydrate – II: geophysical evidence. In: Cox, J.L. (Ed.), Natural Gas Hydrates: Properties, Occurrence and Recovery. Butterworth Publishers, Woburn, MA, pp. 73–90.

Frye, M., 2008. Preliminary evaluation of in-place gas hydrate resources: Gulf of Mexico outer continental shelf. OCS Report MMS 2008-004. Minerals Management Service, Resource Evaluation Division, Herndon, VA.

Fujii, T., Saeki, T., Kobayashi, T., Inamori, T., Hayashi, M., Takano, O., Takayama, T., Kawasaki, T., Nagakubo, S., Nakamizu, M., Yokoi, K., 2008. Resource assessment of methane hydrate in the eastern Nankai Trough, Japan. In: Englezos, P., Ripmeester, J. (Eds.), Proceedings of the 6th International Conference on Gas Hydrates (ICGH 2008), Vancouver, BC, Canada, July 6–10.

Fuller, R.B., 1954. Building construction. U.S. patent 2682235, issued June 29, 1954.

Gornitz, V., Fung, I., 1994. Potential distribution of methane hydrates in the world's oceans. Global Biogeochem. Cycles 8, 335–347.

Hammerschmidt, E.G., 1934. Formation of gas hydrates in natural gas transmission lines. Ind. Eng. Chem. 26, 851–855.

Johnson, A.H., 2011. Global resource potential of gas hydrate – a new calculation. In: Proceedings of the 7th International Conference on Gas Hydrates (ICGH 2011), Edinburgh, Scotland, UK, July 17–21.

Kastner, M., Kvenvolden, K.A., Lorenson, T.D., 1998. Chemistry, isotopic composition, and origin of a methane-hydrogen sulfide hydrate at the Cascadia subduction zone. Earth Planetary Sci. Lett. 156, 173–183.

Klauda, J., Sandler, S., 2003. Predictions of gas hydrate phase equilibria and amounts in natural sediment porous media. J. Mar. Pet. Geol. 20 (5), 459–470.

Klauda, J.B., Sandler, S.I., 2005. Global distribution of methane hydrate in ocean sediment. Energy Fuels 19, 459–470.

Krason, J., Finley, P., Rudloff, B., 1985. Basin Analysis, Formation, and Stability of Gas hydrates in the Western Gulf of Mexico. DOE-NETL, DE-AC21-84MC21181, 168 pp.

Kurotobi, K., Murata, Y., 2011. A single molecule of water encapsulated in fullerene C_{60}. Science 333, 613–616.

Kvenvolden, K.A., 1988. Methane hydrate – a major reservoir of carbon in the shallow geosphere? Chem. Geol. 71, 41–51.

Kvenvolden, K.A., Claypool, G.E., 1988. Gas hydrates in oceanic sediment. USGS Open-File Report 88-216, 50 pp.

Liss, W., 2012. Demand outlook: a golden age of natural gas. Chem. Eng. Prog. 108 (8), 35–40.

Lorenson, T.D., Claypool, G.E., Dougherty, J.A., 2005. Natural gas geochemistry of sediments drilled on the 2005 Gulf of Mexico JIP cruise. Mar. Pet. Geol. 25, 873–883.

Lu, H., Moudrakovski, I.L., Ripmeester, J.A., Ratcliffe, C.I., Matsumoto, R., Tani, A., 2011. The characteristics of gas hydrate recovered from Joetsu Basin, Eastern Margin of the Sea of Japan. In: Proceedings of the 7th International Conference on Gas Hydrates (ICGH 2011), Edinburgh, Scotland, UK, July 17–21. Paper 355.

MacDonald, G.J., 1990. The future of methane as an energy resource. Annu. Rev. Energy 15, 53–83.

Makogon, Y.F., 1981. Hydrates of Natural Gas. Pennwell Publishing Co., Tulsa, OK.

Masters, C.D., Root, D.H., Attansi, E.D., 1991. Resource constraints in petroleum production potential. Science 253 (5016), 146–152.

Max, M.D., Lowrie, A., 1996. Oceanic methane hydrates: a frontier gas resource. J. Pet. Geol. 19 (1), 41–56.

Max, M.D., Johnson, A., Dillon, W.P., 2006. Economic Geology of Natural Gas Hydrate. Springer, Berlin, Dordrecht, 341 pp.

McIver, R.D., 1981. Gas hydrates. In: Meyer, R.G., Olson, J.C. (Eds.), Long-Term Energy Resources. Pitman, Boston, MA, pp. 713–726.

Milkov, A.V., 2004. Global estimates of hydrate-bound gas in marine sediments: how much is really out there? Earth Sci. Rev. 66 (3–4), 183–197.

Milkov, A.V., Sassen, R., 2000. Thickness of the gas hydrate stability zone, Gulf of Mexico continental slope. Mar. Pet. Geol. 17, 981–991.

Milkov, A.V., Sassen, R., 2001. Estimate of gas hydrate resource, northwestern Gulf of Mexico continental slope. Mar. Geol. 179 (1–2), 71–83.

Popescu, I., De Batist, M., Lericolais, G., Nouzé, H., Poort, J., Panin, W., Versteeg, W., Gillet, H., 2006. Multiple bottom-simulating reflections in the Black Sea: potential proxies of past climate conditions. Mar. Geol. 227, 163–176.

Ripmeester, J.A., Tse, J.S., Ratcliffe, C.I., Powell, B.M., 1987. A new clathrate hydrate structure. Nature 325, 135–136.

Rotenberg, A., 2014. Truncated icosahedron similar to a soccer ball. <http://commons.wikimedia.org/wiki/File:Comparison_of_truncated_icosahedron_and_soccer_ball.png>. Licensed under the Creative Commons Attribution-Share Alike 4.0 International.

Shakirov, R., Obzhirov, A., 2011. Western Pacific gas hydrate belt. In: Proceedings of the 7th International Conference on Gas Hydrates (ICGH 2011), Edinburgh, Scotland, UK, July 17–21. Paper 781.

Shelander, D., Dai, J., Bunge, G., Boswell, R., Collett, T.S., Balczewski, J., Jones, E., 2011. Predictions of gas hydrate using pre-stack seismic data, deepwater Gulf of Mexico. In: Proceedings of the 7th International Conference on Gas Hydrates (ICGH 2011), Edinburgh, Scotland, UK, July 17–21. Paper 131.

Sloan, Jr., E.D., 1998. Clathrate Hydrates of Natural Gases, 2nd ed Marcel Dekker, Inc., New York, NY.

Stewart, I., 2011. Mathematics of Life. Basic Books, Philadelphia, PA, pp. 138–157.

Ströeck, M., 2006. A 3D model of a en:C60 molecule, also called a "buckyball". Mstroeck at en.wikipedia. Later versions were uploaded by C. Bryn at en.wikipedia. GFC HTML. <http://commons.wikimedia.org/wiki/File%3AC60a.png>.

Suess, E., Torres, M.E., Bohrmann, G., Collier, R.W., Greinert, J., Linke, P., Rehder, G., Trehu, A., Wallman, K., Winckler, G., Zuleger, E., 1999. Gas hydrate destabilization: enhanced dewatering, benthic material turnover and large methane plumes at the Cascadia Convergent Margin. Earth Planetary Sci. Lett. 170, 1–15.

Swart, P.K., Wortmann, U.G., Mitterer, R.M., Malone, M.J., Smart, P.L., Feary, D.A., Hine, A.C., 2000. Hydrogen sulfide-rich hydrates and saline fluids in the continental margin of South Australia. Geology 28, 1039–1042.

Tanahashi, M., 2011. Present status of Japanese methane gas hydrate research and development program. In: PETRAD-CCOP-PETROVIETNAM-VASI Workshop on Gas Hydrates, HaLong, Vietnam, March 1–3, 44 pp.

Taylor, M.H., Dillon, W.P., Pecher, I.A., 2000. Trapping and migration of methane associated with the gas hydrate stability zone at the Blake Ridge Diapir: new insights from seismic data. Mar. Geol. 164, 79–89.

Trehu, A.M., Torres, M.E., Moore, G.F., Suess, E., Bohrmann, G., 1999. Temporal and spatial evolution of a gas hydrate-bearing accretionary ridge on the Oregon continental margin. Geology 27 (10), 939–942.

Trofimuk, A.A., Cherskiy, N.V., Tsarev, V.P., 1973. Accumulation of natural gases in zones of hydrate-formation in the hydrosphere. Dokl. Akademic Nauk SSSR 212, 931–934, (in Russian).

Waseda, A., Uchida, T., 2002. Origin of methane in natural gas hydrates from the Mackenzie Delta and Nankai Trough. In: Proceedings of the 4th International Conference of Gas Hydrates, Yokohama, May 19–23, pp. 169–174.

Wood, W.T., Jung, W.Y., 2008. Modeling the extent of earth's marine methane hydrate cryosphere. In: Proceedings of the 6th International Conference on Gas Hydrates (ICGH 2008), Vancouver, BC, Canada, July 6–10.

Ziegenhain, W.T., 1931. Every precaution taken to eliminate clogging of new Chicago gas line. Oil Gas J. 30 (19), 34.

CHAPTER TWO

Deep Ocean Sediment–Hydrate Relationships

A seismic wipeout zone is a fairway above bottom of gas hydrate stability (BGHS) where natural gases, having come from forceful disruption of a low-permeability interface at BGHS, enrich but disrupt sediment continuity while permeating to the seafloor via faults, fractures, and chimney-like structures. Hydrates, accessible carbon, and microbes abound in these fairways.

Early efforts to systematically locate seafloor gas hydrates focused on finding bottom-simulating reflectors (BSRs), which manifest seismic reflections outlining floors of hydrate occurrence. Envisioned was a means to economically explore ocean floors for hydrate reservoirs. Although useful, BSRs proved to have limitations. Accomplishments and shortcomings of seismic BSRs in hydrate exploration are discussed in this chapter, as well as well logging advancements that help profile gas hydrates qualitatively and quantitatively.

Gas hydrate morphologies in fine-particle ocean sediments and hydrate morphology in coarse-sand reservoirs are discussed in this chapter. Studies of the hydrate morphologies are aided by noninvasive core analyses utilizing computed tomography (CT) scans.

Especially important for near-surface hydrate studies are fracture propagations in fine sediments, polygonal faulting, and the filling of prevailing fissures with gas hydrates. Overall, the sediment–hydrate matrix relationships involving particle size, permeability, porosity, hydrate saturation, thermal conductivity, geothermal gradient, and heat flux are complex, but empirical bases are presented in this chapter for later determinations of their consequential effect on hydrate gas production.

2.1 DETERMINING ORIGIN OF HYDRATE-OCCLUDED GASES

2.1.1 Carbon Isotope Analysis

Carbon isotope analyses of fairway methane and methane/ethane molecular ratios help determine origins of the hydrate-occluded gas. Carbon

Offshore Gas Hydrates. http://dx.doi.org/10.1016/B978-0-12-802319-8.00002-4

Table 2.1 Carbon isotope concentrations in nature

Carbon isotope	Terrestrial content (%)	Comments	References
Carbon-12	98.89	IUPAC specifies its molecular weight as basis of all elements. Stable	Rounick and Winterbourn (1986)
Carbon-13	1.11	Allows distinguishing microbial source. Stable	Rounick and Winterbourn (1986)
Carbon-14	Trace	Radioactive, half-life 5730 years; dates wood. Unstable	Burdige (2006)

IUPAC, International Union of Pure and Applied Chemistry.

has two stable isotopes, carbon-12 and carbon-13, which find use in determining origins of carbon-containing gases associated with hydrates. An unstable isotope, carbon-14, has traditionally been used for archaeological dating. These three isotopes occur throughout nature in the characteristic amounts and are used in primary applications, as presented in Table 2.1.

Because of weaker bond energies, compounds with the lighter carbon-12 isotope are preferentially processed by bacteria. An enriched substrate of carbon-13 is left behind, depleting that isotope in the gaseous bioproduct. To distinguish microbial sources of hydrate carbon compounds from thermogenic sources, delta values of the stable isotope $\delta^{13}C$ comprising methane are calculated and compared with a standard. The standard is Pee Dee Belemnite (PDB), a calcium carbonate marine fossil from Cretaceous times found in the Pee Dee formation of South Carolina (Rounick and Winterbourn, 1986; Whiticar, 1999). Extremely depleted in the carbon-12 isotope, the delta value of the fossil was arbitrarily set as zero by the National Bureau of Standards.

Delta values ($\delta^{13}C$) may be calculated from Equation 2.1:

$$\delta^{13}C_{PDB} = \left[\frac{(^{13}C/^{12}C)_{sample}}{(^{13}C/^{12}C)_{PDB\ standard}} - 1\right] \times 10^3 \tag{2.1}$$

The isotopic ratio $\delta^{13}C/\delta^{12}C_{PDB}$ effectively determines whether the subject carbon has been fractionated by microbial activity. If $\delta^{13}C_{PDB}$ values of hydrate-occluded methane calculated by Equation 2.1 are more negative than −60‰ (in units of parts per thousand), the gases are considered to

Table 2.2 Representative $\delta^{13}C$ values of methane in seafloor gas hydrates

Location	$\delta^{13}C$ (‰)	References	Comments
Indonesia	−70.6 to −52.6	Sassen and Curiale (2006)	0–6 m <seafloor 1396–1989 m water depth
MC-118, Gulf of Mexico	−45.7	Sassen et al. (2006)	Vent gas 0.5 m above seafloor Below 840 m water column
Bush Hill, Gulf of Mexico	−44.1	Sassen et al. (1999)	Vent gas
Sea of Okhotsk	−49.5 to −65.8 −31.7 to −77.5	Cho et al. (2005) Shakirov and Obzhirov (2011)	Seeps Both thermogenic and biogenic
Nankai Trough	−96 to −63	Waseda and Uchida (2002)	Upper 300 m of sediments
Nankai Trough	−48 to −35	Waseda and Uchida (2002)	Gases deeper than 1500 mbsf
Cascadia margin	−71.5 to −62.4	Suess et al. (1999)	Methane released from gas hydrates
Eastern margin of the Sea of Japan	−36.2 to −40.33	Lu et al. (2011a)	Carbon isotope content of methane in retrieved gas hydrates

be of microbial origin, but $\delta^{13}C_{PDB}$ values of methane more positive than −50‰ are considered to be of thermogenic origin (Paull et al., 2005; Bernard et al., 1978).

Some representative $\delta^{13}C$ values of methane associated with seafloor gas hydrates are presented in Table 2.2.

2.1.2 Molecular Structure Ratios

Further assistance in determining whether a hydrate gas source is biogenic or thermogenic is given by the criterion in Equation 2.2 of the molecular structure ratio [C_1 (methane)]/[C_2 (ethane) + C_3 (propane)]. A ratio greater than 1000 defines a predominantly biogenic origin of the gas mixture because of the relatively large amounts of methane (Bernard et al., 1978):

$$\frac{C_1}{C_2 + C_3} > 1000 \tag{2.2}$$

For example, sediment gas samples from the Nankai Trough exhibiting molecular structure ratios greater than 4000 are decidedly of biogenic origin (Waseda and Uchida, 2002).

Combinations of isotopic ratio and molecular structure ratio become more informative, as well as reliable, in designating gas source. In the preceding Nankai sample, a $\delta^{13}C$ analysis and use of Equation 2.1 yields delta values −71 to −66‰, well beyond the minimum −60‰ criterion for biogenic origin. Then, combination of results from Equations 2.1 and 2.2 firmly establishes hydrate gas origin as biogenic (Bernard et al., 1976; Whiticar, 1999).

Thermogenic origins are indicated when the molecular structure ratios are less than 100, as given by Equation 2.3:

$$\frac{C_1}{C_2 + C_3} < 100 \tag{2.3}$$

For example, a Green Canyon vent gas exemplifies thermogenic sources (Sassen et al., 1999). Composed of 90.4% methane, 4.5% ethane, 3.7% propane, and 1.4% of heavier hydrocarbons, the 11.0 molecular structure ratio is well below the 100 upper limit value specified by Equation 2.3.

Molecular ratios between 1000 and 100 represent mixed gases from both biogenic and thermogenic sources. For example, ratios from 6.1 to 283 in the Xisha Trough of the northern South China Sea show the presence of thermogenic gases, and in some samples show mixtures of thermogenic–biogenic gases (Jiang et al., 2008).

Bernard et al. (1978) first analyzed about 18 cold seeps and vents in the Gulf of Mexico (GOM), finding that the pattern of molecular structure and isotope ratio indicated gas origin. Whiticar (1999) set boundaries and applied the criteria to give a more definitive resolution of hydrate gas origin. Pohlman et al. (2009), as well as others in current standard practice, use the criteria of combined molecular ratio and isotope ratio summarized in Table 2.3 to estimate hydrate gas origins in the seafloor.

Table 2.3 Criteria to establish hydrate gas source

Source	**Molecular structure criterion** $[C_1 / (C_2 + C_3)]$	**Isotope criterion ($\delta^{13}C$)**
Microbial	>1000	More negative than −60‰
Mixed	100–1000	−50 to −60 ‰
Thermogenic	<100	Less negative than −50‰

A general graphical depiction of combined criteria makes readily apparent spreads of biogenic, mixed, and thermogenic gas origins. The graphical depiction of the criteria given in Table 2.3 is now commonly used to define and illustrate hydrate gas origins of cores extracted from hydrate zones.

2.2 WIPEOUT ZONES

2.2.1 Description of Wipeout Zones

Free gas tends to accumulate below sediments made impermeable by hydrates at the BGHS interface. If discontinuities such as faults, mud volcanoes, fractures, or shifting salt columns pierce the interface, the pressured gases break through. Then, tortuous paths of gas flow develop throughout the overburden volume and create shifting vents at the seafloor surface. Seismic reflections from such areas outline irregular patterns of disrupted strata – i.e., a seismic wipeout zone. These blanks within hydrate equilibrium boundaries show clathrates to form in coarse sands, fine-grained muds, slits of fractures, and even occasional mounds protruding from seafloor surfaces – wherever space affords or can be generated for hydrate expansion.

The wipeout-zone gas flux, possibly warmer than surrounding materials because of flow from deeper sources, has marked consequences: (1) in the case of high gas flux, pockmarks may appear at the surface of the seafloor; (2) extensive microbial activity occurs along the tortuous path; (3) extensive authigenic carbonates identify current venting sites as well as historic venting sites; (4) seismic reflections are not coherent; (5) BSR becomes discontinuous at the interfacial breakthrough point; (6) hydrates are scattered throughout the volumetric wipeout zone, although changing water salinity and localized temperatures create a spatially nonuniform phase equilibrium curve.

Figure 2.1 presents finite element modeling of a profile revealed by seismic data from Cascadia margin between Ocean Drilling Program (ODP) Site 889 and Site 890 (Wood et al., 2002). Note the wipeout zone resembling a chimney of gas flow, being wider at the base where warmer gases accompanied by more saline waters dissociate base hydrates near thermodynamically stable temperatures. It represents typical wipeout zones found in gas hydrate provinces.

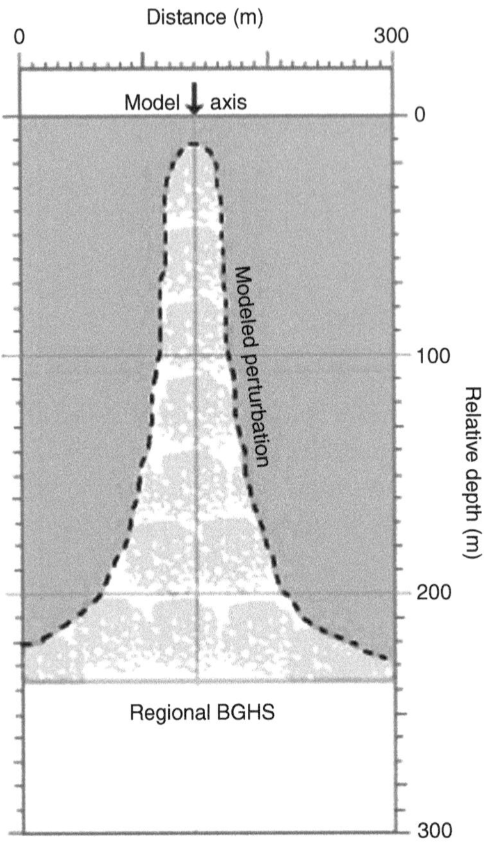

Figure 2.1 ***Finite element modeling of seismic profile of the wipeout zone (Wood et al., 2002).***

2.2.2 Venting Variability within Acoustic Wipeout Zones

Analyses show that methane fluxes in wipeout zones exceed those outside the zones (Lapham et al., 2008). Exemplifying offshore oil- and gas-producing areas, the North Sea exhibits natural seeps of hydrocarbons in acoustic wipeout zones where chaotic reflections characterize seismic waves. The seismic profile sketch presented in Figure 2.2 for the Tommeliten Field, North Sea, illustrates seepage generation and flow mechanism typifying a wipeout zone (Hovland, 2002).

The unordered reflections of seismic waves in the chimney-like center of Figure 2.2 represent breaks and discontinuities in the strata through which gas seeps. In the Tommeliten Field case, fractures originate and spread above piercement of an underlying salt diapir, allowing unhindered

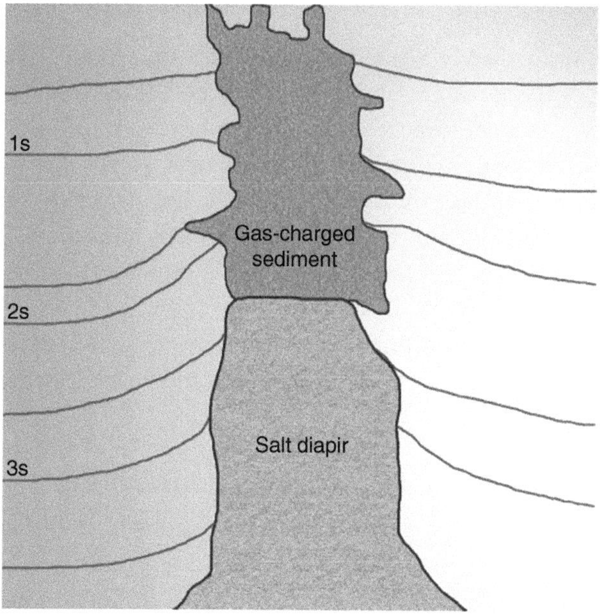

Figure 2.2 ***Sketch of wipeout zone associated with cold seep at Tommeliten (Hovland, 2002).***

hydrocarbon gas flow to the seafloor. Over the entire areal extent of the Tommeliten Field's wipeout feature of low-seismic-amplitude reflections, seeps penetrate the seafloor.

Throughout wipeout zones, vent sites change periodically. Gas seepage varies with time, location, and rate (Lapham et al., 2008). Within hydrate equilibrium boundaries, these changes occur for three primary reasons. First, carbonates as anaerobic oxidation of methane (AOM) reaction products accumulate in the fracture conduits, reducing fracture widths and increasing resistance to gas flow. Second, hydrates form within the conduits, further increasing flow resistance. Incremental resistances from both carbonate accumulations and hydrate formations eventually force changes to alternate flow paths (Hovland, 2002). Third, seismic events redirect flow paths and relocate vents at the seafloor.

The wide distribution over wipeout zones of carbonate precipitates signifies changing vent sites, where carbonates mark active venting sites as well as sites long deactivated.

Radich (2009) established further evidence of the ubiquitous nature of vents in wipeout zones by culturing indigenous microbes in sediment samples taken near an active venting site and comparing with those from an adjacent inactive site at MC-118 in the GOM. Microbes from currently venting sites immediately began producing biosurfactants on imposing the hydrocarbon gas source in laboratory cultures. Within sediment samples extracted from currently inactive sites, microbes were in a dormant state; longer times were required to initiate biosurfactants from the culture. The lag represents time to reactivate spores from the dormant state in those sediments long deprived of carbon because of gas-vent relocation.

Over a several-year span between scientific cruises, MC-118 seeps were observed to relocate. In the days immediately following seismic events several hundred miles to the southeast, observations from a manned submersible noted new active vents having opened and vent relocations having developed.

2.2.3 Gas Flares Emanating from Wipeout Zones

Subsea gas discharges from wipeout zones within hydrate *P–T* equilibrium envelopes enter the overlying water column. In the water column, shells of hydrate form around each gas bubble to retard gas dissipation until the mass reaches a sufficiently shallow depth where the water temperature exceeds hydrate stability. High concentrations of hydrate and gas create densities within plumes less than densities of surrounding seawater. The acoustic differences afford means of monitoring seafloor gas plumes.

Hydroacoustic images are recorded in the Sea of Okhotsk, where gas flares increase with seismic events. Such flares are common in this region above large-scale, periodic gas emissions (Obzhirov et al., 2004).

Generally, voluminous emissions in deepwater gas hydrate zones are consequences of mud volcano eruptions, seismic events, or drilling blowouts. In each case, hydrate shells surround gas bubbles to delay gas dissipation, allowing plumes to rise hundreds of meters above emission depth and sustain far-reaching drift with ocean currents.

The relationship of subsea gas flares with the upper boundary of the gas hydrate stability zone, when that boundary extends into the overlying water column, may be seen from acoustic images terminating near the hydrate equilibrium limit.

2.3 MORPHOLOGIES OF SEAFLOOR HYDRATES

Hydrate morphology determination improves by preserving *in situ* pressures and temperatures during coring. Then, improved morphology characterizations from the pressure cores lead to more accurate estimates of in-place hydrate gas since morphology affects physical properties, seismic velocities, and electrical resistances of sediments from which seismic and logging evaluations derive (Holland et al., 2008).

Early studies of seafloor hydrates concentrated on fine near-surface sediments, for example, mud and clays of grain diameters less than or equal to 3.9 μm. In the quest for potentially producible methane hydrate reservoirs, deeper pore-filling hydrates in coarser sands of high permeability near the BGHS now dominate interest, for example, the grades of sand from 62.5 to 2000 μm. Each of these two major classifications may overlap into silts of 3.9–62.5 μm. Hydrate morphology varies in the two environments where, along with other factors, sediment grain sizes and effective stress differ.

With these considerations, Lu et al. (2011a) suggest classification of seafloor hydrate morphology based on visibility. The classification of Lu et al. simply depends on whether hydrates are pore filling, massive, veins, or nodular. Visibility occurs when: (1) hydrates fill open spaces such as fractures and faults; (2) hydrates agglomerate in veins and small fractures as allowed, for example, by effective stresses of relatively shallow fine-grained sediments; (3) massive hydrates expand above seafloor/ocean interfaces. The invisible category refers to hydrates dispersed within pore spaces of fine-particle sediments.

Another practical classification of seafloor hydrate morphology, influenced by prospects for future gas production, is proposed by Boswell et al. (2012). A similar approach is preferred by Holland et al. (2008). These morphology types are summarized in Table 2.4.

2.3.1 Fine Sediment Morphology

Seafloor hydrate morphology is influenced by gas flux rate, sediment porosity, fracturing–faulting network of sediments, *in situ* bioproducts, sediment mineral content, *in situ* stresses, depths below seafloor, and sediment particle sizes. The opportunity for hydrates to expand or aggregate in free spaces (fractures, seafloor surface, porosity, veins) during crystallization plays an important role in final hydrate morphology. Therefore, hydrates in fine sediments appear as nodular, massive, stratified as lens, veins, or even dispersed.

2.3.1.1 Nodular

Fine-grained, unconsolidated sediments containing clay platelets typify near-surface sediments that harbor low-energy-density hydrates. Percolation of hydrocarbon seep gases through muds of less than 1–3.9 μm has significance for hydrate morphology.

Table 2.4 Morphology classification seafloor hydrates and their production prospect (Boswell et al., 2012)

Morphology type	Grain size (μm)	S_{gh} (%)	Producibility	Drill hazard
Near-surface mounds, masses	Solid mass		Relatively small gas quantity No production means yet	Drilling avoids; minimum hazard
Pore filling				
a. Fine grains	<3.9	≤10	No production means yet	Manageable
b. Coarse grains	3.9–2000	50–90	Excellent. Conventional means	High if free gas present
Grain displacement				
a. Nodular	<3.9	<5	No production means yet	Low
b. Veins		≥40		
c. Fractures				

The unconsolidated muds perform as elastic solids and consequently linear elastic fracture mechanics describe their behavior (Johnson et al., 2002). The elastic behavior facilitates gas bubbles forming particles of hydrates, aggregating, expanding, and consolidating to form nodules. Consolidation of hydrate particles into nodules may occur after initially forming as dispersed or small-fracture-filled hydrates.

Sediment sampling with unpressurized cores, where at least some hydrates decompose during retrieval, may lead to mistakenly identifying hydrates as dispersed, whereas the decomposed gas may actually have come from hydrates filling small fractures or nodules embedded in the elastic muds (Holland et al., 2008).

Equation 2.4 helps explain fracture initiation and propagation in muds according to linear elastic fracture mechanics (Johnson et al., 2002):

$$P_c = \left(\frac{\pi^4 K_{1C}^6}{24EV} \right)^{1/5} \tag{2.4}$$

where P_c, critical value of pressure for fracture initiation; E, Young's modulus; K_{1C}, critical stress intensity factor; V, bubble volume.

Observe in Equation 2.4 that a fracture initiates in the mud when the pressure P_c is reached or exceeded.

Boudreau et al. (2005) injected gas bubbles into a fine mud and then followed bubble movement with the aid of high-resolution CT scan. The exercise helps demonstrate the process described by Equation 2.4. A fracture forms from the injection of a single gas bubble at a specific angle in elastic mud; numerous parallel fractures may develop from streams of gas bubbles and even coalesce to form nodular hydrates (Boudreau et al., 2005).

Three examples are representative of nodular hydrate occurrences.

First, Suess et al. (1999) found that nodular hydrates exist 4–17 m below seafloor in the Cascadia margin (Suess et al., 1999). By CT scans, Abegg et al. (2006) evaluated 57 Cascadia sediment samples and found hydrate nodules approximately spherical with 1–5 cm diameters, where hydrate appears as gray matter in the scans, the lightest tints represent sediment, and the darkest spots shown within the cores are free gas. The noninvasive CT scans are especially helpful for verifying hydrate morphology in fine-grained sediments when cores are retrieved while pressurized, chilled, and then analyzed with original hydrates remaining intact.

As a second example, in sediments beneath 1396–1989 m deep waters near Borneo, Sassen and Curiale (2006) report nodular hydrates retrieved in push cores from 6 m below seafloor surface. The 5-cm-diameter hydrate nodules contained biogenic methane and ethane in fine-grained sediments. Developing nodules had pushed aside surrounding mud to make room for the volume increase required for hydrate growth. Sassen et al. (2001) also reported similar appearances of nodules in fine sediments near GOM gas vents.

A third interesting example relates to nodular clathrates formed in the hydrate laboratory at Mississippi State University in which the nodules were promoted by water solutions of anionic biosurfactants saturating packed media of sand/smectite clay. The hydrate nodules may be seen in Figure 8.8 where a water solution of emulsan, an anionic emulsifier produced by the bacterium *Acinetobacter calcoaceticus*, saturated packed sand/smectite clay media, and nodular hydrates preferentially formed above the smectite (Rogers et al., 2004). Anionic biosurfactants and smectite clay platelets are common in near-surface, fine-grained sediments, and it is suggested that they influence hydrate nucleation as well as ultimate morphology.

2.3.1.2 Massive Hydrates

Abegg et al. (2006) in their study of cores from Cascadia margin define massive hydrates to be greater than 10 cm thick and having less than 25% intercalated sediment. It becomes apparent in observing the CT scans that one route to massive hydrates is the agglomeration of the smaller nodular hydrates into the massive morphology.

If hydrates form, they must have room to expand into pores of coarse packed sands, fractures, or stress-enlarged volumes of fine-grain sediments. If effective stresses plus tensile strength of overburden in fine-grained muds exceed pressures of phase change at hydrate formation, then space cannot by enlarged for massive hydrates to accumulate at that depth. This threshold depth was calculated by Abegg et al. (2007) to be 19 mbsf in fine-grained muds at Hydrate Ridge conditions. In fact, massive hydrates were not found there below 11 mbsf (Abegg et al., 2007).

Consider the growth of massive gas hydrate in a laboratory test cell in which packed media of Ottawa sand and sodium montmorillonite were saturated with seawater and rhamnolipid, a biosurfactant from the bacterium *Pseudomonas aeruginosa* identified to be around gas hydrates on the floor of the GOM (Lanoil et al., 2001). Massive hydrates develop around the smectite, drawing water from the shrinking clays and using clay platelets/adsorbed biosurfactant as crystal nucleation centers. Capillary diffusion allows massive hydrates to develop in test samples, expanding into available space but especially building in unlimited expansion spaces outside the perforated sample container, as can be observed in Figure 8.7. Since both smectite clay platelets and anionic biosurfactants populate gas hydrate zones, an influence of bioproducts on hydrate morphology is probable in seafloors.

Laboratory preparations of gas hydrate in packed porous media saturated with biosurfactant/water solutions (both seawater and distilled water) indicate a stepwise process of some hydrates going from disseminated to nodular to massive as time proceeds and hydrates grow. This stepwise progression of hydrates seems to agree with field data (Holland et al., 2008).

2.3.1.3 Fracture Propagation in Fine Sediments

Seven deep hydrate wells were drilled in the GOM by the Joint Industry Project (JIP). Well log analyses revealed gas hydrate–containing vertical fractures throughout shales in four of the seven wells (Lee and Collett, 2011).

In a wipeout zone hosting gas hydrates as well as free gas, deformable fine-grained sediments saturated with water must accommodate high gas fluxes. In order to do so in the absence of natural fractures and faults, gas percolation develops through imposed fingering or fracturing. A model developed by Holtzman and Juanes (2011) predicts that critical diameters of sediment particles determine when fracturing develops; these critical diameters decrease in size with increasing confining stresses. Thus, there is a

Table 2.5 Tensile strengths of sediments, samples of northeastern Nankai Trough (Suzuki and Narita, 2011)

Material	Tensile strength (kPa)
Sand	90
Sandy silt	161
Mud	347

transition from fingering to fracturing in a constant stress field when particle diameter decreases to a critical size.

Consider the case of fine-grained, unconsolidated sediments within the gas hydrate zone. In the absence of preexisting faults or fractures as conduits for gas movement to the surface, fractures propagate when the difference in gas pressure minus water pressure exceeds the prevalent horizontal stress and interparticle cohesion (Jain and Juanes, 2009).

The northeastern segment of Nankai Trough has unconsolidated sand layers sandwiched between mud and silty layers. Cores were extracted and tensile strengths evaluated with the aid of a mercury porosimeter (Suzuki and Narita, 2011). Discrete, dispersed hydrate particles occupy pores of the sand pack. From multiple core samples, Suzuki and Narita estimated the tensile strengths given in Table 2.5.

The high tensile strengths of the mud and silt–clay layers explain the absence of hydrate-filling fractures. Capillary entrance pressures of the mud are too high for fracturing networks of small fractures to have developed in the mud. In eventual production of gas from hydrates by the depressurization method, the existence of nonpermeable boundaries becomes important, so the silt and mud characterized in Table 2.5 could serve that purpose. That is, the sand itself might be fractured to improve permeability during production while leaving the bounding silt and mud intact as an impermeable boundary.

2.3.1.4 Fracture Orientations

In ODP Leg 204, Abegg et al. (2007) retrieved 57 samples from pressurized cores that covered all sediment depths from surface to BGHS at 130 mbsf. Their CT-scan analyses gave an undisturbed view of sediments within the cores. Numerous fractures were observed in these Cascadia margin cores, giving insight into hydrate morphology and fracture orientation as a function of depth.

Horizontal fractures parallel to bedding planes were found near seafloor. By contrast, fractures at depths approaching BGHS were almost exclusively vertical. The analyses of Abegg et al. showed fractures dipping 0–30° above 45 mbsf, dipping predominantly 30–60° in intervals between 45 and 85 mbsf, and angling 60–90° at depths beyond 85 mbsf. See Figure 2.3.

Analogously, operator-imposed hydraulic fracturing in land-based coal-bed methane production or in conventional onshore gas production results in fractures trending toward the horizontal for shallow formations and trending toward the vertical for deeper formations. For example, mine-throughs in coal seams reveal horizontal fractures less than 229 m below surface and vertical fractures greater than 610 m deep, with inclined fractures between these two depths (Rogers, 1994).

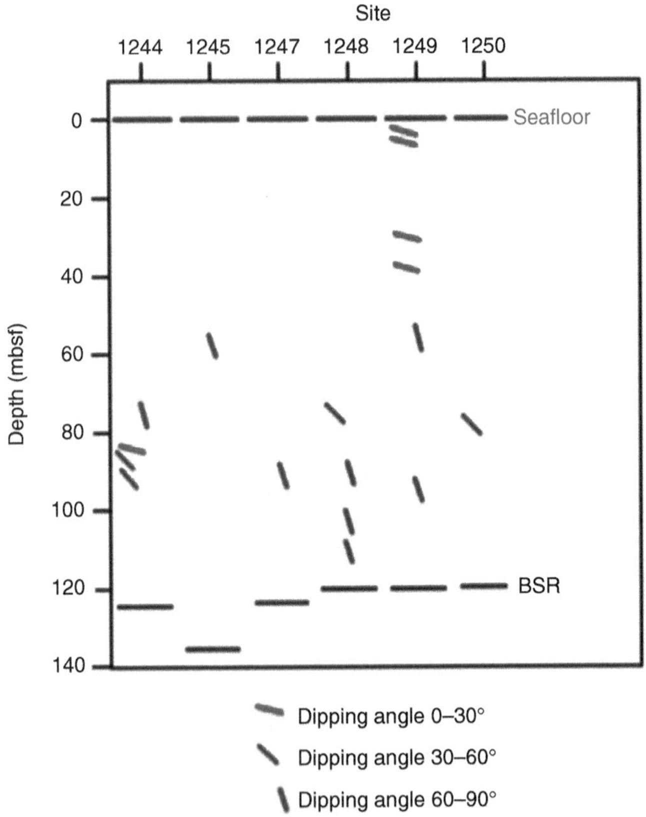

Figure 2.3 ***Plot of measured angles of dip in samples of Sites 204-1244 to 1250.*** Sample 204-1249C-1H-1 is hardly visible because it is only 0.25 mbsf. (© *Springer-Verlag 2007. With kind permission from Springer Science and Business Media; Abegg et al., 2007, Figure 6.)*

Thus, hydrates fill veins within fine sediments of gas hydrate zones and conform to the vein's shape and size (Abegg et al., 2007), as Suess et al. (1999) verify in the Cascadia margin.

2.3.1.5 CT Scans of Fracture-Filled Hydrates

With noninvasive CT scans coupled with pressurized core retrieval, fine hydrate-filled fractures or hydrate particles dispersed throughout porous sand can be accurately detailed, as well as unreacted free gas or water isolated within the core. A few of the following results illustrate the power of the CT scans.

Holland et al. (2008) analyzed pressurized cores retrieved from Cascadia margin, GOM, East Sea, Ulleung Basin, and offshore India, using X-ray CT-scan analysis, P-wave velocity measurements, and gamma ray density. With their retrieval and analytical technique, they were able to distinguish between hydrates displacing grains and those filling pores – important for establishing physical properties of sediments and improving estimates of hydrate gas.

Hydrate reservoirs composed of numerous small fractures filled with gas hydrates have been discovered offshore Korea and India (Kurihara et al., 2011b). In 2007 pressure cores were retrieved from South Korea's Ulleung Basin, East Sea; pressures and temperatures were maintained on the cores within aluminum chambers while CT scans were made. Complex networks of hydrate-filled fractures were observed (Park et al., 2009).

2.3.1.6 Polygonal Faults

Polygonal faults are normal fractures found in fine-grained, low-density sediments. These faults terminate at intersections with coarser-grained sediments, thus leaving unique geometric patterns (Dewhurst et al., 1999). Further, the numerous fissures per sediment volume are identified by their patterns in passive margins of minimized horizontal stresses (Cartwright, 2011). Three-dimensional seismic technique is needed for their discovery and for fully detailing their patterns (McGee et al., 2009).

Shiguo et al. (2011) estimate that polygonal faults exist in about 50 marine basins around the world. Most of these polygonal systems seem to supply important pathways for hydrocarbon fluid movements. In addition, the faults become significant by providing storage space for gas hydrate accumulations.

In the Lower Congo Basin, polygonal fault systems at 0–800 mbsf connect to the center of seafloor pockmarks, providing pathways for thermogenic hydrocarbon migrations (Gay et al., 2006). Larter et al. (2000)

report vertical migration of hydrocarbon fluids through a polygonal fault system within the Heimdal field of the North Sea. Camerlenghi et al. (2009) report existence of a polygonal fault system, although probable because detection was by 2D seismic technique, acting as fluid conduits in the Western Mediterranean Sea.

In three basins on the northern slope of the South China Sea, a polygonal faulting network provides conduits for gas to flow from deep sources below BSR. Detected by 3D seismic technique, the 150–1500 m long faults in fine-grained sediments penetrate the bottom of the hydrate zone, causing discontinuities in the BSR and ultimately creating gas hydrate fields between BSR and seafloor. Wipeout zones of acoustic blanking above BSR and pockmarks at the seafloor surface attest to functioning of the fault system as conduits for gas flow (Shiguo et al., 2011).

At sufficient depths within fine sediments of a gas hydrate stability zone, porosity created by fractures becomes important to provide the necessary space for expansion when hydrates form and permeability created by fractures provides the necessary fluid transport (Abegg et al., 2007). Therefore, polygonal faults may provide general, and possibly universal, fracture networks amid low-permeability environments of fine-grained sediments in which gas hydrates accumulate (McGee et al., 2009).

Since polygonal faults are not confined to near-surface sediments and 3D seismic technique was lacking, data at the beginning of offshore hydrate research on polygonal fault/hydrate systems are scant (Allen et al., 2013; McGee et al., 2009).

Two-dimensional seismic data indicate probable polygonal faulting at MC-798 block, GOM. The closely aligned fractures, resembling a broom, progress through the upper 100 m of fine-grained sediments. Core samples from the sediments show a high content of smectite clay, and laboratory tests of their sediments establish a trend of decreasing hydrate induction time, possibly as a function of increasing fractures containing smectites. See Figure 2.4 (McGee et al., 2009).

Others report polygonal faults associated with colloidal smectite particle fillings (Burst, 1965; Dewhurst et al., 1999). Based on these field observations and Mississippi State University (MSU) Laboratory test results, gas hydrates should form rapidly in polygonal fault conduits of hydrate zones. Hydrates would be promoted by *in situ* clay platelets/bioproduct nucleation sites.

Gas hydrates are verified within polygonal fault patterns, for example, as seen in CT-scanned cores extracted from offshore-India hydrate zones (Holland et al., 2008).

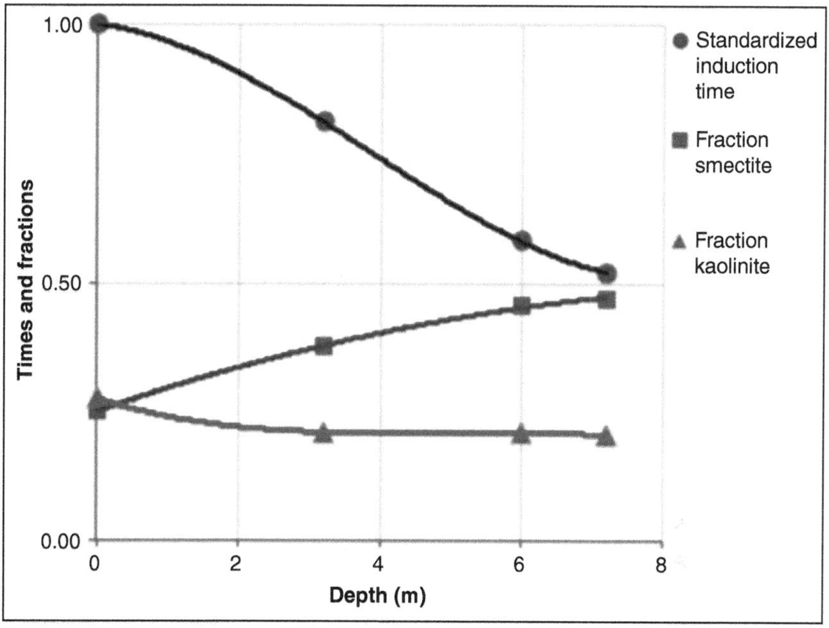

Figure 2.4 ***Hydrate formation catalyzed by smectite particles (McGee et al., 2009).***

Other discoveries of interest include the identification from 3D seismic data by Andreassen et al. (2007) of a polygonal fault system in the hydrate zone of the Barents Sea continental margin.

2.3.2 Dispersed Gas Hydrates

2.3.2.1 Coarse Sands; Porous, Permeable Hydrate Reservoirs

Dispersed gas hydrates in sand reservoirs near BGHS, representing high-density energy sources, have the greatest potential for economical methane gas production. Among other attributes, coarse-sand reservoirs offer exceptional permeability to have initiated original hydrates as well as to transport dissociated-hydrate gas during future production (Boswell and Collett, 2006; Zhang and McConnell, 2011).

A mechanism whereby sand reservoirs of favorable hydrate saturation, porosity, and permeability develop is proposed by Jain and Juanes (2009). Initially, when gas pressures entering pores exceed water pressures within the pores – that is, the capillary entrance pressure differential – gas slowly invades the pore space. This pressure differential is inversely proportional to grain diameter of the sediment. Therefore, pores of large grains of sand more easily become saturated with gas hydrates. In the model of Jain and Juanes the large sand grains remain stationary during gas invasion and subsequent hydrate formation. Gas hydrates volumetrically expand the *in situ* water precursor. When effective stress of confining sediment exceeds hydrate expansive forces, or when insufficient water or gas within pores limits hydrate formation, dispersed hydrate becomes confined to a fraction of available pore spaces. Alternatively, if effective stresses are less than the hydrate phase pressure, hydrate expansion displaces the porous media, and dispersed hydrate particles consolidate to form nodular, massive, or stratified hydrates (Holland et al., 2008; Abegg et al., 2006).

Coarser-grained sediments at Blake Ridge predominantly contain dispersed, pore-filling hydrates (Lorenson, 2000) formed from low gas flux primarily of biogenic origin (Davie and Buffett, 2002). This is also the case in the Nankai Trough.

2.3.2.2 Fine-Grain Sands and Muds with Dispersed Gas Hydrates

Large surface area/volume ratios of dispersed hydrate particles, as in hydrates occupying small pores of a fine sand, increase heat transfer rates to accelerate hydrate dissociation during unpressurized core retrieval.

Rapid dissociations of dispersed hydrates are reported by Abegg et al. (2006) while retrieving fine-grained cores from the Cascadia margin. Holland et al. (2008) point out that unless the samples are retrieved in pressure cores and subsequently analyzed by noninvasive means such as X-ray CT scan, decomposition of hydrates during retrieval may mask veined hydrate presence from the true dispersed hydrate state. This is verified in cores from the Ulleung Basin, South Korea, where Bahk et al. (2011b) report difficulties distinguishing dispersed hydrate morphology from veined hydrates in mud samples.

Dispersed, pore-filling hydrates are reported in the small-grain volcanic ash of the Andaman Islands in the Bay of Bengal, off India's eastern coast; beds of the ash are surrounded by an even more finely grained matrix. The pore spaces of the 6 μm mean diameter ash, having a 64% median porosity, are filled with gas hydrates to a maximum saturation greater than 50% (Winters, 2011).

2.4 PHYSICAL PROPERTIES OF SEDIMENT MATRIX INFLUENCE HYDRATES

2.4.1 Particle Size Effects on Permeability

In computer simulations, high intrinsic permeability is shown to be important for successful hydrate gas production. As are other important physical

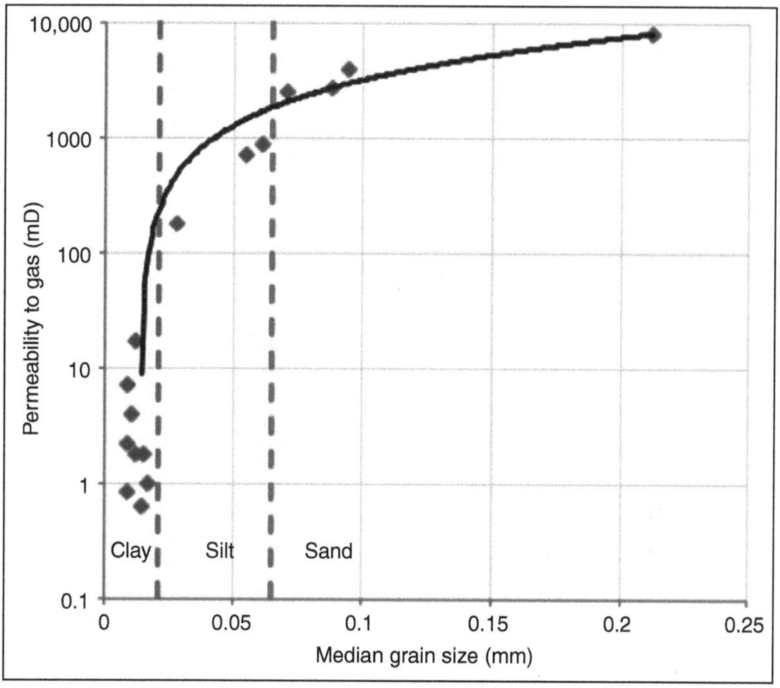

Figure 2.5 ***High permeability extends into silt grain sizes (Winters et al., 2011).***

properties of sediments, intrinsic permeability is a function of matrix grain size – permeability increases with grain size. The boundary between categories of very fine sand and smaller-sized silt is taken as a grain size of about 62.5 μm. However, it is seen from Figure 2.5 that intrinsic permeability to gas may be adequate down to grain sizes comprising silts before a sudden drop-off in permeability steepens in midrange of the silt category (Winters et al., 2011).

A minimum permeability floor must be exceeded to accumulate hydrates in silty sands. Based on data of Lee and Collett (2011) from wells at Mallik and Mount Elbert in which the sands contain about 10% silt, that threshold permeability is 10–40 mD.

Another way to judge hydrate production viability is to consider impacts of sediment particle sizes on porosity and permeability. Kurihara et al. (2011a) relate permeability to clay, silt, and sand content of cores taken from Eastern Nankai Trough. Sediments of 50/50 sand/silt gave permeabilities between 1 and 100 mD, whereas 100% sand exceeded 100 mD for favorable production potential. Their correlations predict increasing production difficulties as silt increases to 100% and further deterioration thereafter with the introduction of clay in the silt.

Particle sizes of the clay, silt, and sand fractions are taken as less than 3.9 μm, 3.9–62.5 μm, and 62.5 μm to 2 mm, respectively.

The permeabilities and porosities of Nankai Trough agree closely with those of Mallik and Mt. Elbert hydrate zones beneath Arctic permafrost.

2.4.2 Particle Size Effects on Hydrate Saturation

2.4.2.1 Core Analyses

Particle size distributions in sandy media correlate well with fractional hydrate saturation if the hydrates contained therein are pore filling. Larger particle sizes in the sediments yield higher hydrate saturations (Ukita et al., 2011; Bahk et al., 2011b). This axiom holds over wide particle size ranges of predominantly sandy porous media. Based on numerous samples, Ukita et al. (2011) verified the trend in the Nankai Trough. Although their compilations showed data scatter, saturation values trended from less than 20% hydrate saturations in particles sizes less than 75 μm to hydrate contents exceeding 40% in cores of average particle diameters above 150 μm. Data scatter is attributed to wide ranges of particle sizes in the cores from which average values were determined for the plot, and the scatter may also be due in part to somewhat random mineral content in the samples, especially charged clay particles and iron-containing minerals (Ukita et al., 2011).

Therefore, important to the phenomenon of larger sand particles accommodating higher hydrate saturations within sediment pore spaces are the water-binding characteristics of surrounding sediment particles. Because increased particle binding restricts available water for hydrate formation within pore spaces, significantly smaller sand particles, or especially charged clay particles, bind higher percentages of connate water and lower S_{gh}.

A substantiating example comes from Ulleung Basin core samples that exhibit higher hydrate saturations accompanying particle size distributions of larger sand particles (Bahk et al., 2011b). However, the occurrence has two important provisos: the core sample comes from a sandy interval and the hydrate morphology is pore filling. See Figure 2.6.

In Figure 2.6 the outlier points may be consequences, among other factors, of methane supply being deficient or excessive, but the primary culprit is probably the distribution of clay platelets throughout the sand matrix. In this regard, random distribution of smectite platelets would be especially decisive.

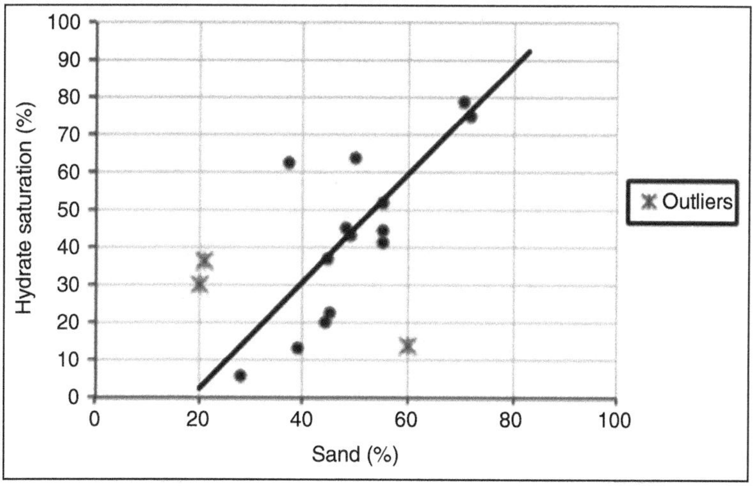

Figure 2.6 ***Core samples from Ulleung Basin.*** Data derived from pore water salinity of squeeze-cake (Bahk et al., 2011a).

Note that the increasing S_{gh} accompanying higher sand percentages in the sediment did not hold for fine-particle muds containing fracture-filled hydrates or fine-particle muds containing dispersed hydrates (Bahk et al., 2011a).

2.4.2.2 Laboratory Analyses, Hydrates Saturating Sands

Lu et al. (2011b) evaluated in the laboratory effects of sand particle sizes on hydrate saturations of packed media. The particles were silica powders of greater than 99% purity. Using size-graded silica powders with 20–1000 μm diameter (i.e., ranging over size classifications from silt to coarse sand) and high-purity methane, Lu et al. created gas hydrates by imposing 12 MPa and 3°C on a perforated sample container holding packed media saturated with 3.5 wt% NaCl–water solution. Hydrate pore saturations S_{gh} increased rapidly as sample particle sizes increased until media diameters approached 200 μm, where approximately 100% hydrate saturation S_{gh} was reached. See Figure 2.7.

It is significant that pore water conversion to hydrate depends on size of sand particles. The data of Lu et al. (2011b) indicate that silica surfaces interact with interstitial waters adjacent to the silica particles, making the affected water unavailable for hydrate formation. Therefore, only unbound bulk waters within the pores are available to form hydrates. In the case of smaller sand grains with higher specific surface areas, unbound waters

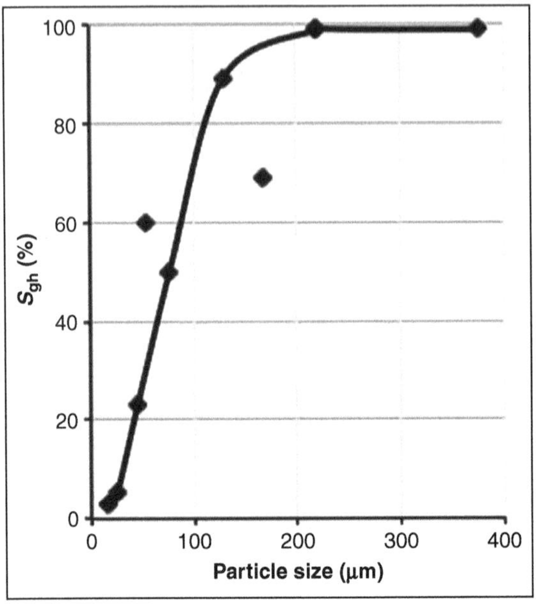

Figure 2.7 ***Hydrate saturations prevalent in coarser sands (Lu et al., 2011b).***

represent lower percentages of total pore water – consequently lower S_{gh} values. See Figure 2.8.

2.4.2.3 CT Scan for Noninvasive Hydrate Analysis of Seafloor Cores

X-ray CT scan, a noninvasive means for evaluating cores with hydrates, was adapted from the medical profession. Differences in bulk densities of mineral-hydrate gas within the core cause corresponding differences of X-ray attenuations (Abegg et al., 2007). Importantly, CT scans allow mineral matrices within extracted cores to be studied without decomposing hydrates or disturbing relative placements of minerals, hydrates, and entrapped gas–water pockets.

In an extraordinary work Klapp et al. (2011) extended the CT-scan analysis to a high resolution capable of a noninvasive 3D view of internal pore water–ice–gas configurations of hydrates within the core. They term their method "synchrotron radiation X-ray cryotomographic microscopy" (SRXCTM). In the procedure of Klapp et al., internal hydrate details are viewed at submicrometer resolutions. Consequently, the hydrate and mineral matrix may be mapped as a composite of extremely small volumes. For the first time, the SRXCTM of Klapp et al. provides valuable insight into gas hydrate decomposition mechanisms, hydrate formation from *in situ*

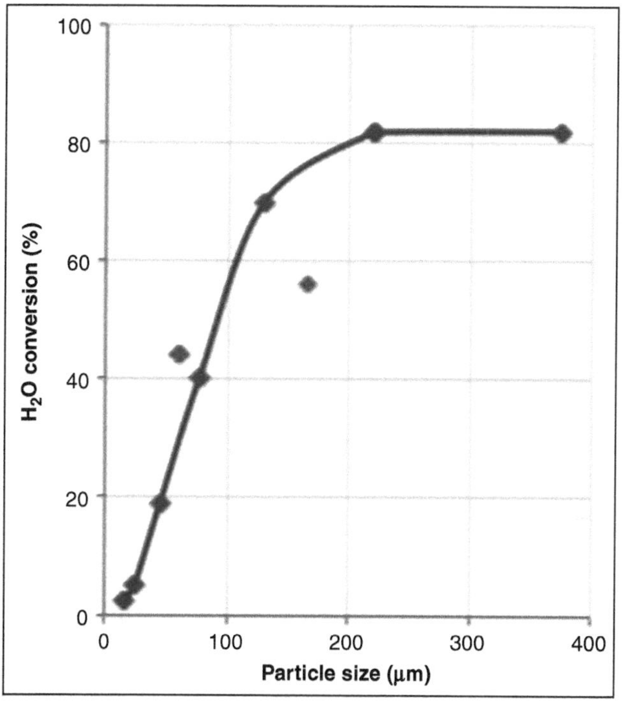

Figure 2.8 ***More unbound waters available to form hydrates in larger-particle sand samples (Lu et al., 2011b).***

gas bubbles, and detection of multiple hydrate structures within massive gas hydrates of cores retrieved from the seafloor.

2.4.3 Thermal Conductivities of Seafloor Hydrates

2.4.3.1 Hydrate Conductivities Compared with Ice

Many physical properties of hydrates approach the values of their ice counterparts, but thermal conductivity is an exception. Thermal conductivity of gas hydrate diverges sharply from that of ice. A weak bonding alliance between nonpolar guest gases and water in the hydrate framework accounts for the dissimilarities. In those cases, hydrate properties become weighted averages of structural water and occluded gas properties, biasing outcomes toward water properties since gas concentrations in hydrates are small relative to water content (Inoue et al., 1996).

When temperatures rise from 100 to 260 K, thermal conductivities of ice I_h decrease over a range of 5.94–2.35 W/(m K) (Inoue et al., 1996;

Table 2.6 Representative thermal conductivities of pure gas hydrates

Material	Temperature (K)	Thermal conductivity [W/(m K)]	References
Xenon hydrate	100	0.36	Inoue et al. (1996)
Xenon hydrate	180	0.44	Inoue et al. (1996)
Xenon hydrate	260	0.47	Inoue et al. (1996)
Xenon hydrate	245	0.36	Handa and Cook (1987)
Methane hydrate	213	0.45	Handa and Cook (1987)
Ethylene oxide hydrate	263	0.49	Handa and Cook (1987)
Propane hydrate		0.394	Stoll and Bryan (1979)

Handa and Cook, 1987). Compare from Table 2.6 thermal conductivities of hydrates over a similar temperature range.

At very low temperatures, thermal conductivities of hexagonal ice I_h can be orders of magnitude more than the thermal conductivities of gas hydrates at the same temperatures. Near melting points, for example, thermal conductivities of ice are typically about 5 times the values of their hydrate counterparts, but at lower temperatures the thermal conductivities of ice may be greater than hydrate values by a factor of 20 (Tse and White, 1988). As seen in Table 2.6, at 100 K conductivity of ice is 16 times larger than the value for xenon hydrate.

Thermal conductivities of pure hydrates characteristically increase as temperatures elevate, while thermal conductivities of ice decrease. However, within the temperature range of interest in seafloor hydrate zones, thermal conductivity changes of hydrates in a practical sense are minimal (Warzinski et al., 2008).

Hydrates are porous. Methane hydrates prepared in the laboratory vary in measured thermal conductivities as a consequence of differences in preparation methods, interstitial water contents, porosities, and measurement techniques (Warzinski et al., 2008). It is interesting that when laboratory-prepared hydrates are compressed to diminish porosity, the measured thermal conductivities increase. This compaction effect can be seen in Table 2.7.

It is informative to extrapolate conductivity to its maximum value on compression. The ultimate thermal conductivity of methane hydrate at zero porosity is estimated to be 0.70 ± 0.04 W/(m K) (Gupta et al., 2006).

Table 2.7 Thermal conductivities of compressed methane hydrates

Formation temperature (K)	Thermal conductivity [W/(m K)] Porous	Compressed	Compaction pressure (MPa)	References
263–278	0.334–0.381	0.564–0.587	2	Huang and Fan (2004)
253–290	0.36–0.34	0.62–0.63	102	Waite et al. (2002a,b, 2007)

2.4.3.2 Composite Conductivities of Hydrate–Sediments

Thermal conductivity values for sand, shale, water, and hydrate of 7.0, 1.9, 0.6, and 0.45 W/(m K), respectively, were used by Henninges et al. (2005) in mixing-law models to estimate composite conductivity of hydrate-containing sediments. To do so, effective conductivity values of the minerals were used along with the true values of water and hydrate.

Important to hydrate exploration and production are the thermal conductivities of the *in situ* hydrate and its encompassing mineral matrices. Some representative composite values are presented in Table 2.8.

From Nankai margin cores, Hyndman et al. (1992) determined sediment thermal conductivity; bottom hole temperature and seafloor temperature

Table 2.8 Representative thermal conductivities of hydrate-containing sediments

Material	Temperature (K)	Thermal conductivity of hydrate–sediment [W/(m K)]	References
Peru subduction zone, DSDP Site 688	275–298	0.89	Hyndman et al. (1992)
Nankai Trough, Site 808	275–299	1.08	Hyndman et al. (1992)
Mallik, NW Territories, Canada	280	2.24	Rutqvist et al. (2009)
Mount Elbert, Alaska	276	3.1	Rutqvist et al. (2009)
Southern Hydrate Ridge, Oregon	284–289	1.0	Liu and Flemings (2006), Trehu et al. (2003)
Typical dispersed hydrate formation	275–288	2-4	Kurihara et al. (2011b)
Salt of salt dome, Gulf of Mexico	*In situ*	3.01	Selig and Wallick (1966)

were measured by thermal probes inserted in an ODP Leg 131 well. Over an approximate 400 m of hydrate zone, the thermal conductivity k decreased linearly with depth according to the empirical Equation 2.5, valid for Site 808:

$$k = 0.91 + 0.0017 D_{\mathrm{bsf}} \tag{2.5}$$

where k, thermal conductivity [W/(m K)]; D_{bsf}, depth below seafloor (mbsf).

Note that Equation 2.5 predicts a thermal conductivity range of about 0.9–1.6 W/(m K) through 400 mbsf of hydrate zone. The mean value of thermal conductivity over the interval is 1.08 W/(m K).

To estimate composite thermal conductivity of hydrate-bearing sediments, the general Equation 2.6 may be used in computer models for production simulations (Moridis et al., 2005):

$$k_{\theta} = \left(\sqrt{S_{\mathrm{H}}} + \sqrt{S_{\mathrm{W}}}\right)\left(k_{\mathrm{SW}} - k_{\mathrm{Sd}}\right) + k_{\mathrm{Sd}} \tag{2.6}$$

where k_{θ}, composite thermal conductivity of hydrate-containing sand; k_{Sd}, thermal conductivity of dry sand; k_{SW}, thermal conductivity of water-saturated sand; S_{W}, liquid water saturation in the pores; S_{H}, hydrate saturation in the pores. Moridis et al. (2005) found the expression to acceptably approximate composite thermal conductivity of a hydrate-bearing porous media.

In eventual hydrate gas production, natural heat flux by conduction into the gas hydrate zone needs to be a maximum for efficient hydrate gas recovery; that is, composite reservoir conductivities are important to process profitability. The heat flux q is a function of thermal conductivity k and geothermal gradient dT/dz (Grevemeyer and Villinger, 2001), as given in Equation 2.7.

2.5 DETERMINING THE HYDRATE ZONE

2.5.1 Hydrate Zones in Seafloor

Gas hydrates form in zones of ocean sediments where hydrocarbon gas flux (thermogenic, biogenic, or combinations) saturates *in situ* water in sediments below a water column deep enough for sufficiently high pressures and low temperatures to stabilize sI or sII hydrates. (Type H hydrates may form, but they have not been shown to be significant.) The lower zone boundary, the deepest hydrate extension below seafloor, depends on gas composition, water salinity, hydrostatic pressure, and temperature as determined by geothermal gradient. The hydrostatic head of a water column provides pressure for hydrate stabilization. Temperature of the water column decreases with depth until it reaches a minimum near seafloor; in deep waters the upper boundary of the hydrate zone extends appreciably into the water column. For example, representative calculations of the hydrate zone at JIP drilling sites, Keathley Canyon and Atwater Valley of approximately 1300 m water depths, indicate that hydrate zones extend into overlying water columns almost 700 m, while the BGHS approaches 500 mbsf (Ruppel et al., 2008).

To what depth in the sediments can hydrates form? As sediments deepen, pressures continue to increase, but that positive influence on hydrate formation is negatively countered by temperature increases from the minimum near seafloor. The sediment depth at which the geotherm intersects the *P–T* hydrate equilibrium curve for prevailing gas composition and water salinity establishes lower boundary of the hydrate zone, BGHS.

Exceptionally high, seafloor thermal gradients decrease depth of hydrate zone. This condition might occur from rising hot magma far below sediments causing unusually high temperature gradients. Radioactive decay or deformation in tectonically active areas may contribute to high temperature gradients. But more commonly, higher thermal conductivities of massive salt deposits elevate the BGHS.

Hydrate stability is possible at sites similar to Keathley Canyon or Atwater Valley approximately 700 m into the water column above seafloor (Ruppel et al., 2008). Any hydrates outcropping the sediments and protruding into the overlying water column are transient because of vagaries in ocean current temperature and methane undersaturation in surrounding waters. Moreover, large emissions of natural gas from the seafloor under these prevailing conditions could create a plume of gas bubbles encased with hydrate shells that could conceivably remain pseudostable until buoyed 700 m above the seafloor.

Geothermal gradients become important for establishing BGHS to estimate extent of hydrates, but the gradients especially impact eventual hydrate gas production.

2.5.2 Geothermal Gradients Influenced by Salt Deposits

The thickness of salt below the hydrate zone affects thermal gradient. A case in point is the GOM, where thermal gradients vary because of underlying salt distributions: western regions have steeper gradients, but eastern regions generally have thermal gradients less steep and cooler (Hutchinson et al., 2008). Thermal gradients increase in the GOM wherever high thermal conductivity and greater thicknesses result from underlying salt (Epp et al., 1970; Nagihara et al., 1992).

As an example of variation in thermal gradients, three proposed hydrate production sites in the GOM – Alaminos Canyon, Green Canyon, and Walker Ridge – show thermal gradients of 44, 32, and 19 mK/m, respectively. In the absence of underlying salt at Walker Ridge, its thermal gradient is lowest of the three sites. As a result, BGHS is a much deeper 841 m at Walker Ridge despite a shallower water column and an accompanying lower hydrostatic pressure.

Milkov and Sassen (2000) estimate a 10–20% error in the calculated BGHS if thermal gradient is misjudged in the calculation by a mere 3.0 mK/m. Therefore, BGHS calculations, and often the in-place gas estimate, are especially vulnerable to the thermal gradient chosen within salt-prevalent GOM.

2.5.3 Thermal Gradients in Diverse Gas Hydrate Zones

Geothermal gradients at various gas hydrate sites are given in Table 2.9.

Microbial activity is closely linked with gas hydrate occurrences in ocean sediments. An optimum temperature range promotes microbial activity and an extreme upper temperature range reduces the activity, so the sediment thermal gradient influences microbial activity throughout the hydrate zone and even beyond. Stetter et al. (1990) estimate microbial activities in deep ocean sediments to be limited by high temperatures of 110–115°C. With ocean sediment temperature gradients generally in the range of 30–100°C/km (Parkes et al., 1994), significant microbial activity could be expected to depths of 1.1–5.0 km.

2.5.4 Seismic Techniques for Hydrate Exploration

Seismic techniques provide an efficient, economical hydrate exploration tool for assistance to determine the following reservoir conditions: (1) hydrate fractional pore saturation; (2) location and extent of high-quality sand reservoirs in the hydrate zone; (3) methane source; (4) methane access to reservoir; (5) reservoir proximity to BGHS; (6) local thermal gradients; and (7) wipeout zones.

Advancements were made coordinating and relating seismic, logging, and coring data as a result of the 2009 JIP drilling program in the GOM.

Table 2.9 Representative geothermal gradients reported in gas hydrate zones

	°C/km	References
Blake Ridge	38.9–49.2	Yamano et al. (1982)
Blake Ridge Sites 995, 997	32, 37	Wellsbury et al. (1997)
Carolina Rise and Blake Ridge	40	Ruppel et al. (1995)
Outer Ridge	39.2	Hyndman et al. (1992)
Western GOM	30.2–33.2–	Yamano et al. (1982)
Western GOM	25–40	Milkov and Sassen (2000)
Bush Hill	25	Milkov and Sassen (2000)
Central Gulf of Mexico	10–25	Milkov and Sassen (2000)
Nankai Trough	41.5–65.0	Yamano et al. (1982), Suess et al. (1999)
Eastern Nankai Slope	35	Foucher et al. (2002)
Nankai margin, Site 808	119	Hyndman et al. (1992)
Cascadia margin	51	Suess et al. (1999), Kastner et al. (1998)
Southern Hydrate Ridge, Oregon	55	Liu and Flemings (2006), Torres et al. (2004)
Peru Trench	52	Hyndman et al. (1992)
Peruvian margin, ODP Site 688	51.6	Grevemeyer and Villinger (2001)
Storegga Slide, Norway	50	Posewang and Mienert (1999)
Storegga Slide, Norway	56	Bouriak et al. (2000)
Krishna–Godavari Basin, India	42	Kundu et al. (2008)
West Siberian permafrost	24	Duxbury et al. (2004)

The JIP wells provide accurate sonic and density logs through the hydrate zone at depths where quality logs from conventional oil and gas wells are not commonly available.

Seismic data were compared with logs from JIP Leg II wells GC-955, WR-313, and AC-21 by Shelander et al. (2011), who found close agreements in estimating hydrate saturations S_{gh} between seismic data and well logs if hydrates fill sediment pores to the extent of 60–80% in 16–40 m thick high-quality sand reservoirs. However, unreliable accuracies resulted in wells having only low hydrate concentrations in thick sand intervals, namely AC-21 and GC-955. Another inconsistency occurred in the case of thin hydrate-filled fractures and veins that do not provide adequate acoustic impedance contrasts with background sediments to be detected precisely by seismic technique (Shelander et al., 2011).

2.5.4.1 BSR Limitations

Although seismic detections of BSRs have been an important tool since the discovery of seafloor gas hydrates, BSRs have unresolved limitations. First, they fail to quantify hydrates in the seafloor. Second, BSRs sometimes fail to indicate hydrate accumulations such as in the Krishna–Godavari Basin, offshore India. Finally, prolific hydrate accumulations located by drilling may not correlate with the direct presence of a BSR at that location (Collett et al., 2008).

Gas hydrates occur extensively in the GOM. Some of these hydrate accumulations exhibit continuous BSRs, but a majority exhibit discontinuous or pluming BSRs. For example, the agency *Bureau of Ocean Energy Management* (BOEM) overseeing oil and gas development in the Gulf reports 145 BSRs in the Gulf which identify BGHS, but more than 50% of these BSRs are discontinuous (Shedd et al., 2012).

2.5.4.2 BSR Principles

When creating impermeable barriers at BGHS, solid-phase hydrates block upward movement of free gas. In the absence of lower-resistance escape routes, unassociated methane gas accumulates below BGHS when concentrations of methane exceed solubility limits in the water. In sound wave reflectance involving a sufficiently extensive gas column, BSRs show as a bright area of recorded wave impedance, distinguishing the interface of hydrate-laced sediments and free gas.

In Figure 2.9, the gas hydrate zone and the regional BSR are superposed on the phase diagram for methane hydrates in Ulleung Basin, East Sea of

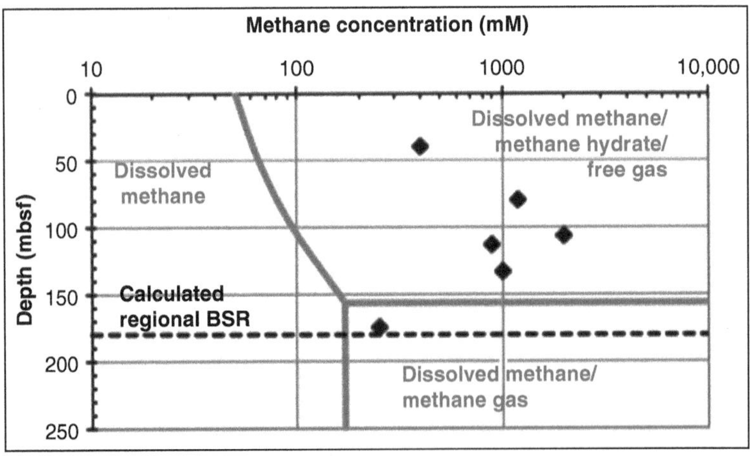

Figure 2.9 ***BSR in Ulleung Basin coincides with phase diagram, core analyses (Bahk et al., 2011a).***

Korea; the calculated phase diagram, which delineates limits of hydrate equilibrium, fits closely to the measured BSR and core-derived analyses (Park et al., 2009). Note that diamond symbols in Figure 2.9 represent gas hydrates as determined from piston cores extracted while maintaining *in situ* pressures.

At these East Sea sites, clathrates occur as vein fillings in the hydrate zone.

2.5.4.3 Acoustic Impedance

Acoustic impedance is the product of porous media density through which the sound wave travels and the velocity of the sound wave (Andreassen et al., 2007). Examples of sound velocity, media density, and impedance relative to seafloor gas hydrates are given in Table 2.10.

Table 2.10 Representative acoustic impedances, hydrate zones

Material	Compressional V_p (km/s)	Density (g/cm³)	Acoustic impedance (kg/s g/cm³)	References
Ice	3.83	0.90	3.45	Winters et al. (2004), Perry (1950)
Pure water	1.50	1.00	1.50	Winters et al. (2004)
Water	1.46	1.00	1.46	Helgerud et al. (1999)
Pure CH_4 hydrates	3.78	CH_4 hydrate, 90%	3.44	Jakobsen et al. (2000)
Pure gas hydrates	3.27	Cage occupancy	3.27	Ghosh et al. (2010)
Pure gas hydrates	3.18	0.926	2.89	Helgerud et al. (1999)
Pure gas hydrates	3.80	0.91	3.50	Winters et al. (2004)
Quartz	6.019	2.65	15.95	Helgerud et al. (1999), Winters et al. (2004), Jakobsen et al. (2000), Ghosh et al. (2010)
Methane	0.415	0.108	0.0448	Helgerud et al. (1999)
Salt	4.6	2.2	10.12	Amery (1969), Selig and Wallick (1966)
Clay	3.40	2.58–2.60	8.84	Jakobsen et al. (2000), Ghosh et al. (2010), Helgerud et al. (1999), Winters et al. (2004)

Compressional velocity V_p ranges between 0.69 and 6.0 m/s for common materials involved in seafloor gas hydrates, while shear velocity V_s ranges up to 4.04 m/s. It is apparent from values in Table 2.10 that methane free-gas impedance in sediments directly below the BGHS interface accounts for a distinct negative transition of impedance.

Acoustic impedance of gas hydrates will be much greater than that of the gas-rich sediments directly below BGHS (Paull et al., 2005). A BSR results when a large acoustic impedance increment is abrupt and negative over the transition from hydrate-containing sediments to free gas in sediments. Therefore, data plots of compressional P-wave reflections versus depth display BGHS as a bright line parallel to the seafloor (Wood et al., 2002; Trehu et al., 1999; Brooks et al., 1984). The bright line parallel to the seafloor is the BSR, identifying extremity of gas hydrate stability at that depth (Suess et al., 2002; Taylor et al., 2000).

However, to establish a defining bright line at the hydrate–sediments and free-gas interface without interference of wave reflections from the bottom of the gas column, gassy sediments below the hydrates must be suitably thick. Andreassen et al. (2007) evaluated, according to the theory of Gregory (1977), sediments from the Barents Sea continental margin having porosity of 0.4 and compressional velocity of 1900 m/s and found that reflections recorded from the top and bottom of the underlying gas column appear distinctly without interference if the thickness of the free gas column is at least 32 m.

Below BGHS at Blake Ridge, free-gas zone thicknesses of 200 m give strong BSRs (Taylor et al., 2000). On the northern margin of the Storegga Slide, a strong BSR exists above a 150 m thick free-gas zone (Westbrook et al., 2008).

Interestingly, Andreassen et al. found a drop in compressional velocity V_p and a corresponding dramatic drop in acoustic impedance as free gas–occupying sediment pores increase from 0 to 4%.

Models show that acoustic impedances from P-wave velocities in hydrate-bearing sediments increase with depth below seafloor as well as with increasing fractional hydrate saturation (Shelander et al., 2010).

2.5.4.4 *Double BSR*

Double BSRs are not common in gas hydrate zones. But in a few instances where they have been identified, interesting observations may be made (Foucher et al., 2002; Posewang and Mienert, 1999). The occurrences provide historical insight into gas hydrates at that location.

Generally, the upper BSR of the couplet, designated BSR1, has stronger amplitude and corresponds in depth to present temperatures and pressures that identify the current gas hydrate stability boundary. Hyndman et al. (1992) found BSR1 to generally be at pressures and temperatures matching current hydrate equilibrium values when methane hydrate is assumed to form from pure water.

Although the reason for existence of a lower BSR2 is not certain, the following three reasons have been offered (Foucher et al., 2002; Posewang and Mienert, 1999): (1) fractionation of natural gas occurs during initial hydrate formation. Residual gas of high methane concentration escapes. (2) AOM at the BGHS interface results in carbonate deposition, affecting seismic wave reflectivity as well as sediment permeability. As an aside, temperatures warmer at BGHS compared with shallower depths would enhance microbial activity. (3) Either seawater temperature increases or water depth lowering forces BGHS shrinkage to shallower sediments. Seawater temperature could increase from atmospheric warming. Water depth alteration might have resulted from either an uplifting tectonic event or from a decrease of the sea level.

A double BSR is located in the eastern Nankai Trough. BSR1 and BSR2 parallel the seafloor for about 10 km and occur at depths with 50–100 m between them. The two lines cross stratigraphic units and terminate on one side of a wipeout zone but reemerge on the opposite side. The upper BSR1 depth corresponds closely to the calculated depth for current methane hydrate stability – based on known temperatures and pressures at the location. Foucher et al. (2002) found the lower BSR2 trace to be modeled closely by assuming either a 1.5°C temperature increase or a 90 m tectonic uplift in its history. See Figure 2.10.

Another double BSR is reported at the Storegga Slide, offshore Norway, where the 285 m depth of the upper BSR coincides with the current temperature and the forming gas composition of 99% methane and 1% ethane (Posewang and Mienert, 1999). Calculations account for the older and 45 m lower BSR by a 1.9°C decrease in BGHS temperature or by a gas composition change to 92% methane and 8% ethane, as might happen in fractionation.

In the Cascadia margin, second BSRs are verified through magnetic properties of the sediments at BGHS and below (Musgrave et al., 2006). Magnetic parameters over the interval 20–60 m below the current BGHS indicate greigite collection in hydrate accumulations during glacial times when the hydrate zone was deeper. Then, below the apparent glacial BGHS, the magnetic signal decreases abruptly to signify H_2S conversion of greigite to pyrite.

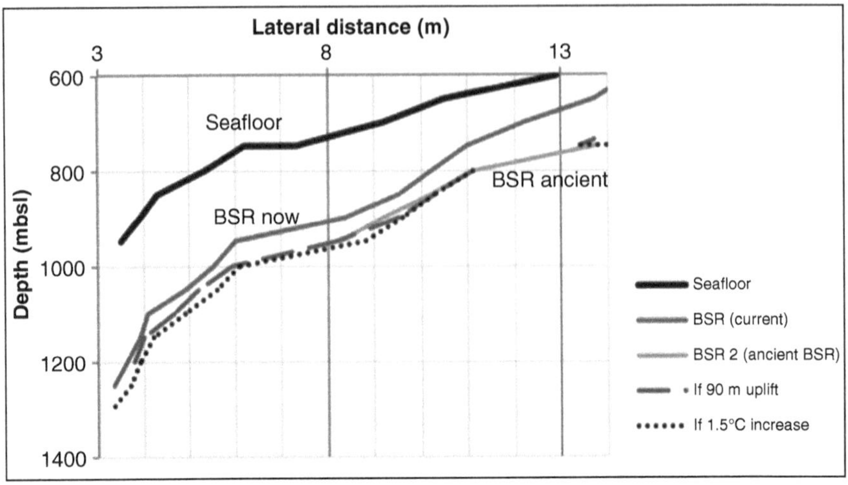

Figure 2.10 ***Double BSR on the Nankai Slope (Foucher et al., 2002).***

2.5.4.5 Determining Thermal Gradients from BSR

Fairly accurate temperature determinations from seismically established BSRs allow mapping thermal gradients in gas hydrate zones that have BSRs (Hyndman et al., 1992). These gradients can be determined using the following steps (Minshull, 2011):

1. Establish the BSR in relationship to the seafloor.
2. Convert seismic transit times to depth interval Δz.
3. Convert depth to pressure by using hydrostatic pressure gradient.
4. Estimate BSR temperature by referring calculated pressure to *P–T* phase equilibrium curve.
5. Calculate thermal gradient $\Delta T/\Delta z$ from sediment surface to BSR.

The thermal gradient calculations of Hyndman et al. (1992) or procedures outlined by Minshull (2011) are accurate within constraints of the following errors: (1) assigning correct sound velocities to interval sediments, (2) assuming linear temperature gradients from seafloor to BSR, (3) selecting correct phase equilibrium for hydrates at BSR, and (4) assigning correct seafloor temperatures.

Establishing an appropriate phase equilibrium curve evokes the largest error because of gas composition and pore water salinity uncertainties at the BSR. To counter the problem, several approximations may be chosen. A composition/salinity combination may be assumed to be either methane/pure water (Hyndman et al., 1992) or methane/seawater (Grevemeyer and Villinger, 2001), where the BSR temperature calculated by the two methods may differ by about 1°C. In each case, hydrostatic pressure accurately establishes pressure at the BSR.

Known downhole temperatures at the BSR for 10 drilling sites were superposed on the values calculated from equilibrium curves (Grevemeyer and Villinger, 2001; Brown and Bangs, 1995; Dickens and Quinby-Hunt, 1994). Relative good agreements between calculated temperatures and downhole measured temperatures were found. On converting temperatures and depths to thermal gradients and comparing, errors ranged from 5 to 25% for both methane/pure water and methane/seawater assumptions, although the seawater assumption seemed slightly superior.

Therefore, errors may exist in temperatures estimated for the seafloor–water column interface, but errors may also be present in estimated gas compositions (Grevemeyer and Villinger, 2001; Minshull, 2011).

While drilling in the Nankai Trough, temperature measurements of Figure 2.11 were made. The temperature calculated for the seismically indicated BSR occurring at 205 mbsf matches well with those accurate values obtained while drilling (Hyndman et al., 1992). BSR temperature was determined by Hyndman et al. (1992) to be 26.2 ± 0.7°C at the subject Nankai margin location.

2.5.4.6 Heat Flux Determined from BSR

To produce methane commercially from seafloor hydrates, the simplest technique reduces pressure on the hydrates to below the equilibrium value and then allows natural, upward heat flux to supply the necessary latent heat to isothermally decompose the hydrates. Usually, the natural heat flux will be insufficient to maintain constant temperature as decomposition progresses, but increasing the magnitudes of makeup energy significantly lowers process profitability.

The heat flux by conduction is given by Equation 2.7:

$$q = k\frac{dT}{dz} \tag{2.7}$$

where q, heat flux; k, thermal conductivity; dT/dz, geothermal gradient with depth.

If a linear temperature gradient between the seafloor and BSR is assumed and that gradient is constrained by conductive heat transfer, then Equation 2.8 is applicable (Shankar et al., 2011):

$$q = 1000k\left(\frac{T_{\mathrm{BSR}} - T_{\mathrm{sf}}}{\Delta z}\right) \tag{2.8}$$

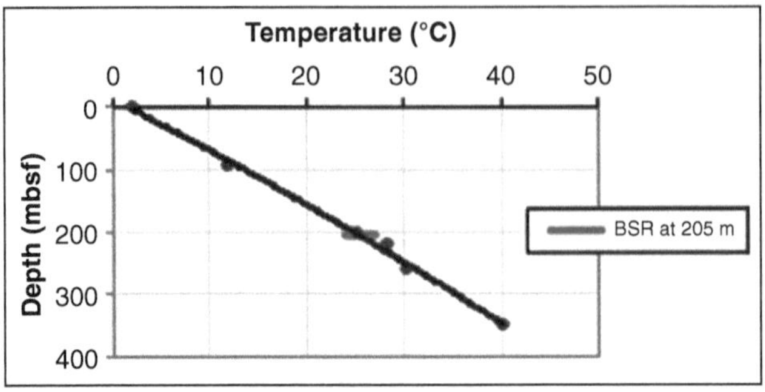

Figure 2.11 ***Thermal gradient, ODP Site 808, Nankai margin (Hyndman et al., 1992).***

where q, heat flux (mW/m^2); k, thermal conductivity [W/(m K)]; T_{BSR}, temperature at BSR; T_{sf}, temperature at seafloor; Δz, depth from seafloor to BSR (m).

Temperatures and thermal conductivities may be measured *in situ*, or thermal conductivity may also be measured in the laboratory using seafloor materials from the subject interval (Grevemeyer and Villinger, 2001).

Shankar et al. (2011) calculated for the Krishna–Godavari Basin, India, a heat flux that ranged from 23 mW/m^2 for the water depths less than 1000 m in the northeast of the basin to 62 mW/m^2 in the deeper regions of the basin – the differences attributed to topography.

Distinct BSRs were found in three gas hydrate zones of Peru margin, Nankai margin, and Blake-Bahama Outer Ridge (Hyndman et al., 1992). Common temperatures at the BGHS, as determined from the respective BSRs, were 25–27°C. These temperatures verify that equilibrium values measured in the laboratory for a hydrate phase formed from methane–pure water.

Calculated heat fluxes from the same three gas hydrate zones are given in Table 2.11.

Table 2.11 Calculated heat flux of representative hydrate zones

Location	**Heat flux (mW/m^2)**	**References**
Peru Trench, DSDP Site 688	46	Hyndman et al. (1992)
Krishna–Godavari Basin, India	23–62	Shankar et al. (2011)
Blake-Bahama, Outer Ridge, DSDP Sites 102, 104, 533	54	Hyndman et al. (1992)
Nankai margin, Site 808	129	Hyndman et al. (1992)

REFERENCES

Abegg, F., Bohrmann, G., Kuhs, W., 2006. Data report: shapes and structures of gas hydrates imaged by computed tomographic analyses, ODP Leg 204, Hydrate Ridge. In: Trehu, A.M., Bohrmann, G., Torres, M.E., Colwell, F.S. (Eds.), Proceedings of the Ocean Drilling Program, Scientific Results, vol. 204.

Abegg, F., Bohrmann, G., Freitag, J., Kuhs, W., 2007. Fabric of gas hydrate in sediments from Hydrate Ridge – results from ODP Leg 204 samples. Geo-Mar. Lett. 27, 269–277.

Allen, C.C., Oehler, D.Z., Etiope, G., Van Rensbergen, P., Baciu, C., Feyzullayev, A., Martinelli, G., Tanaka, K., Van Rooij, D., 2013. Fluid expulsion in terrestrial sedimentary basins: a process providing potential analogs for giant polygons and mounds in the Martian lowlands. Icarus 224 (2), 424–432.

Amery, G.B., 1969. Structure of Sigsbee Scarp, Gulf of Mexico. Am. Assoc. Pet. Geol. Bull. 58 (12), 2480–2482.

Andreassen, K., Nilssen, E.G., Odegaard, C.M., 2007. Analysis of shallow gas and fluid migration within the Plio-Pleistocene sedimentary succession of the SW Barents Sea continental margin using 3D seismic data. Geo-Mar. Lett. 27, 155–171.

Bahk, J.-J., Um, I.-K., Holland, M., 2011a. Core lithologies and their constraints on gas-hydrate occurrence in the East Sea, offshore Korea: results from the site UBGH1-9. Mar. Pet. Geol. 28, 1943–1952.

Bahk, J.-J., Kim, D.-H., Chun, J.-H., Son, B.K., Kim, J.-H., Ryu, B.-J., Torres, M., Schultheiss, P., Collett, T., Riedel, M., Lee, S.R., 2011b. Gas hydrate occurrences and their relation to hosting sediment properties: results from UBG2, East Sea. In: Proceedings of the 7th International Conference on Gas Hydrates (ICGH 2011), Edinburgh, Scotland, UK, July 17–21. Paper 567.

Bernard, B.B., Brooks, J.M., Sackett, W.M., 1976. Natural gas seepage in the Gulf of Mexico. Earth Planetary Sci. Lett. 31 (1), 48–54.

Bernard, B.B., Brooks, J.M. Sackett, W.M. 1978. Light hydrocarbons in recent Texas continental shelf and slope sediments. J. Geophys. Res. 83, 4053–4061.

Boswell, R., Collett, T., 2006. The gas hydrates resource pyramid. In: Fire in the Ice, DOE NETL Newsletter. Fall Issue, pp. 5–7.

Boswell, R., Collett, T.S., Frye, M., Shedd, W., McConnell, D.R., Shelander, D., 2012. Subsurface gas hydrates in the northern Gulf of Mexico. Mar. Pet. Geol. 34, 4–30.

Boudreau, B.P., Algar, C., Johnson, B.D., Croudace, I., Reed, A., Furukawa, Y., Kelley, M.D., Jumars, P.A., Grader, A.S., Gardiner, B.S., 2005. Bubble growth and rise in soft sediments. Geology 33 (6), 517–520.

Bouriak, S., Vanneste, M., Saoutkine, A., 2000. Inferred gas hydrates and clay diapirs near the Storegga Slide on the southern edge of the Voring Plateau, offshore Norway. Mar. Geol. 163, 125–148.

Brooks, J.M., Kennicutt, II, M.C., Fay, R.R., McDonald, T.J., 1984. Thermogenic gas hydrates in the Gulf of Mexico. Science 225 (460), 409–411.

Brown, K.M., Bangs, N.L., 1995. Thermal regime of the Chile triple junction: constraints provided by downhole temperature measurements and distribution of gas hydrate. In: Proceedings of the ODP, Initial Reports, vol. 141, pp. 259–275.

Burdige, D.J., 2006. Geochemistry of Marine Sediments. Princeton University Press, Princeton, Oxford.

Burst, J.F., 1965. Subaqueously formed shrinkage cracks in clay. J. Sediment. Petrol. 35, 348–353.

Camerlenghi, A., Accetella, D., Costa, S., Lastras, G., Acosta, J., Canals, M., Wardell, N., 2009. Morphogenesis of the SW Balearic continental slope and adjacent abyssal plain, Western Mediterranean Sea. Int. J. Earth Sci. (Geol. Rundsch.) 98, 735–750.

Cartwright, J.A., 2011. Diagenetically induced shear failure of fine-grained sediments and the development of polygonal fault systems. Mar. Pet. Geol. 28, 1593–1610.
Cho, J.H., Jeong, K.S., Kim, K.H., Obzhirov, A., Jin, Y.K., Shoji, H., 2005. Migration patterns of sulfate reduction–methane oxidation interface in hydrate-bearing and free-gas saturated sediments in the Sea of Okhotsk. In: Proceedings of the Fifth International Conference on Gas Hydrates, Trondheim, Norway.
Collett, T.S., Riedel, M., Cochran, J., Boswell, R., Presley, J., Kumar, P., Sathe, A., Sethi, A., Lall, M., Siball, V., 2008. The NGHP Expedition 01 Scientists. The National Gas Hydrate Program Expedition 01 Reports.
Davie, M.K., Buffett, B.A., 2002. A comparison of methane sources using numerical models for hydrate formation. In: Fourth International Conference on Gas Hydrates, Yokohama, pp. 25–30.
Dewhurst, D.N., Cartwright, J.A., Lonergan, L., 1999. The development of polygonal fault systems by syneresis of colloidal sediments. Mar. Pet. Geol. 16 (8), 793–810.
Dickens, G.R., Quinby-Hunt, M.S., 1994. Methane hydrate stability in seawater. Geophys. Res. Lett. 21, 2115–2118.
Duxbury, N.S., Abyzov, S.S., Romanovskh, V.E., Yoshikawa, K., 2004. A combination of radar and thermal approaches to search for methane clathrate in the Martian subsurface. Planet. Space Sci. 52, 109–115.
Epp, D., Grim, P.J., Langseth, Jr., M.G., 1970. Heat flow in the Caribbean and Gulf of Mexico. J. Geophys. Res. 75 (29), 5655–5668.
Foucher, J.-P., Nouzé, H., Henry, P., 2002. Observation and tentative interpretation of a double BSR on the Nankai slope. Mar. Geol. 189, 161–175.
Gay, A., Lopez, M., Cochonat, P., Levache, D., Sermondadaz, G., Seranne, M., 2006. Evidences of early to late fluid migration from an upper Miocene turbiditic channel revealed by 3D seismic coupled to geochemical sampling within seafloor pockmarks, Lower Congo Basin. Mar. Pet. Geol. 23 (3), 387–399.
Ghosh, R., Sain, K., Ojha, M., 2010. Estimating the amount of gas-hydrate using effective medium theory: a case study in the Blake Ridge. Mar. Geophys. Res. 31, 29–37.
Gregory, A.R., 1977. Fluid saturation effects on dynamic properties of sedimentary rocks. Geophysics 41, 895–921.
Grevemeyer, I., Villinger, H., 2001. Gas hydrate stability and the assessment of heat flow through continental margins. Geophys. J. Int. 145 (3), 647–660.
Gupta, A., Kneafsey, T.J., Moridis, G.J., Seol, Y., Kowalsky, M.B., Sloan, E.D., 2006. Composite thermal conductivity in a large heterogeneous porous methane hydrate sample. J. Phys. Chem. B 110 (33), 16384–16392.
Handa, Y.P., Cook, J.G., 1987. Thermal conductivity of xenon hydrate. J. Phys. Chem. 91 (25), 6327–6328.
Helgerud, M., Dvorkin, J., Nur, A., Sakai, A., Collett, T., 1999. Elastic wave velocity in marine sediments with gas hydrates: effective medium modeling. Geophys. Res. Lett. 26, 2021–2024.
Henninges, J., Huenges, E., Burkhardt, H., 2005. In-situ thermal conductivity of gas-hydrate-bearing sediments of the JAPEX/JNOC/GSC, et al. Mallik 5L-38 Well. In: Proceedings of the Fifth International Conference on Gas Hydrates 3, Trondheim, Norway, June 13–16. Paper 3034.
Holland, M., Schultheiss, P., Roberts, J., Druce, M., 2008. Observed gas hydrate morphologies in marine sediments. In: Proceedings of the 6th International Conference on Gas Hydrates, Vancouver, British Columbia, Canada, July 6–10.
Holtzman, R., Juanes, R., 2011. Mechanisms of gas transport in sediments: crossover from fingering to fracturing. In: Proceedings of the 7th International Conference on Gas Hydrates (ICGH 2011), Edinburgh, Scotland, UK, July 17–21. Paper 303.

Hovland, M., 2002. On the self-sealing nature of marine seeps. Continental Shelf Res. 22, 2387–2394.
Huang, D., Fan, S., 2004. Thermal conductivity of methane hydrate formed from sodium dodecyl sulfate solution. J. Chem. Eng. Data 49 (5), 1479–1482.
Hutchinson, D.R., Shelander, D., Dai, J., McConnell, D., Shedd, W., Frye, M., Ruppel, C., Boswell, R., Jones, E., Collett, T., Rose, K., Dugan, B., Wood, W., Latham, T., 2008. Site selection for DOE/JIP gas hydrate drilling in the northern Gulf of Mexico. In: Proceedings of the 6th Conference on Gas Hydrates, Vancouver, British Columbia, Canada, July 6–10. Paper 5506.
Hyndman, R.D., Foucher, J.P., Yamano, M., Fisher, A., Scientific Team of Ocean Drilling Program Leg 131, 1992. Deep-sea bottom-simulating-reflectors-calibration of the base of the hydrate stability field as used for heat-flow estimates. Earth Planetary Sci. Lett. 109, 289–301.
Inoue, R., Tanaka, H., Nakanishi, K., 1996. Molecular dynamics simulation study of the anomalous thermal conductivity of clathrate hydrates. J. Chem. Phys. 104 (23), 9569–9577.
Jain, A.K., Juanes, R., 2009. Preferential mode of gas invasion in sediments: grain-scale mechanistic model of coupled multiphase fluid flow and sediment mechanics. J. Geophys. Res. 114, B08101, 19 pp.
Jakobsen, M., Hudson, J.A., Minshull, T.A., Singh, S.C., 2000. Elastic properties of hydrate-bearing sediments using effective medium theory. J. Geophys. Res. 105, 561–577.
Jiang, S.-Y., Yang, T., Ge, L., Yang, J.-H., Ling, H.-F., Wu, N.-Y., Liu, J., Chen, D.-H., 2008. Geochemistry of pore waters from the Xisha Trough, northern South China Sea and their implications for gas hydrates. J. Oceanogr. 64, 459–470.
Johnson, B.D., Boudreau, B.P., Gardiner, B.S., Maass, R., 2002. Mechanical response of sediments to bubble growth. Mar. Geol. 187, 347–363.
Kastner, M., Kvenvolden, K.A., Lorenson, T.D., 1998. Chemistry, isotopic composition, and origin of a methane-hydrogen sulfide hydrate at the Cascadia subduction zone. Earth Planetary Sci. Lett. 156, 173–183.
Klapp, S.A., Enzmann, F., Walz, P., Huthwelker, T., Tuckermann, J., Schwarz, J.-O, Pape, T., Peltzer, E.T., Hester, K.C., Zhang, X., Mokso, R., Wangner, D., Marone, F., Kersten, M., Bohrmann, G., Kuhs, W.F., Stampanoni, M., Brewer, P.G., 2011. Fluid flow in natural gas hydrates revealed by X-ray tomographic microscopy. In: Proceedings of the 7th International Conference on Gas Hydrates (ICGH 2011), Edinburgh, Scotland, UK, July 17–21. Paper 248.
Kundu, N., Pal, N., Sinha, N., Budhiraja, I.L., 2008. Do paleo hydrates play a major role in deepwater biogenic gas reservoirs in Krishna–Godavari Basin? In: Methane Hydrate Newsletter. Fall Issue. National Energy Technology Laboratory, pp. 13–15.
Kurihara, M., Ouchi, H., Sato, A., Yamamoto, K., Noguchi, S., Narita, H., Nagao, J., Masuda, Y., 2011a. Prediction of performances of methane hydrate production tests in the Eastern Nankai Trough. In: Proceedings of the 7th International Conference on Gas Hydrates (ICGH 2011), Edinburgh, Scotland, UK, July 17–21. Paper 450.
Kurihara, M., Ouchi, H., Narita, H., Masuda, Y., 2011b. Gas production from methane hydrate reservoirs. In: Proceedings of the 7th International Conference on Gas Hydrates (ICGH 2011), Edinburgh, Scotland, UK, July 17–21. Paper 792.
Lanoil, B.D., Sassen, R., La Duc, M.T., Sweet, S.T., Nealson, K.H., 2001. Bacteria and archaea physically associated with Gulf of Mexico gas hydrates. Appl. Environ. Microbiol. 67 (11), 5143–5153.
Lapham, L.L., Chanton, J.P., Martens, C.S., Sleeper, K., Woolsey, J.R. 2008. Microbial activity in surficial sediments overlying acoustic wipeout zones at a Gulf of Mexico cold seep. Geochem. Geophys. Geosyst. 9(6), doi: 10.1029/2008GC001944.
Larter, S., Aplin, A.C., Bowler, B., Lloyd, R., Zwach, C., Hansen, S., Telnaes, N., Sylta, O., Yardley, G., Childs, C., 2000. A drain in my graben: an integrated study of the Heimdal area petroleum system. J. Geochem. Exploration 69–70, 619–622.

Lee, M.W., Collett, T.S., 2011. Three types of gas hydrate reservoirs in the Gulf of Mexico identified in LWD data. In: Proceedings of the 7th International Conference on Gas Hydrates (ICGH 2011), Edinburgh, Scotland, UK, July 17–21. Paper 39.

Liu, X., Flemings, P.B., 2006. Passing gas through the hydrate stability zone at southern Hydrate Ridge, offshore Oregon. Earth Planetary Sci. Lett. 241, 211–226.

Lorenson, T.D. (Ed.), 2000. Microscopic character of marine sediments containing disseminated gas hydrate. Examples from the Blake Ridge and the Middle America Trench. Ann. N.Y. Acad. Sci. 912, 189–194.

Lu, H., Moudrakovski, I.L., Ripmeester, J.A., Ratcliffe, C.I., Matsumoto, R., Tani, A., 2011a. The characteristics of gas hydrate recovered from Joetsu Basin, eastern margin of the Sea of Japan. In: Proceedings of the 7th International Conference on Gas Hydrates (ICGH 2011), Edinburgh, Scotland, UK, July 17–21. Paper 355.

Lu, H., Kawasaki, T., Ukita, T., Moudrakovski, I., Fujii, T., Noguchi, S., Shimada, T., Nakamizu, M., Ripmeester, J., Ratcliffe, C., 2011b. Particle size effect on the saturation of methane hydrate in sediments – constrained from experimental results. Mar. Pet. Geol. 28, 1801–1805.

McGee, T., Lutken, C., Rogers, R., Brunner, C., Dearman, J., Lynch, F.L., Woolsey, J.R., 2009. Can fractures in soft sediments host significant quantities of gas hydrates? In: Collett, T., Johnson, A., Knapp, C., Boswell, R. (Eds.), Natural Gas Hydrates – Energy Resource Potential and Associated Geologic Hazards: AAPG Memoir 89, pp. 1–11.

Milkov, A.V., Sassen, R., 2000. Thickness of the gas hydrate stability zone, Gulf of Mexico continental slope. Mar. Pet. Geol. 17, 981–991.

Minshull, T.A., 2011. Some comments on the estimation of geothermal gradients from the depths of bottom simulating reflectors. In: Proceedings of the 7th International Conference on Gas Hydrates (ICGH 2011), Edinburgh, Scotland, UK, July 17–21. Paper 731.

Moridis, G.J., Seol, Y., Kneafsey, T.J., 2005. Studies of reaction kinetics of methane hydrate dissociation in porous media. In: Proceedings of the 5th International Conference on Gas Hydrates, Trondheim, Norway, June 13–16, pp. 21–30. Paper 1004.

Musgrave, R.J., Bangs, N.L., Larrasoaña, J.C., Gracia, E., Hollamby, J.A., Vega, M.E., 2006. Rise of the base of the gas hydrate zone since the last glacial recorded by rock magnetism. Geology 34 (2), 117–120.

Nagihara, S., Schlater, J.G., Beckley, L.M., Behrens, E.W., Lawver, L.A., 1992. High heat flow anomalies over salt structures on the Texas continental slope, Gulf of Mexico. Geophys. Res. Lett. 19 (16), 1687–1690.

Obzhirov, A., Shakirov, R., Salyuk, A., Suess, E., Biebow, N., Salomatin, A., 2004. Relations between methane venting, geological structure and seismo-tectonics in the Okhotsk Sea. Geo-Mar. Lett. 24, 135–139.

Park, K.P., Bahk, J.J., Holland, M., Yun, T.S., Schultheiss, P.J., Santamarina, C., 2009. Improved pressure core analysis provides detailed look at Korean cores. In: Fire in the Ice, DOE NETL Newsletter. Winter Issue, pp. 18–20.

Parkes, R.J., Cragg, B.A., Bale, S.J., Getliff, J.M., Goodman, K., Rochelle, P.A., Fry, J.C., Weightman, A.J., Harvey, S.M., 1994. Deep bacterial biosphere in Pacific Ocean sediments. Nature 371, 410–413.

Paull, C.K., Ussler, III, W., Lorenson, T., Winters, W., Dougherty, J., 2005. Geochemical constraints on the distribution of gas hydrates in the Gulf of Mexico. Geo-Mar. Lett. 25, 273–280.

Perry, J.H. (Ed.), 1950. Chemical Engineers' Handbook, third ed. McGraw-Hill Book Company, Inc., New York, Toronto, London.

Pohlman, J.W., Bauer, J.E., Canuel, E.A., Grabowski, K.S., Knies, D.L., Mitchell, C.S., Whiticar, M.J., Coffin, R.B., 2009. Methane sources in gas hydrate-bearing cold seeps: evidence from radiocarbon and stable isotopes. Mar. Chem. 115, 102–109.

Posewang, J., Mienert, J., 1999. The enigma of double BSRs: indicators for changes in the hydrate stability field? Geo-Mar. Lett. 19, 157–163.
Radich, J.G., 2009. Laboratory and theoretical investigations of direct and indirect microbial influences on seafloor gas hydrates. A thesis in partial fulfillment of the requirements for the degree of Master of Science in Chemical Engineering in the Dave C. Swalm School of Chemical Engineering, Mississippi State University, Mississippi State, MS.
Rogers, R.E., Zhang, G., Kothapalli, C., French, W.T., 2004. Laboratory evidence of microbial-sediment–gas hydrate synergistic interactions in ocean sediments. ISOPE, Paper 2004-JSC-218. Presented at Toulon, France.
Rogers, R.E., 1994. Coalbed Methane: Principles and Practices. Prentice-Hall, Inc., Englewood Cliffs, NJ, pp. 301–302.
Rounick, J.S., Winterbourn, M.J., 1986. Stable carbon isotopes and carbon flow in ecosystems. BioScience 36 (3), 171–177.
Ruppel, C., Von Herzen, R.P., Bonneville, A., 1995. Heat flux through an old (≈175 Ma) passive margin, offshore southeastern U.S.A. J. Geophys. Res. 100, 20037–20057.
Ruppel, C., Boswell, R., Jones, E., 2008. Scientific results from Gulf of Mexico Gas Hydrates Joint Industry Project Leg 1 drilling: introduction and overview. Mar. Pet. Geol. 25, 819–829.
Rutqvist, J., Moridis, G.J., Grover, T., Collett, T., 2009. Geomechanical response of permafrost-associated hydrate deposits to depressurization-induced gas production. J. Pet. Sci. Eng. 67, 1–12.
Sassen, R., Joye, S., Sweet, S.T., DeFreitas, D.A., Milkov, A.V., MacDonald, I.R., 1999. Thermogenic gas hydrates and hydrocarbon gases in complex chemosynthetic communities, Gulf of Mexico continental slope. Org. Geochem. 30 (7), 485–497.
Sassen, R., Losh, S.L., Cathles, II, L., Roberts, H., Whelan, J.K., Milkov, A.V., Sweet, S.T., DeFreitas, D.A., 2001. Massive vein-filling gas hydrate: relation to ongoing gas migration from the deep subsurface of the Gulf of Mexico. Mar. Pet. Geol. 18, 551–560.
Sassen, R., Curiale, J.A., 2006. Microbial methane and ethane from gas hydrate nodules of the Makassar Strait, Indonesia. Org. Geochem. 37, 977–980.
Sassen, R., Roberts, H.H., Jung, W., Lutken, C.B., DeFreitas, D.A., Sweet, S.T., Guinasso Jr., N.L., 2006. The Mississippi Canyon 118 gas hydrate site: a complex natural system. In: 2006 Offshore Technology Conference, OTC 18132, 10 pp.
Selig, F., Wallick, G., 1966. Temperature distribution in salt domes and surrounding sediments. Geophysics XXXI (2), 346–361.
Shakirov, R., Obzhirov, A., 2011. Western Pacific gas hydrate belt. In: Proceedings of the 7th International Conference on Gas Hydrates (ICGH 2011), Edinburgh, Scotland, UK, July 17–21. Paper 781.
Shankar, U., Sain, K., Riedel, M., 2011. Assessment of heat flow and gas-hydrate in Krishna Godavari Basin, constrained from 2D/3D seismic and well log data. In: Proceedings of the 7th International Conference on Gas Hydrates (ICGH 2011), Edinburgh, Scotland, UK, July 17–21. Paper 451.
Shedd, W., Boswell, R., Frye, M., Godfriaux, P., Kramer, K., 2012. Occurrence and nature of "bottom simulating reflectors" in the northern Gulf of Mexico. J. Mar. Pet. Geol. 34, 31–41.
Shelander, D., Dai, J., Bunge, G., 2010. Predicting saturation of gas hydrates using pre-stack seismic data, Gulf of Mexico. Mar. Geophys. Res. 31, 39–57.
Shelander, D., Dai, J., Bunge, G., 2011. Predictions of gas hydrates using pre-stack seismic data, deepwater Gulf of Mexico. In: Proceedings of the 7th International Conference on Gas Hydrates (ICGH 2011), Edinburgh, Scotland, UK, July 17–21. Paper 131.
Shiguo, W., Xiujuan, W., Duanxin, C., Qiliang, S., 2011. Polygonal fault, pockmark and gas hydrate occurrence in the northern slope of the northern South China Sea. In: Proceedings of the 7th International Conference on Gas Hydrates (ICGH 2011), Edinburgh, Scotland, UK, July 17–21. Paper 137.

Stetter, K.O., Fiala, G., Huber, G., Huber, R., Segerer, A., 1990. Hyperthermophilic microorganisms. FEMS Microbiol. Rev. 75, 117–124.
Stoll, R.D., Bryan, G.M., 1979. Physical properties of sediments containing gas hydrates. J. Geophys. Res. 84, 1629–1634.
Suess, E., Bohrmann, G., Greinert, J., Lausch, E., 1999. Flammable ice. Sci. Am. 281 (5), 77–83.
Suess, E., Bohrmann, G., Rickert, D., Kuhs, W.F., Torres, M.E., Trehu, A., Linke, P., 2002. Properties and fabric of near-surface methane hydrates at Hydrate Ridge, Cascadia margin. In: Proceedings of the Fourth International Conference on Gas Hydrates, Yokohama, May 19–23, pp. 740–744.
Suzuki, K., Narita, H., 2011. Tensile strengths of marine sediments sampled from methane hydrate bearing zone near the northeast Nankai Trough. In: Proceedings of the 7th International Conference on Gas Hydrates (ICGH 2011), Edinburgh, Scotland, UK, July 17–21. Paper 586.
Taylor, M.H., Dillon, W.P., Pecher, I.A., 2000. Trapping and migration of methane associated with the gas hydrate stability zone at the Blake Ridge Diapir: new insights from seismic data. Mar. Geol. 164, 79–89.
Torres, M.E., Wallmann, K., Trehu, A.M., Bohrmann, G., Borowski, W.S., Tomaru, H., 2004. Gas hydrate growth, methane transport, and chloride enrichment at the southern summit of Hydrate Ridge, Cascadia margin off Oregon. Earth Planetary Sci. Lett. 226, 225–241.
Trehu, A.M., Torres, M.E., Moore, G.F., Suess, E., Bohrmann, G., 1999. Temporal and spatial evolution of a gas hydrate-bearing accretionary ridge on the Oregon continental margin. Geology 27 (10), 939–942.
Trehu, A.M., Bohrmann, G., Rack, F.R., Torres, M. E., 2003. Proceedings of the ODP, Initial Reports, vol. 204.
Tse, J.S., White, M.A., 1988. Origin of glassy crystalline behavior in the thermal properties of clathrate hydrates: a thermal conductivity study of tetrahydrofuran hydrate. J. Phys. Chem. 92 (17), 5006–5011.
Ukita, T., Noguchi, S., Shimada, T., Lu, H., Moudrakovski, I., Ripmeester, J., Ratcliffe, C., 2011. Experimental study on the possible factors that affect the saturation of gas hydrate in natural sediments. In: Proceedings of the 7th International Conference on Gas Hydrates (ICGH 2011), Edinburgh, Scotland, UK, July 17–21. Paper 329.
Waite, W.F., deMartin, B.J., Kirby, S.H., Pinkston, J., Ruppel, C.D., 2002a. Thermal conductivity measurements in porous mixtures of methane hydrate and quartz sand. Geophys. Res. Lett. 29 (24), 82-1–82-4.
Waite, W.F., Pinkston, J., Kirby, S.H., 2002b. Preliminary laboratory thermal conductivity measurements in pure methane hydrate and methane hydrate–sediment mixtures: a progress report. In: Fourth International Conference on Gas Hydrates, Yokohama, Japan.
Waite, W.F., Stern, L.A., Kirby, S.H., Winters, W.J., Mason, D.H., 2007. Simultaneous determination of thermal conductivity, thermal diffusivity and specific heat in sI methane hydrate. Geophys. J. Int. 169, 767–774.
Warzinski, R.P., Gamwo, I.K., Rosenbaum, E.J., Myshakin, E.M., Jiang, H., Jordan, K.D., English, N.J., Shaw, D.W., 2008. Thermal properties of methane hydrate by experiment and modeling and impacts upon technology. In: Proceedings of the 6th International Conference on Gas Hydrates (ICGH 2008), Vancouver, British Colombia, Canada, July 6–10.
Waseda, A., Uchida, T., 2002. Origin of methane in natural gas hydrates from the Mackenzie Delta and Nankai Trough. In: Proceedings of the 4th International Conference on Gas Hydrates, Yokohama, May 19–23.
Wellsbury, P., Goodman, K., Barth, T., Cragg, B.A., Barnes, S.P., Parkes, R.J., 1997. Deep marine biosphere fueled by increasing organic matter availability during burial and heating. Nature 388, 573–576.

Westbrook, G.K., Chand, S., Rossi, G., Long, C., Bunz, S., Camerlenghi, A., Carcione, J.M., Dean, S., Foucher, J.P., Flueh, E., Gei, D., Haacke, R.R., Madrussani, G., Mienert, J., Minshull, T.A., Nouze, H., Peacock, S., Reston, T.J., Vanneste, M., Zillmer, M., 2008. Estimation of gas hydrate concentration from multi-component seismic data at sites on the continental margins of NW Svalbard and the Storegga region of Norway. Mar. Pet. Geol. 25 (8), 744–758.

Whiticar, M.J., 1999. Carbon and hydrogen isotope systematics of bacterial formation and oxidation of methane. Chem. Geol. 161, 291–314.

Winters, W.J., Pecher, I.A., Waite, W.F., Mason, D.H., 2004. Physical properties and rock physics models of sediment containing natural and laboratory-formed methane gas hydrate. Am. Mineral. 89, 1221–1227.

Winters, W.J., 2011. Physical and geotechnical properties of gas-hydrate-bearing sediment from offshore India and the Northern Cascadia margin compared to other hydrate reservoirs. In: Proceedings of the 7th International Conference on Gas Hydrates (ICGH 2011), Edinburgh, Scotland, UK. Paper 408.

Winters, W., Walker, M., Hunter, R., Collett, T., Boswell, R., Rose, K., Waite, W., Torres, M., Patil, S., Dandekar, A., 2011. Physical properties of sediment from the Mount Elbert Gas Hydrate Stratigraphic Test Well, Alaska North Slope. J. Mar. Pet. Geol. 28, 361–380.

Wood, W.T., Gettrust, J.F., Chapman, N.R., Spence, G.D., Hyndman, R.D., 2002. Decreased stability of methane hydrates in marine sediments owing to phase-boundary roughness. Nature 420, 656–660.

Yamano, M., Uyeda, S., Aoki, Y., Shipley, T.H., 1982. Estimates of heat flow derived from gas hydrates. Geology 10, 339–343.

Zhang, Z., McConnell, D., 2011. Detailed seismic amplitude analysis and characterization of a gas hydrate and free gas mixed system in Green Canyon 955, Gulf of Mexico. In: Proceedings of the 7th International Conference on Gas Hydrates (ICGH 2011), Edinburgh, Scotland, UK, July 17–21. Paper 683.

CHAPTER THREE

Gulf of Mexico, Thermobiogenic Hydrates

Hydrates occlude biogenic and thermogenic gas mixtures in the Gulf of Mexico (GOM) gas hydrate province as a consequence of a series of interesting geologic and contemporary events. River systems draining much of the North American continent over geologic time deposited organic matter on a thick base of Louann salt. Multiple river reroutings left deep canyons fingering seaward over the continental shelf and the continental slope. The prodigious and rapid sediment deposits initiated underlying salt movement to create mini-basins, salt domes, and structural folding within deepwater hydrate zones, which in turn facilitates hydrocarbon flows, gas trapping, and gas hydrate generation – that is, creation of hydrate reservoirs. The extensive faulting and fracturing above and below bottom gas hydrate stability (BGHS) provide conduits for oil and gas movement manifested at the seafloor as outcroppings of hydrate mounds, massive carbonate deposits, hydrocarbon seeps, and chemosynthetic communities. Within the many acoustic wipeout zones, gas hydrates of all morphologies are widespread. Among significant outcomes of these features is the modulation of hydrocarbon gas escape into the overlying water column.

Early gas hydrate studies focused on near surface, so GOM data accrued for hydrate formation, fluid flow, chemosynthetic occurrences, and anaerobic oxidation of methane in fine sediments within about 6 m of the seafloor surface. But recent studies center on deep waters of the continental slope, where gas hydrate wells drilled in mini-basins and near the forward edge of the Sigsbee Escarpment contribute data for gas hydrate behavior in conventional reservoir settings – especially pertinent to eventual commercial hydrate-gas production.

3.1 GEOLOGIC ORIGINS SIGNIFICANT TO GAS HYDRATE ACCUMULATIONS

3.1.1 Outline of Geologic Events

The GOM originated as a rift in late Triassic to late Jurassic Periods (Kirkland and Gerhard, 1971; Roberts et al., 2010a). Isolated without outlets to the

Offshore Gas Hydrates http://dx.doi.org/10.1016/B978-0-12-802319-8.00003-6

oceans, evaporation in the northern shallow sea left massive deposits of Louann salt – on the order of 3000 m thick (Humphris, 1979).

Counterparts of the contemporary Mississippi River drainage system deposited massive sediments/organic residue into the basin. These masses increased as glaciation products were swept into the basin by the continental drainage system during Pleistocene times. The prodigious river-deposited sediments moved the shelf edge of the Gulf hundreds of kilometers seaward. Rapid and enormous sediment deposits steepened the shelf's slope, leading to failures at the leading edge (Roberts et al., 2010a; Hutchinson et al., 2008a,b).

Cumulative overburden weight of depositions forced movement and folding of the thick salt base, initiating faulting in the northern GOM that continues today as sediments deposit on the lower continental slope (Hutchinson et al., 2008a).

An array of canyons, some now partially obscured, developed serially to distribute the river and its products at changing locations across the northern Gulf. Mississippi Canyon accommodates current river depositions; older canyons feature Keathley Canyon (KC), Alaminos Canyon (AC), and Green Canyon (GC) – all of them are now Joint Industry Program (JIP) exploration targets for the initial production of gas from hydrates (Hutchinson et al., 2008b).

Therefore, although the GOM is a passive plate-tectonic margin (Peel et al., 1995), the voluminous sediments and organic materials deposited at high rates are underlain by salt domes and a shifting salt base that impose dynamic trapping and faulting networks. The complex geology creates conditions for gas hydrate accumulations in hundreds of meters thick sediments segmented by networks of faults, fractures, and gas chimneys. Not inconsequential, oil/gas cold seeps populate the seafloor where chemosynthetic communities thrive. The many flow conduits afford release of gas trapped below the BGHS, as well as communication with deeper hydrocarbon reservoirs that may be associated with conventional production.

3.1.2 Salt Withdrawal Mini-Basins

As mentioned in the preceding discussion, high sediment deposition rates in the GOM compressed a Louann salt base. Subsidence followed from displaced salt dissolution, leaving concavities eventually filled with sediments – the metamorphosis of a mini-basin. Attendant peripheral fracturing and faulting provided gas access as an integral step in gas hydrate accumulations.

Table 3.1 Mini-basins evaluated by JIP for hydrate gas production

Mini-basin	JIP drill site	Comments	References
Casey	KC-151	Drilled 2005–2006; 1320–1330 m water depth; fine-grain muds, silts, and clays; hydrates fill vertical fractures; faults and seafloor vents; sand interval 95–110 mbsf	Hutchinson et al. (2008c), Winters et al. (2008)
Terrebonne	WR-313	Target near BGHS; rapid sedimentation; coarse-grained sands; channel, levee, fans	Boswell et al. (2012a)
Diana	AC-21	Reservoir midway gas hydrate stability (GHS) zone. Areal extent of gas hydrates = 3×10^7 m^3	Boswell et al. (2012a)

Thus, numerous mini-basins influenced by salt tectonics dot the northern GOM seafloor landward of the Sigsbee Escarpment, the forward front of salt movement (Ruppel et al., 2008). For example, the Mississippi Canyon on the continental slope crosses a series of the mini-basins (Hart et al., 2008).

In their search for gas hydrate prospects, JIP targeted a few of these miniature basins because of their extraordinary organic matter deposits and their associated fans, channels, and levee systems. Three of the mini-basin hydrate sites chosen by JIP are listed in Table 3.1.

Orca, on the continental slope offshore Louisiana, is a salt withdrawal mini-basin not on the short list of JIP for initial hydrate exploration, although gas hydrates were retrieved in cores from Orca during Leg 96 of the Deep Sea Drilling Program. Covering the area around the basin's low-point center is a lake 220 m deep of hypersalinity that Pilcher and Blumstein (2007) determine to come from dissolution of 3.62×10^9 tons of base salt. The lake's oxygen-depleted waters helped preserve deposited organic matter, allowing Meckler et al. (2008) to distinguish terrestrial material as the dominant volume deposited by river flow during glacial melt. Such deposits logically represent other GOM intraslope mini-basins.

Features of the mini-basins relate to quality hydrate reservoirs. Shelander et al. (2011) further hypothesize on features developed during mini-basin formations: (1) deposits rich in sands and clays displaced base salt to form Tertiary Age mini-basins; (2) sand interbeddings of shales developed above thinned salt bases; (3) extruded salt formed surrounding ridges; (4) migrating sands formed channels between mini-basins; (5) leading salt edges collapsed from increasing overburdens, redirecting distribution of sediments.

3.2 SALT TECTONIC EFFECTS

3.2.1 Faulting

Fracturing networks are important to the accumulation, discovery, and production of oil and conventional gas in the GOM. A shifting salt base in the northern Gulf exerts domineering influences on gas hydrate accumulations. The salt deformation model of Humphris (1979) predicts salt-dome growth to be dependent on the weight of at least 1220–1525 m of incremental sediment height above the adjacent structural low in order to deform the salt base.

The Sigsbee Escarpment represents the current forward wall, the southern extremity, of the salt mass that has progressed seaward into the lower continental slope of the northwestern GOM (Humphris, 1979; Shelander et al., 2011).

Accommodating oil and gas migration from deep hydrocarbon reservoirs, fault networks are the basis of mixed thermogenic/biogenic gas hydrates so widespread in the GOM (Milkov and Sassen, 2002; Sassen et al., 2001). The networks are extensive above massive salt-forming sheets, diapirs, ridges, and basin edges (Milkov and Sassen, 2000). Based on different geometric traits, faults have been grouped into nine extension families, two contraction families, one strike-slip group, and six fault weld families – all of which have been observed in the northern GOM (Rowan et al., 1999).

3.2.2 Thermal Gradients

Additional heat flux enters hydrate zones when there are underlying salt masses and imparts variability to GOM thermal gradients. Consequently, higher thermal conductivities of the salt decrease hydrate zone depths (Milkov and Sassen, 2000).

Generally, thermal conductivity of salt is greater by a factor of approximately 6 over that of adjacent sediments (Epp et al., 1970). Nagihara et al. (1992) report typical thermal conductivities of salt to be greater by a factor of 4 than the adjacent GOM sediments, leading them to use salt and sediment conductivities of 6.0 and 1.5 W/(m K), respectively, in computer simulations.

Selig and Wallick (1966) calculated an approximate geothermal gradient in sediments above a hypothetical salt body versus a gradient through similar sediments in the absence of salt. With a thermal conductivity of the salt assigned a value five times greater than surrounding sediments, the geothermal gradient in the sediments above the salt dome in their model was larger by a factor of 2 than thermal gradients of similar sediments in the absence of salt.

A salt diapir lies several hundred meters below hydrate mounds protruding from the seafloor at Atwater Valley Blocks 13/14 in the Mississippi Canyon. Partly because of the higher thermal conductivity of the salt body (Wood et al., 2008) and partly because of higher salinity of surrounding waters (Kastner et al., 2008), an elevated BGHS exists above the salt diapir. Here, heat flow measurements by Coffin et al. (2008) show flux of 160 and 132 mW/m^2 at Mounds F and D, respectively. This compares with background values away from the mounds of 40–50 mW/m^2.

In another study of the Texas continental slope, Nagihara et al. (1992) made 74 measurements within ±10% accuracy of heat flows over 2 irregularly shaped salt domes, spacing samples 1–3 m apart. Nagihara et al. found that heat flow ranged from 25 to 40 mW/m^2 beyond salt influence, while ranging from 50 to 100 mW/m^2 directly over the salt. The latter measurements may vary due to thickness and distribution of the underlying salt.

Milkov and Sassen (2001) derived Equation 3.1 from heat flux and bottom-hole temperatures of oil and gas wells in the GOM. It is considered a general relationship for temperature gradients not in close proximity to salt deposits:

$$G = -9.6 \ln D_W + 88.4 \tag{3.1}$$

where G, geothermal gradient (°C/km); D_W, water depth (mbsf).

Milkov and Sassen (2001) arrived at the empirical Equation 3.2 using data of Wash et al. (1998) for water temperatures T_W at the sea floor:

$$T_W = 295.1 D_W^{-0.6} \tag{3.2}$$

A linear geothermal gradient to the sediment depth D_S is assumed in Equation 3.3:

$$G = \frac{T_S - T_W}{D_S} \tag{3.3}$$

Equations 3.1–3.3 are combined to estimate in Equation 3.4 the sediment temperature at desired depth:

$$T_S = (-9.6 \ln D_W + 88.4) D_S + 295.1 D_W^{-0.6} \tag{3.4}$$

where T_S, sediment temperature (°C); D_S, sediment depth (mbsf).

The Milkov and Sassen's Equation 3.4 may then be used to estimate BGHS on substituting the pressure gradient for sediment depth and using a known *T–P* equilibrium relationship for the site's gas composition. Thus, depths of hydrate zones are diminished by higher bottom water temperatures, high-salinity effects of salt domes, and higher heat flux above salt domes (Francisca et al., 2005). Estimates of BGHS may be optimistic when based on erroneous thermal properties near salt domes, leading to errors in quantifying GOM gas hydrates.

3.3 LOOP CURRENTS AND TEMPERATURES AT WATER COLUMN BOTTOM

In the GOM with its numerous seafloor hydrocarbon seeps, hydrates were first discovered, detailed, and studied in near-surface sediments (Dillon and Max, 2000).

How do bottom water temperature fluctuations affect near-surface hydrate stabilities? To what depth within sediments do bottom water temperature fluctuations influence formation and decomposition of hydrates?

Based on computer simulations, Reagan and Moridis (2008) suggest that a ±1–3 K change of bottom water temperatures could significantly affect gas hydrate formation and dissociation in near-surface sediments.

In a study to determine temperature variations on near-surface Bush Hill hydrates, Solomon et al. (2008) monitored bottom water temperatures over a period of 15 months. Temperatures ranged between 6.9 and 9.6°C – the mean being 7.9°C. They found slow, periodic thermal cycling of the waters during tidal changes. An independent study by MacDonald et al. (2005) verified the mean 7.9°C bottom water temperature as presented in Figure 3.1.

At Bush Hill water depths, bottom water temperature variations influence hydrate stability within 2 m of the subsurface. Note in Figure 3.1 that only the maximum temperature spikes of bottom waters of domineering influence are included. Hydrate temperatures lag bottom water temperature by 0.5 days, while sediment temperature lags bottom water temperature by 3 days. The low-amplitude, high-frequency temperature variations of bottom waters are dampened by the hydrate and to a greater extent by the sediments (Walker et al., 1993).

MacDonald et al. (2005) report the largest bottom water temperature fluctuation at the Bush Hill site to be within the range 6.64–9.73°C. During the time period of the maximum temperature change, a protruding

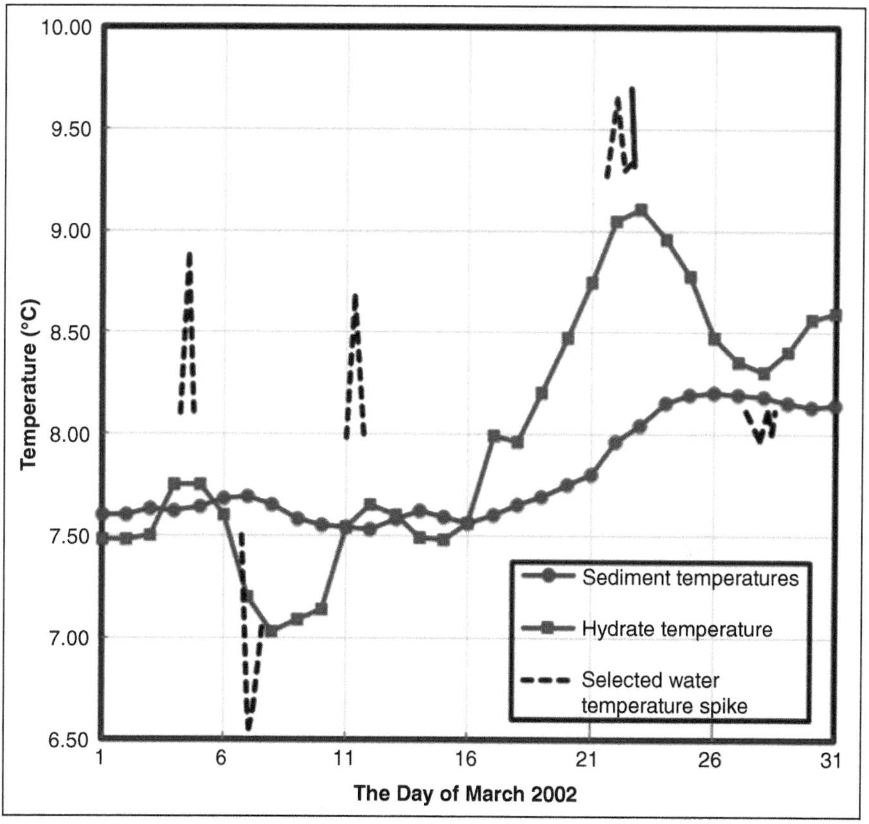

Figure 3.1 ***A month-long episode of hydrate and sediment temperatures influenced by bottom water temperatures at Bush Hill (MacDonald et al., 2005).***

hydrate mound was monitored; significant changes to the mound could not be seen over this short term. However, during a subsequent 5-year interval between submersible visits, hydrate outcrops on the seafloor at Bush Hill were noted to dissipate (MacDonald, 2009).

Mounds at Bush Hill contain type sII hydrates. Program CSMHYD (Sloan, 1998) analyses indicate an equilibrium sII temperature range of 14–16.5°C for prevalent seafloor pressures and gas compositions. The temperatures to maintain sII hydrate equilibrium reside safely above the locale's short-term peak temperature fluctuations outlined in Figure 3.1. However, sI hydrates under like conditions are only marginally stable. Methane hydrates, type sI of biogenic gas origin, exhibit equilibrium temperatures calculated to range from 8.2 to 7.5°C at Bush Hill conditions – values often exceeded during seafloor temperature fluctuations, thus dictating predominantly sII hydrates of this site's protruding mounds (Solomon et al., 2008).

Fluctuations in GOM bottom water temperatures are attributed primarily to warm eddies breaking from the Loop Current (MacDonald et al., 2005; Walker et al., 1993). A large rogue eddy might impose a temperature difference of 14°C with usual bottom water temperature and occur approximately once every 11 months. Satellite infrared sensors recorded a large eddy with dimensions 546 × 300 km separating from the Loop Current at a velocity of 1.25 km/h (Walker et al., 1993; MacDonald et al., 1994).

3.4 GULF OF MEXICO CONVENTIONAL DRILLING AND INFRASTRUCTURE

The GOM is a remarkable region to study gas hydrates. Oil and gas industry activity throughout the northern Gulf archives extensive seismic and logging data, geologic interpretations, block leasing information, and environmental monitoring. The most complete offshore oil and gas production infrastructure in the world exists there at water depths supporting gas hydrates. For example, in a recent 14-year span, about 0.36 tcm (12.7 tcf) of natural gas and 2.9 × 10^9 barrels of oil were produced from deep Gulf waters (Boswell et al., 2012b). Another 0.85 tcm (30 tcf) of conventional gas reserves is proven in the northern GOM, yet possibly 7.08 tcm (250 tcf) of conventional natural gas remains to be found (Savidge and Adewumi, 1995/1996). There is an unprecedented comprehensive accumulation of scientific data from this region that is important to the study of offshore hydrates and the eventual production of hydrate gas.

Figure 3.2 presents the number of conventional oil and gas wells drilled in GOM at water depths conducive to hydrate formation, the shallowest depth being 500 m. Thereafter, each data point refers to the number of wells drilled within a 500 m interval, using in the figure the median depth of each interval (Boswell et al., 2012b).

Near-surface hydrates in the northwestern GOM were first observed in shallow piston cores retrieved near oil seeps (Brooks et al., 1984). Hydrocarbons escaping deep reservoirs through networks of fractures, faults, gas chimneys, mud volcanoes, and other conduits form gas hydrates during transit through the hydrate zone. Fracture-filled hydrates and hydrate mounds populate the seafloor near hydrocarbon seeps.

Deep GOM gas hydrates were first sought in seven wells drilled by JIP in 2009. Drilling these deepwater wells revealed three characteristic hydrate

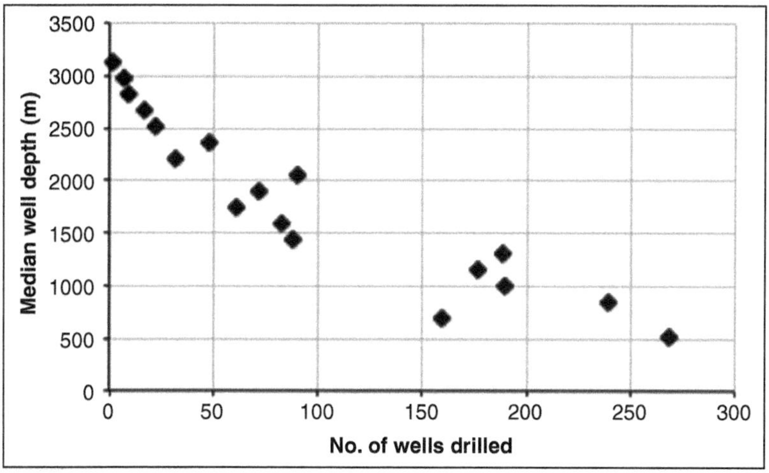

Figure 3.2 ***Conventional oil and gas wells drilled in GOM water depths supporting known gas hydrate accumulations.*** *(Data from Boswell et al., 2012b.)*

accumulations: (1) filling sand pores, (2) filling vertical shale fractures, and (3) layering between silt, shale, and water-saturated sediments (Lee and Collett, 2011).

About greater than 400 seafloor cold seeps in the northern GOM support chemosynthetic communities, tube worms, ice worms, bacterial mats, and gas hydrates (Sassen et al., 2001). At these cold seeps, hydrate mounds 1–3 m diameter (Sassen et al., 1999) and carbonate rubble (Lorensen et al., 2002) are products of a long-term ecological balance. Depths of the sulfate–methane interface (*SMI*) in the sediments depend on methane flux (Paull et al., 2005). All entities concentrate near current or past faults and flow conduits of hydrocarbon seeps.

3.5 ORIGINS OF GULF OF MEXICO HYDRATE GAS

The Bureau of Ocean Energy Management analyzed cold seeps in the northern GOM, finding vented gas composed equally of thermogenic and biogenic origins (Boswell et al., 2012b). The mixture derives from the following sources: (1) biogenic gas from rich carbon deposited by continental drain, (2) thermogenic gas from deeply buried organic matter, (3) thermogenic gas of deep reservoirs from migration, and (4) biogenic gas from action of microbes on reservoir gases in transit.

To determine gas sources in the near-surface sediments, Paull et al. (2005) analyzed cores from the northern GOM hydrate zone that included

Table 3.2 Molecular structure criterion for hydrocarbon gas source (Paull et al., 2005)

$C_1/(C_2+C_3)$	No. of samples	Percentage of samples	Source
>10,000	9	7.4	Heavily biogenic
1000–10,000	21	17.2	Moderately biogenic
<1000	92	75.4	Mixed to thermogenic

hydrate mounds, gas vents, and basinal areas. The $C_1/(C_2 + C_3)$ ratios of 122 core samples taken by Paull et al. are presented in Table 3.2.

Analyses of the ratios of methane to ethane plus propane gases in the cores reveal that only 7.4% of the samples, primarily those near gas vents, were heavily biogenic. Overall, the 122 core samples were dominated by mixtures of the biogenic and the thermogenic gases (Paull et al., 2005). In 43 of the near-surface cores, a $\delta^{13}C$ isotope analysis indicated gas source distributions as presented in Table 3.3 (Paull et al., 2005).

The data indicate that biogenic gases prevail in basinal areas and thermogenic gases prevail around gas vents and mounds, while mixtures of the two gas sources dominate overall (Paull et al., 2005).

3.6 BSRs IN GULF OF MEXICO

In most gas hydrate provinces, continuous bottom-simulating reflectors (BSRs) frequently define bottom of hydrate stability as they parallel the seafloor (Cooper and Hart, 2003). Blake Ridge is a positive example of continuous BSRs; GOM is an exception. In early gas hydrate explorations, a scarcity of continuous BSRs in the complex geology of the GOM made it more difficult to locate hydrate boundaries (Paull et al., 2005). Extensive faulting, fracturing, and salt-dome disruption help dissipate BSR markers of free gas accumulations, resulting in fewer classic continuous BSRs (Milkov and Sassen, 2000). Additionally, gas composition variations and uneven thermal gradients involving GOM salt masses disrupt BSR continuities and symmetries relative to the seafloor.

Table 3.3 Carbon-13 criterion for hydrocarbon gas source (Paull et al., 2005)

$\delta^{13}C$ (‰)	No. of samples	%	Source
−60 to −94.6	9	20.9	Biogenic
−50 to −60	33	76.7	Mixed
−45.7 to −50	1	2.4	Thermogenic

From 3D GOM seismic records, Shedd et al. (2009, 2012) identified 145 BSRs in the northern GOM, which led them to 3 BSR categorizations: (1) continuous, (2) discontinuous or segmented, and (3) pluming.

Shedd et al. found 24% of the BSRs to be continuous, usually occurring in relatively homogeneous fine-particle sediments of uniform gas distribution and diffusion.

The group also found 58.6% of the Gulf BSRs to be discontinuous segmented. As explanation of the discontinuity mechanism, a pattern of disrupting faults may develop (Cooper and Hart, 2003; Losh et al., 1999). The faults allow conduit fluid flow to divert into adjoining permeable sands bounded by shale, thus promoting isolated seismic high-reflectivity zones. With judicious judgment, these isolated segments of high reflectivity can be connected as missing pieces of the discontinuous BSR puzzle. Complete analysis leading to a reliable composite depends on supplementing with other data. Dai et al. (2008a) relied heavily on such data to pinpoint acoustic wipeout zones, gas chimneys, and gas plumes.

Finally, plume BSRs represent 17.2% of all BSRs found by Shedd et al. These seismic bright spots have a continuous profile but are incongruent with seafloor topography because they are caused by ascending fluid flows having localized higher temperatures and greater salinities (Shedd et al., 2012; Wood et al., 2002).

3.7 SATELLITE LOCATING GAS HYDRATES

One hundred and twenty-four oil slicks from natural cold seeps were documented by MacDonald et al. (1993) from a satellite survey of 15,000 km^2 of northern GOM sea surfaces above hydrate-forming water depths of 500–2000 m. Independently, images from a second satellite revealed 66 oil slicks in 8200 km^2 of sea surface. Some of the areas covered by the two surveys overlapped. Follow-ups by surface vessels allowed MacDonald et al. to verify that the slicks came from natural seepages. Based on slick sizes, they estimated the barrels of oil per unit area of sea surface as follows: (1) oil slicks originate at point sources; (2) seepage rates from the point sources range from 1.4 to 30 m^3/1000 km^2/day; (3) annual rate of oil seepage into this satellite-monitored area amounts to 20,000 m^3/year (120,000 bbls/year) – estimates not including natural gas leakage.

Additionally, gas emissions in approximately 400 plumes occurring naturally from GOM waters were estimated to be 0.5 Tg/year of methane (Leifer and MacDonald, 2003).

Satellite views guided the team to the recurring oil slicks at the GC-600 block. Site exploration of the seafloor with remotely operated vehicles revealed hydrate outcrops on the seafloor. MacDonald et al. (2000) found a direct correlation between surface oil slicks and gas hydrates in the associated seafloor.

3.8 NEAR-SURFACE GAS HYDRATES

3.8.1 Bush Hill; Green Canyon Block 185

Seafloor gas hydrate accumulations in the GOM, as well as chemosynthetic communities around surface hydrates and cold seeps, were first discovered at Bush Hill, near the boundary of Green Canyon lease Blocks 184/185. With the assistance of research submersibles, Bush Hill became the most extensively studied of seafloor hydrate sites (Brooks et al., 1984). Although the data cover the infancy of near-surface hydrate exploration in relatively shallow waters, Bush Hill has characteristics widely relevant to hydrate formation.

Bush Hill is a mud mound of area 101,300 m^2 (Neurauter and Bryant, 1990), residing above a fault connecting to a large hydrocarbon field 1–3 km below seafloor. Its thermogenic oil and gas source is nearby Jolliet Field (McConnell et al., 2012). Gases venting around the hydrate mound at Bush Hill have the signature composition of the Jolliet Field.

Bush Hill hydrates are limited in structures to sII because of bottom water temperature fluctuations between 5.4 and 12.5°C and shallow water depths. The site's overlying water column establishes a hydrostatic pressure only marginal for sI methane hydrate stability, but Milkov and Sassen (2000) show that sII hydrates containing propane and *i*-butane readily stabilize at the prevalent 540 m water depths.

Vent gas compositions of Bush Hill contain primarily methane as well as heavier alkanes including 4.5% ethane and 3.7% propane that help stabilize sII hydrates (Milkov and Sassen, 2001; Sassen et al., 1999).

3.8.2 Mississippi Canyon Block 118

Understandably, Mississippi Canyon is the dominant feature on the continental slope of northern GOM, accommodating one of the largest continental drainage basins in the world (Hart et al., 2008; Roberts et al., 2010a). Extending 150 km in length and 10–15 km in width, its entrance has 300–400 m high walls that decline as the canyon progresses seaward. Goodwin and Prior (1989) estimate that 600–700 m of sediments has been dumped into the Canyon since late Pleistocene.

The Gas Hydrate Observatory, approved by the US Department of the Interior for hydrate studies under Research Reserve status and directed by the Mississippi Mineral Resources Institute at the University of Mississippi, was established with early support from a consortium of universities, industry, and government scientists. Funded by Department of Energy, National Oceanic and Atmospheric Administration, and the Department of Interior, the Observatory has produced comprehensive data of this near-surface, hydrate-bearing site.

The academe, industry, and government consortia at the Observatory evaluate: (1) gas venting, fracturing, and compositions of water columns and seafloor as factors influencing hydrate formation/dissociation; (2) rate of seafloor change; (3) seafloor stability effects of hydrates; (4) oil and gas venting relevant to gas hydrate formation/dissociation and carbonate deposition; (5) synergy of hydrates, microbes, and sediment minerals; (6) microorganisms in the sediments; and (7) dynamic ecology of chemosynthetic communities, hydrocarbon flow rates, and shifting gas vents (Lutken et al., 2011).

A photograph of a massive hydrate/carbonate outcrop at the Observatory is presented in Figure 3.3.

As seen in Figure 3.4, the Observatory is situated in Mississippi Canyon Block 118. At inception, the Observatory was situated on an open lease in the northern GOM centered at Woolsey Mound.

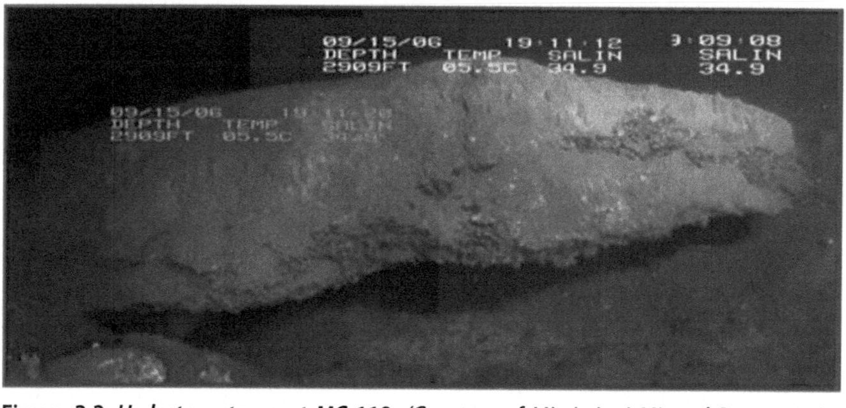

Figure 3.3 ***Hydrate outcrop at MC-118.*** *(Courtesy of Mississippi Mineral Resources Institute, University of Mississippi. MMRI overseer of Gulf Coast Gas Hydrate Consortium supported by U.S. Department of Interior, Department of Energy, and NOAA.)*

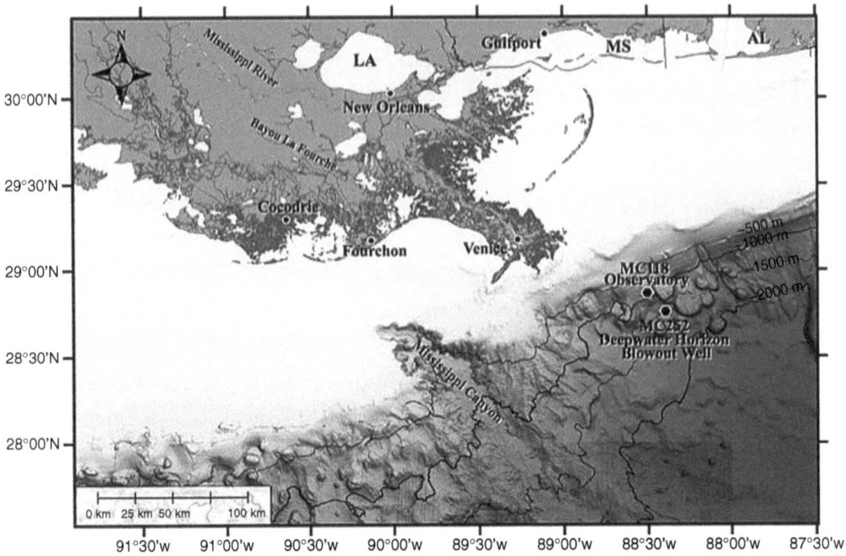

Figure 3.4 ***Map of the MC-118 site of Gas Hydrate Observatory.*** *(Drawing credit: Marco D'Emidio, Graphic Information Systems Research Analyst, MMRI.)*

At 890 m water depths, the Observatory's area of interest covers 1 km^2 of seafloor. Large craters from episodic gas expulsion (three major gas vents and crater complexes remain active) mark the seafloor. Outcrops of massive gas hydrates, hydrate-filled fractures and veins, natural gas vents, and carbonate rubble dot the area (Weitemeyer et al., 2009; Sassen et al., 2006). Chemosynthetic communities and bacterial mats are located at the site.

Proximity of the Observatory to the Deepwater Horizon blowout well, also in Mississippi Canyon, may be seen in Figure 3.4.

The Observatory is within an acoustic wipeout zone above a salt dome, typical of near-surface hydrate occurrences in the GOM. Fractures emanating from the salt diapir direct gas and crude oil through vents at the seafloor. Massive gas hydrates reside and even protrude at the surface. Hydrates fill inactive fractures in the fine-grained sediments (Lutken et al., 2011).

Microbial activity is prolific within the wipeout zone. Cores extracted near gas vents exhibit anaerobic methane oxidation sufficiently intensive to reduce SMI to 50 cmbsf, whereas SMI outside the wipeout zone extends to below coring depths (Lapham et al., 2008).

Carbonates litter the seafloor at the Observatory, verifying long-term and voluminous gas flux through shifting vent sites. Sassen et al. (2006)

estimate that 3 billion cubic meters (bcm) of CO_2 from anaerobic oxidation of methane (AOM) was necessary to form the carbonates observed at the site. Based on microbial activity and dormancy, laboratory studies of sediment samples taken at MC-118 substantiate site movement of the vents.

A comparison of vented gas and hydrate-occluded gas compositions at MC-118 demonstrates the degree of fractionation as a result of hydrate formation. The vent gas reported by Sassen et al. (2006) contains hydrocarbon gases heavier than methane, especially ethane/propane/*i*-butane of 3.1/1.4/0.3 proportions, respectively. Preferentially extracting ethane, propane, and *i*-butane, the hydrates become richer in those gases – decreasing occluded-gas composition to 70.0% methane from the rich 94.6% methane in vent gases (Sassen et al., 2006).

3.8.3 Mississippi Canyon Blocks 852/853

Mississippi Canyon Blocks 852/853 were initially explored for gas hydrates primarily with piston coring from surface research vessels. The northern GOM site is located in the middle of the continental slope near two producing oil and gas fields. Water column depth above the near-surface hydrates is 1060–1070 m, reaching a bottom water temperature of 4.8°C – conditions that allow both sI and sII hydrates to form. Milkov and Sassen (2000) report a hydrate mound 1.5 km approximate diameter that extends into the water column above a shallow salt protrusion disrupting strata below and influencing the site's 20°C/km thermal gradient (Milkov and Sassen, 2000).

3.8.4 Mississippi Canyon Block 252; Deepwater Horizon Well

The Deepwater Horizon Well, lease Block 252 of the Mississippi Canyon, is 10 miles (16.1 km) from the MC-118 Gas Hydrate Observatory. From the Observatory, instruments monitored drifting methane plumes 1.5 months after the blowout. Mass spectrometer data were taken of water column samples of plumes that drifted by the Observatory during uncontrolled flow from the well. Data analyses of three water column samples show methane-concentration spikes at 600 and 800 m water depths, indicating methane plumes near the limiting water depth for hydrate stability (Lutken et al., 2011).

Gas chromatograph analyses of samples from the same water depths substantiated the mass spectrometer results. A pronounced methane-concentration spike is observed at about 600 m water depth along with a much weaker concentration spike at the approximate 800 m depth, possibly signifying hydrate stability depths of sI and sII, respectively.

3.9 JOINT INDUSTRY PROGRAM

The JIP oversees a joint hydrate research program focusing on the GOM. Primarily sponsored by US Department of Energy, the program is directed by Chevron/Texaco with academia as secondary participants. The program seeks to further all aspects of offshore gas hydrates pertinent to eventual gas productivity. Especially important are the hydrate data gathered near BGHS, well logging applications to gas hydrates, and resource discovery and characterization with seismic techniques. Prospects of hydrate-gas production have improved by JIP drilling endeavors in deep waters of the northern GOM. JIP drilling locations in the GOM are shown in Figure 3.5 (Roberts et al., 2010b).

Hydrate wells drilled by JIP in these deep waters for the first time acquired data near BGHS showing high S_{gh} content in quality reservoir sands.

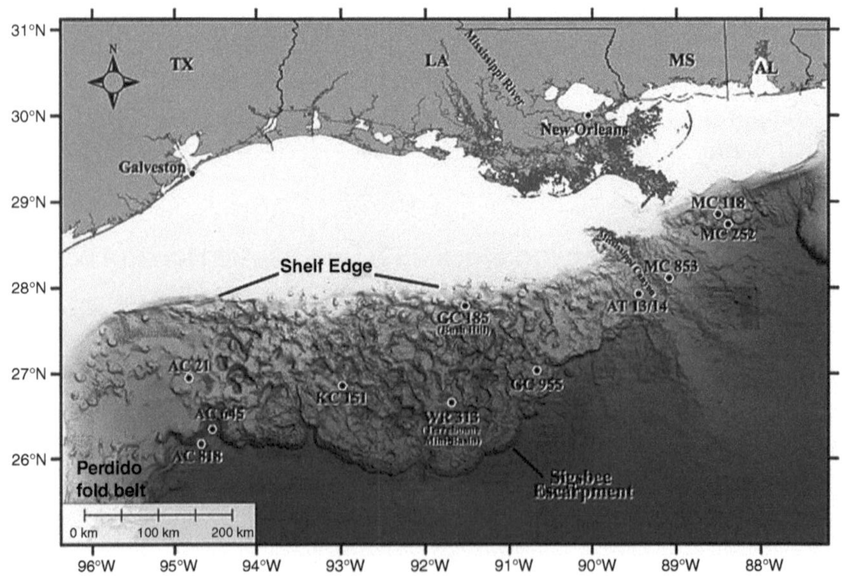

Figure 3.5 ***Location of JIP study area on mid-to-lower slope, northern GOM.*** Drawn by Marco D'Emidio (Roberts et al., 2010b).

The overall JIP test plan and general test results are summarized in Table 3.4 (Balczewski et al., 2011).

Table 3.4 JIP plan and test results (Balczewski et al., 2011)

Phase	Leg	Date	Location	Activity	Results
1 and 2	1	Early in program	Deep-water GOM	Advance seismic data Collect samples Well stability Develop data capability	*J. Mar. Pet. Tech.* publications (Ruppel et al., 2008) New tools, ship tested for on-site analyses Lab tests: physical, mechanical properties of fine-grained sediments
1 and 2	1	May 2005	AT-13/14 KC-151	Drill and log five wells Deep core expedition	Well logs, borehole seismic data of GHZ 200 m core; 60 m good S_H Safe drilling through GHZ
3	2	2006–2008	Deepwater GOM	Characterize coarse-grained sand reservoirs Seismic predictive capability	Published results of Leg 1; workshops to select Leg 2 and Leg 3 sites. Laboratory studies
3	2	April, May 2009	WR-313 GC-955 AC-21	No coring 21-Day drilling expedition 7 wells drilled Logging-while-drilling data Access industry 3D seismic data	Acquired 17,000 ft logging while drilling (LWD) data 7 wells drilled and logged High S_{hyd} in GC-955 and WR-313 sands. Verified predictive techniques. 3D data of hydrate sediments
3	3	Begin: August 2009 End:	GC-955 WR-313	Improved testing tools. Pressure coring, wireline logging, site characterization	1. New tools fabricated 2. TBD

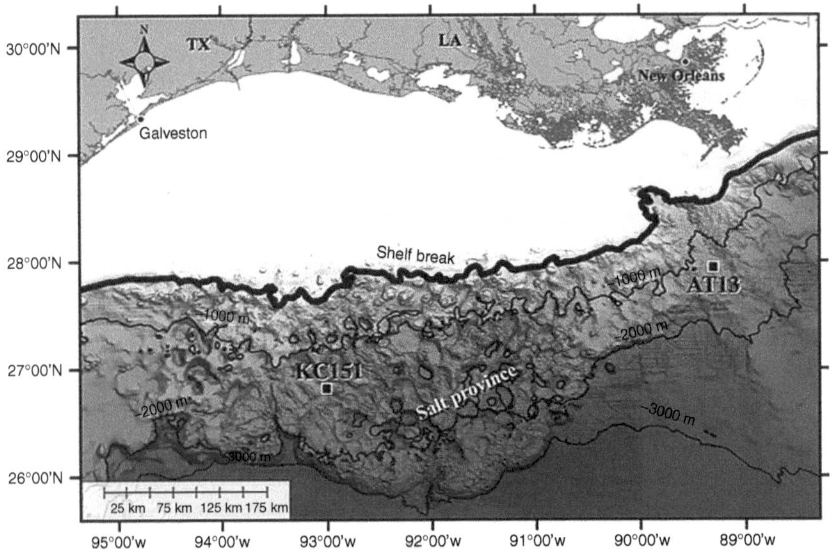

Figure 3.6 ***Location of KC-151 and AT-13 in JIP evaluation of prospects.*** Drawn by Marco D'Emidio (Lee et al., 2008).

3.9.1 Green Canyon, GC-955

Green Canyon Block 955 extends into the abyssal plain of the GOM, seaward of the Sigsbee Escarpment boundary and in front of the Sigsbee's northeastern to southwestern movement. See Figure 3.6. Its deep reservoir sands near BGHS within the hydrate stability zone overlie a Louann salt base (Hutchinson et al., 2008b).

The geologic establishment of a GC-955 hydrate-bearing system to a large extent typifies hydrate reservoir developments throughout the northern GOM. Since Pleistocene time, channel-levee-type sands began accumulating what now serves as quality hydrate reservoirs 480 mbsf (McConnell, 2000). To complete the GC-955 hydrate system development, reservoirs became linked to gas sources by faults emanating from salt diapirs below (McConnell et al., 2009; Boswell et al., 2012a).

Before JIP drilling, seismic data had revealed geologic features conducive to hydrate accumulations at GC-955. Salt uplifts near the Sigsbee Escarpment caused folding to disrupt strata at BGHS, generating widespread fault systems and numerous active gas chimneys that convey gas throughout the basin. The seismic data also revealed a mud volcano on the GC-955 seafloor, resulting in a 40 m diameter mound extending about 10 m above the seafloor (Heggland, 2004; McConnell and Zhang, 2011).

Table 3.5 Results from LWD data, GC-955H (McConnell and Zhang, 2011)

Depth interval (mbsf)	Log	Value	Results
410–450	Resistivity	1–200 Ω m	Concentrated hydrates, >70% saturation
410–450	Sonic velocity	1600–2800 m/s	Concentrated hydrates
375–490	Gamma ray	Low	Concentrated hydrates in sand
375–490	Density	Low	Hydrates

Industry drilled two GC-955 conventional hydrocarbon wells, which supplied historical data for hydrate endeavors, prior to JIP involvement. The first well was drilled in the block's center that verified thick reservoir sands. Later, JIP drilled three GC-955 wells at water depths of 2000–2200 m to test, among other things, hydrate content of the porous sands. The first well, designated *I* well, showed only minor hydrate accumulations with their thick sands seemingly apparently isolated from gas sources (McConnell et al., 2012).

Quality logs from the second well, designated the *H* well, taken while drilling were evaluated by Zhang and McConnell (2011). They found over the interval 410–450 mbsf sonic velocity and resistance logs highlighting hydrates saturating sands to an extent of 70%. To a lesser extent, gamma ray and density logs defined hydrates in the larger interval 375–490 mbsf. The significant results of the Zhang and McConnell logging data interpretations are presented in Table 3.5.

Structural traps provide reservoirs within sands sealed by shale above (Hutchinson et al., 2008a,b). Apparently, random impermeability of gas hydrates traps free gas within the channel sands (McConnell and Zhang, 2011). The numerous interbedded thin sand beds of high S_{gh} observed in GC-955H give similar structures to Nankai Trough (Boswell et al., 2012a).

At the GC-955 *I* and *H* sites, hydrate reservoirs exist deep in thick channel sands. Being near the BGHS, having relatively high 32 mK/m thermal gradients as a consequence of underlying salt bases, and demonstrating 50–89% hydrate saturation, these reservoirs have high production potential (Hutchinson et al., 2008b).

The GC-955 data converge on multiple desirable features for successful gas production from its hydrate reservoirs. Volumetric calculations substantiate the data. For example, Boswell et al. (2012a) estimate in-place gas in the hydrate sands intersected by *I* and *H* wells based on the following data: 40% porosity, 80% hydrate saturation, and average net sand thickness

30 m for the *H* zone and 21 m for the *I* zone in their calculations. Overall in-place gas for *I* and *H* hydrate sands is 5.5×10^8 m^3 (19.8 bcf).

3.9.2 Keathley Canyon and Atwater Valley Drill Sites

Gas hydrate GOM sites with most favorable features for initial natural gas production were sought by JIP early in the program according to the schedule of Table 3.4. Two sites of early interest were Keathley Canyon and Atwater Valley. Located north of the Sigsbee Escarpment on the continental slope, Atwater Valley has thick Pleistocene sands and shales underlain by salt (Birchwood and Noeth, 2012; Hart et al., 2008; Lee et al., 2008).

Locations of KC-151 and AT-13 are pinpointed in Figure 3.6.

Although KC-151 and AT-13 eventually were not chosen for the final seven wells to be further developed, it is informative to review data acquired for future reference in prospective hydrate reservoir evaluations.

Although its interpretation was clouded by extensive faulting in the area, high-resolution 3D and 2D multichannel seismic data taken over a 1000 km path (621 miles) at the Keathley Canyon site indicate a truncated BSR. However, additional coring, logging, and seismic data substantiated the interpretations (Hutchinson and Hart, 2003; Hutchinson et al., 2003).

Surface mounds follow a linear NW to SE fault line at 1300 m water depth (Dai et al., 2008a). A salt diapir located several hundred meters below the mounds creates fracture networks that transport warmer fluids to vent at the mounds (Hart et al., 2008; Wood et al., 2008; Kastner et al., 2008). Characteristics of two of these mounds, Mound D and Mound F, are given in Table 3.6 (Hart et al., 2008).

Figure 3.6 pinpoints sites where JIP extracted deep cores from Keathley Canyon, lease Block 151, and from Atwater Valley, lease Blocks 13 and 14 (Winters et al., 2008).

Table 3.6 Characteristics of Mound D and Mound F, Atwater Valley Blocks 13/14 (Hart et al., 2008)

	Mound dimensions	Chemosynthetic presence	Mound surfaces	Fluid venting
Mound D	250 m diameter 7–10 m height	*Beggiatoa* mats, mussels	Carbonate exposures	Slow–moderate
Mound F	400 m diameter 15 m height	*Beggiatoa* mats	Flocculate mud flows	Moderate–rapid

Table 3.7 Property similarities of KC-151 and AV-13/14

Property	Keathley Canyon 151 and Atwater Valley 13/14	References
Water depth	Atwater: 1295 m Keathley: 1322 m	Winters et al. (2008), Winters (2011)
Water salinity (ppt)	≈35 at surface, ≈55 at 380 mbsf	Kastner et al. (2008)
Gas composition	96–99.9% microbial methane; CO_2, C_2–C_6 minor amounts	Lorenson et al. (2008)
$\delta^{13}C$ isotopic CH_4	−84.3 to −71.5‰	Lorenson et al. (2008)

In Figure 3.6 contours show similar 1300 m water depths of the two sites near, although the sites are about 387 km apart (Winters et al., 2008). Sites KC-151 and AT-13/14 reveal a common gas source of biogenic methane according to $\delta^{13}C$ values.

Gas samples from cores at KC-151 and AT-13/14 also contain small amounts of C_2–C_5 gases (Lorenson et al., 2008), but not enough to alter the conclusion of primarily biogenic sources. Further confirmation of a primarily biogenic base comes from a Bernard chart (Bernard et al., 1978), where C_1/C_2 and $\delta^{13}C$ isotopic CH_4 values are superposed to fall in the region of biogenic gas dominancy.

Some similarities of the two sites are summarized in Table 3.7.

Seismic anomalies, mounds at the surface, and a relatively shallow sulfate–methane transition zone verify a significant upward flux of the biogenic gases (Kastner et al., 2008; Lorenson et al., 2008).

Dissimilarities between KC-151 and AT-13/14 sites are presented in Table 3.8, where the main contrasts are depths at which hydrates are found, attendant chemosynthetic communities, and seafloor surface features.

A histogram of methane $\delta^{13}C$ from 160 high-flux seeps sampled in the northern GOM was developed by Sassen et al. (2002) that showed about 65 of the seeps peaking at methane $\delta^{13}C$ values −65 to −80‰, well within a biogenic methane definition. When $\delta^{13}C$ values from KC-151 and AT-13/14 having −84.3 to −71.5‰ are superposed on the Sassen histogram, a significant match is evident with the shallower seeps of the GOM (Lorenson et al., 2008).

Although the sulfate–methane transitions are shallow in both Keathley Canyon and Atwater Valley study sites, the transition depth reduces further to only less than or equal to 10 cm atop the Atwater Valley seep mounds. With reduced gas flux, the SMI increases in depth with distance from the mounds (Kastner et al., 2008).

Table 3.8 Property dissimilarities of KC-151 and AT-13/14

Property	Keathley Canyon 151	Atwater Valley 13/14	References
Surface features		Fluid venting; mounds; chemosynthetics at mounds	Hart et al. (2008), Wood et al. (2008)
BSR	A clear BSR	No BSR	Hutchinson et al. (2008a)
Sulfate–methane transition (mbsf)	9	8	Kastner et al. (2008)
Seafloor mounds	No	Yes	Hart et al. (2008)
Hydrate location (mbsf)	In fractured silty–clayey; 220–230 and 264–298 mbsf in near-vertical fractures	Near seafloor surface, but not outcropping	Kastner et al. (2008), Winters et al. (2008)

Coffin et al. (2008) studied geochemical data from 15 piston cores taken along a line beginning atop Mound F, AT-13/14, and extending 3.5 km beyond the mound. A summary of their analysis is given in Table 3.9.

A peak of methane flux at the mound apex is evident, which reduces localized depth of the SMI correspondingly. At the top of Mound F, the higher temperatures and higher salt contents of the water indicate that more stringent hydrate equilibrium conditions retard localized hydrate formation – somewhat substantiated by the higher methane concentrations (Coffin et al., 2008).

3.9.3 Walker Ridge, WR-313; Terrebonne Mini-Basin

Walker Ridge, Block 313, offers an interesting study of gas hydrates associated with mini-basins on the lower slope of the northern GOM. Primary hydrate interest resides with the Terrebonne mini-basin in the lease block's

Table 3.9 Atwater Valley, Mound F, geochemical data (Coffin et al., 2008)

Location relative to Mound F	Concentration CH_4 (mM)	Cl^- concentration (mM)	Heat flux (mW/m^2)	SMI depth (cmbsf)
Top	17.4–13.9	608–983	160	≈52 average
Base	Very low	560–902	–	110 average
Off mound	Very low	560*	40–50	361 average

*Salinity of background seawater.

western region. Refer to Figure 3.5 to locate the area 50–90 km north of the Sigsbee Escarpment (Shelander et al., 2011; Weitemeyer et al., 2009; Birchwood and Noeth, 2012).

Features that distinguish Terrebonne as a hydrate prospect are as follows: at the entrance to the mini-basin, channels of coarse sands become bounded to the south–southeast by salt ridges. A salt diapir at the northeastern limit of the ridges accounts for nearby faults where gas expulsion has resulted in large surface mounds. Hutchinson et al. (2008b) suggest gas accesses the mini-basin interior along stratigraphic units. Because of high hydrostatic pressures from 1917 m water depths and a moderate 19.6 mK/m thermal gradient, BGHS in the Terrebonne mini-basin reaches a deep 949 mbsf (Shelander et al., 2011; Frye et al., 2012). Aligned bright spots of seismic data in the Terrebonne mini-basin form a discontinuous BSR that can be pieced together for a composite. Thick gas hydrate accumulations bottom at 841 mbsf, close to hydrate stability limits – a factor noted by JIP as propitious for potential hydrate-gas production to improve ease and economy of supplying decomposition energy (McConnell and Kendall, 2002; McConnell et al., 2012).

With favorable exploration data, two other hydrate wells were drilled at WR-313 in 2009. Each of the two wells penetrated BGHS, bottoming at 870 mbsf. Designated WR-313-G and -H, the JIP wells reside near an earlier conventional well drilled by industry from which industry had retrieved poor logging data through the hydrate zone. However, advanced logging suites of density, gamma ray, acoustic, and resistivity (Collett et al., 2012) as well as 3D seismic data came from the later JIP wells (Lee and Collett, 2011; Boswell et al., 2012a,b).

Boswell et al. (2012a) found near BGHS the three WR-313 adjacent wells, *G*, *H*, and industry's conventional #001, to have parallel sand intervals trending upward into the basin at 8–12° angles. The three sand units are named according to increasing depositional age as Blue, Orange, and Green, respectively. The hydrate reservoirs are bounded by low-permeability silt/clay above and below. For example, the younger Blue Sands are underlain by Orange Sands with 105–140 m of low-permeability silts and clays intervening (Boswell et al., 2012a).

There are favorable indicators for hydrate-gas production from the three reservoir sands: (1) porosities range from 33 to 40%; (2) S_{gh} is a high 60–90%; (3) cumulative pay zones are thick – 11 m of Blue Sands to 8.2 m of hydrate-containing Orange Sands; (4) the channel sands have an areal extent of 300,000 m^2; (5) impermeable shales between sand units provide flow barriers at top, bottom, and two sides; (6) the hydrate reservoirs reside near BGHS (Boswell et al., 2012a).

Interestingly, above these three reservoir sands, the three wells of the Terrebonne mini-basin penetrate a thick 165 m shale interval of gas hydrates having fractures throughout oriented approximately vertical, giving an average S_{gh} of 3.6%. The WR-313-G well encountered the interval of fine sediments between 237 and 402 mbsf (Frye et al., 2012). Additional hydrates are reported to be in thin sand lenses of high S_{gh} below the fractured shale interval but above the Blue Sands reservoir. Finally, below the Green Sands a permeable interval termed Pink Sands trends upward into the gas hydrate zone (GHZ), but of the three wells only industry's conventional well intersects the Pink Sand unit (Frye et al., 2012; Boswell et al., 2012a).

3.9.3.1 Volumetric Calculations of Terrebonne Hydrate Gas

Data retrieved from JIP wells in the Terrebonne mini-basin (Frye et al., 2012) establish higher standards than previously available for accurate hydrate-gas estimates: (1) multiple hydrate reservoir forms, (2) high-quality logging while drilling, (3) 3D seismic data, (4) multiple wells drilled through BGHS, (5) borehole penetration and monitoring through fractured fine sediments and thin sand lenses in the hydrate zone above target sands, and (6) potential hydrate gas production directly comparable to nearby conventional gas production.

The volumetric estimates of Frye et al. are summarized in Table 3.10.

Cumulative in-place hydrate gas for Blue, Orange, and Green Sands of the Terrebonne mini-basin is estimated to be 3.8 bcm. The two thinner sands of less areal extent above Orange and Green reservoirs contain 0.58 bcm of hydrate gas. Despite a substantial 4.38 bcm of hydrate gas in the coarse reservoir sands, much larger volumes of hydrate gas fill the fractures

Table 3.10 Volumetric estimates* of in-place gas, Terrebonne mini-basin (Frye et al., 2012)

Reservoir	Areal extent (km^2)	Thickness (m)	In-place gas (bcm)
Blue Sands	10.481	14.325	1.627
Orange Sands	15.970	19.812	0.745
Green Sands	4.382	22.860	1.456
Pink Sands	3.784	9.144	0.411
Thin sands	2.023	3.048	0.170
Fractured fines	51.395	152.4	16.638

*Estimates do not include any free gas.

of fine sediments above because the unit of shale has extraordinary thickness (Frye et al., 2012).

Note that the petroleum system technique of evaluating reserves corroborates these relative amounts of hydrate gas worldwide in fine-sediment fractures versus quality reservoir sands (Boswell and Collett, 2006; Johnson, 2011; Max et al., 2006). Implications are tremendous if this hydrate gas mix is representative of a worldwide mix. Specifically, the fractured fines above target contain 79% of the total hydrate gas in the Terrebonne mini-basin. Yet hydrates in the high-quality sands of the principal target near BGHS contain producible gas quantities comparable to conventional gas reservoirs being produced economically in the vicinity (Frye et al., 2012).

3.9.3.2 Myshakin Numerical Simulation, Terrebonne Mini-Basin

Myshakin et al. (2012) simulated the depressurization method (DPM) to produce gas from Blue and Orange Sands at WR-313 wells *G* and *H*. Use of advanced logging data in a 3D simulation accounted for reservoir heterogeneities; seismic data provided reservoir extent. Assuming a bottom-hole pressure of 2.7 MPa, Myshakin et al. predict high percentages of in-place gas recoverable at high rates. Simulation predicts 45 million scfd production rates to be feasible from the Blue Sands of well G and 35 million scfd from the Blue and Orange Sands of well *H* (Myshakin et al., 2012). Moreover, their results predict cumulative extraction of 34–38 bcf of hydrate gas over a 4-year period by the DPM.

Better heat and mass transfer because of heterogeneities of interbedded sands promotes hydrate-gas production according to the Myshakin simulation. There would be less secondary hydrate formation from released gases because of the heterogeneities. However, three red flags waved from their depressurization simulation results: (1) hydrate reformation near the wellbore increases with reservoir size; (2) lag time before gas recovery lengthens with reservoir size; (3) even small increases in permeability of barrier silts and clays have large deleterious effects on gas production.

3.10 ALAMINOS CANYON

AC resides in the northwestern region of the GOM. See Figure 3.5. On the AC seafloor, oil and gas seeps support tube worms, mussel beds, and bacterial mats (Guan et al., 2010). In early hydrate exploration, AC 645 chemosynthetic communities were observed with a manned submersible assistance; samples were retrieved from the seafloor at 2200 m water depth.

Later, deep wells drilled in Blocks AC-21 and AC-818 offered valuable insight into the potential of hydrate-gas production.

3.10.1 Alaminos Canyon Block 21

JIP drilled two wells, designated AC-21A and AC-21B, in the sands of the Diana mini-basin near the northern boundary of AC. Unlike the program's five wells at GC-955 and WR-313, which were near the BGHS, hydrate-bearing sands of AC-21 reside midway between seafloor and BGHS but well within the hydrate stability zone (Miller et al., 2012). Note that BGHS at AC-21 is 300–450 mbsf and is distinguished by BSRs (Shedd et al., 2012).

Muds of low permeability bound the thick sands, restricting gas accessibilities to the sands via numerous, polygonal-like, vertical faults throughout the muds (Miller et al., 2012). Logging and seismic data indicate only moderate 8–28% S_{gh} in the sands, a situation verifying possible limited gas access.

Volumetric calculations of in-place hydrate gas by Boswell et al. (2012a) were partially based on the following: 38% porosity, 12% S_{gh}, 30 m sand thickness, and 7500 acres areal extent of the Diana mini-basin reservoir. Their results predict 3.5×10^{10} m^3 (1200 bcf) of in-place hydrate gas.

3.10.2 Perdido Fold Belt

In the southwestern part of the northern GOM in US waters of 2300–3000 m depth, the Perdido fold belt extends over a 3000 km^2 area (Fiduk et al., 1999). A dominating geologic feature, the Perdido fold belt derives from underlying Louann salt weighted by thick sediment overburden as the northeast to southwest trending Sigsbee Escarpment of salt progresses to the lower continental slope. Under influences of prodigious sediment deposits, hundreds of folds have developed (Trudgill et al., 1999).

The extensive faulting and trapping within the Perdido fold belt generated strong interest in conventional oil and gas exploration when the deepwater area became a frontier for exploration. Numerous expensive blocks were auctioned. The first exploratory conventional well (AC 600) was drilled to 1096 mbsf (Fiduk et al., 1999). Afterwards, an exploratory well for conventional oil and gas was drilled in the Perdido fold belt at AC-818 in 2744 m deep water. Subsequently, this exploratory well designated Tigershark now provides information concerning gas hydrates within Frio Sands of Oligocene origin (Boswell et al., 2009).

3.10.3 Hydrate Reservoir – AC-818, Perdido Fold Belt

The Tigershark well encountered reservoir quality sands 20 m thick with high saturations of gas hydrates (Weitemeyer et al., 2009). The oil and gas exploratory well provided quality logs within the hydrate stability envelope where quality logs are not usually sought during conventional drilling. Logging data combined with seismic data give comprehensive information on a deep gas hydrate reservoir near a fold crest in permeable sands. Hydrate saturations reach 80% with resistivity and sonic logs indicating surprising uniformity of hydrate content (Boswell et al., 2009; Dai et al., 2008b). Resistivity logs show 13 m of hydrate saturated sands in the interval 439–452 mbsf.

Near BGHS several hundred meters below seafloor, the permeable Frio Sands have attractive properties for hydrate gas production: permeability, porosity, trapping mechanism, and water and gas sources of the hydrate reservoir contact permeable conduits (Boswell et al., 2009). See Table 3.11.

A 42% porosity of unconsolidated Frio Sands affords space for hydrate forming and expanding. High permeability, approximately 1 darcy, provides fluid transport capability, as well as a conduit to transmit pressure reductions during potential hydrate-gas production by the DPM. This gas hydrate Class 2 reservoir has on the east and south sides an aquifer, effectively infinite, in communication with Frio Sands.

Frio Sands have been structurally uplifted by the Perdido fold belt and in doing so protrude into the hydrate stability zone. A short distance above BGHS, the sand continuity is terminated by a piercing fault that creates a trapping mechanism. While the upper end of the reservoir extends into the hydrate stability zone, the sand's lower extremity dips below BGHS (Boswell et al., 2009).

3.10.4 In-Place Hydrate Gas Calculations, Tigershark Well

Data for volumetric calculations of hydrate gas content in Block 818 Frio Sands come from published enhanced well log data of the parent company

Table 3.11 Properties of Frio Sand (Boswell et al., 2009)

Sand grain diameter (μm)	Porosity (%)	Permeability (mD)	Temperature (°C)	Age
40–60*	42†	600–1500‡	20§	Oligocene

*Volcanoclastic sand, highly unconsolidated, high porosity.
† From log data.
‡ Retrieved core analysis without hydrates.
§ Temperature near BGHZ where hydrates occupy sand voids.

while drilling the original Tigershark well and from 3D seismic data encompassing the surrounding 23 km^2 area (Boswell et al., 2009; Weitemeyer et al., 2009; Dai et al., 2008b). The volumetric calculations are outlined as follows.

First, seismic data establish areal extent A of 0.8 km^2 for the hydrate reservoir. From this area and an average pay zone depth h of 18 m, Equation 3.5 leads to a bulk volume in the hydrate zone of 14.4 million m^3:

$$V_{hz} = Ah \tag{3.5}$$

Second, Frio Sand porosity ϕ derived from density logs is 0.42. Found from Equation 3.6, the total void volume V_v available for filling is 6.048 million m^3:

$$V_v = Ah\phi \tag{3.6}$$

Third, the fractional hydrate saturation S_{gh} is uniformly near 0.80, yielding by Equation 3.7 gas hydrate volume V_{gh} in the reservoir of 4.84 million m^3:

$$V_{gh} = Ah\phi S_{gh} \tag{3.7}$$

Fourth, methane content n in hydrate at a pressure near 30 MPa may be taken as 164 m^3 of gas at standard temperature–pressure per 1 m^3 of hydrate volume, leading to a volume of in-place gas V_{gSTP} from Equation 3.8 of 793.5 million m^3 (Boswell et al., 2009):

$$V_{gSTP} = nV_{gh} \tag{3.8}$$

3.10.5 Simulated Alaminos Canyon Hydrate-Gas Production

Three independent simulations help determine economic feasibility of AC hydrate-gas production.

3.10.5.1 Moridis and Reagan Simulation

Moridis and Reagan (2007) assumed depressurization supplemented by warm water injection near the wellbore in their production simulation, employing AC-818 reservoir characteristics: (1) 466.5 m (1530 ft) below seabed; (2) 2744 m (9000 ft) water depth; (3) 18.25 m (60 ft) thickness of hydrate; (4) 30% porosity; (5) 60–80% gas hydrate saturation; (6) permeability in the darcy range; (7) 33 MPa pressure; (8) 21°C temperature; (9) near BGHS.

Table 3.12 Assumptions for simulation (Boswell et al., 2009)

Porosity (%)	Permeability (darcy)	Upper 18 m of sand	Lower 30 m of sand	Initial pressure (MPa)
42	1.0	75% saturated with hydrate	100% water-saturated sands on lower 30 m; connected to infinite aquifer	30

By assuming either a Class 2 reservoir with an underlying aquifer or a Class 3 reservoir with an underlying low-permeability mud, Moridis and Reagan found production rate and cumulative production promising to be economical (Dai et al., 2008b). However, high removal rates of water along with initially low gas production rates would reduce profitability as measured by discounted rate of return.

3.10.5.2 Boswell Simulation

Boswell et al. (2009) ran a simulation of hydrate production based on the log and seismic data of AC-818. Some assumptions of the simulation are listed in Table 3.12.

Pressure reduction was to be accomplished by removing 30,000 bwpd. A single horizontal well of 1100 m length was assumed to be drilled into an elliptical reservoir.

The important results of Boswell et al. are summarized as follows:

1. With strong water drives, difficult pressure reduction hinders hydrate decomposition.
2. Contact with an infinite aquifer lying to the east and south makes impractical the DPM acting alone.
3. A closed system without aquifer intrusion is necessary unless supplementary techniques act with depressurization.
4. Depressurization alone cannot economically produce AC-818 hydrate gas.

3.10.5.3 Phirani and Mohanty Simulation

Phirani and Mohanty (2011) considered only horizontal wells, injector and producer, in the Frio Sands of the AC-818 reservoir. An infinite aquifer was assumed to occupy one side. Dual cases of horizontal reservoir and dipping reservoir were considered. Other assumptions in the simulation are given in Table 3.13.

Table 3.13 AC-818 simulation properties of Phirani and Mohanty (2011)

Reservoir dimensions (m)	Hydrate zone (m)	S_H (%)	S_W (%)	Initial* pressure (MPa)	Initial* temperature (K)	Simulator description
1200 (length) × 500 (width)	18	75	25	31	294	3D finite volume Considers heat transfer, multiphase fluid flow, hydrate equilibrium thermodynamics

*Bottom of reservoir.

Significant conclusions of Phirani and Mohanty (2011) are the following:

1. Warm water injection must supplement depressurization to obtain practical production rates.
2. High reservoir and aquifer permeabilities effectively distribute warm waters.
3. Distance between injector and production wells dictates whether water channels develop that bypass hydrates. The simulator should be used to determine well spacing.
4. Best production occurs in dipping reservoirs if producing wells are at the reservoir ceiling and injector wells are adjacent to aquifer.

REFERENCES

Balczewski, J.T., Boswell, R., Collett, T.S., Baker, R., 2011. International collaboration on deepwater natural gas hydrate continues – Gulf of Mexico gas hydrate JIP: past, present, and future. In: Proceedings of the 7th International Conference on Gas Hydrates (ICGH 2011), Edinburgh, Scotland, UK, July 17–21. Paper 739.

Bernard, B.B., Brooks, J.M., Sackett, W.M., 1978. Light hydrocarbons in recent Texas continental shelf and slope sediments. J. Geophys. Res. 83, 4053–4061.

Birchwood, R., Noeth, S., 2012. Horizontal stress contrast in the shallow marine sediments of the Gulf of Mexico sites Walker Ridge 313 and Atwater Valley 13 and 14 – geological observations, effects on wellbore stability, and implications for drilling. Mar. Pet. Geol. 34, 186–208.

Boswell, R., Collett, T.S., 2006. The gas hydrates resource pyramid. In: Fire in the Ice, Methane Hydrate Newsletter. Fall Issue. U.S. Department of Energy, Office of Fossil Energy, National Energy Technology Laboratory, pp. 5–7.

Boswell, R., Shelander, D., Lee, M., Latham, T., Collett, T., Guerin, G., Moridis, G., Reagan, M., 2009. Occurrence of gas hydrate in Oligocene Frio sand: Alaminos Canyon Block 818: Northern Gulf of Mexico. Mar. Pet. Geol. 26, 1499–1512.

Boswell, R., Frye, M., Shelander, D., Shedd, W., McConnell, D.R., Cook, A., 2012a. Architecture of gas-hydrate-bearing sands from Walker Ridge 313, Green Canyon 955, and Alaminos Canyon 21: Northern Gulf of Mexico. Mar. Pet. Geol. 34, 134–149.

Boswell, R., Collett, T., Frye, M., Shedd, W., McConnell, D.R., Shelander, D., 2012b. Subsurface gas hydrates in the northern Gulf of Mexico. Mar. Pet. Geol. 34, 4–30.

Brooks, J.M., Kennicutt, II, M.C., Fay, R.R., McDonald, T.J., Sassen, R., 1984. Thermogenic gas hydrates in the Gulf of Mexico. Science 225, 409–411.

Coffin, R., Hamdan, L., Plummer, R., Smith, J., Gardner, J., Hagen, R., Wood, W., 2008. Analysis of methane and sulfate flux in methane-charged sediments from the Mississippi Canyon, Gulf of Mexico. Mar. Pet. Geol. 25, 977–987.

Collett, T., Lee, M., Lewis, R., Zyrianova, M., Mrozewski, S., Guerin, G., Cook, A., Goldberg, D., 2012. Gulf of Mexico gas hydrate joint industry Project Leg II logging-while-drilling data acquisition and analysis. J. Mar. Pet. Geol. 34, 41–61.

Cooper, A.K., Hart, P.E., 2003. High-resolution seismic-reflection investigation of the northern Gulf of Mexico gas-hydrate-stability zone. Mar. Pet. Geol. 19, 1275–1293.

Dai, J., Snyder, F., Gillespie, D., Koesoemadinata, A., Dutta, N., 2008a. Exploration for gas hydrates in the deepwater, northern Gulf of Mexico: part I. A seismic approach based on geologic model, inversion, and rock physics principles. Mar. Pet. Geol. 25, 830–844.

Dai, J., Banik, N., Shelander, D., Bunge, G., Dutta, N., 2008b. Seismic detection and quantification of gas hydrates in Alaminos Canyon, Gulf of Mexico. In: Proceedings of the 6th International Conference on Gas Hydrates, Vancouver, British Columbia, Canada, July 6–10.

Dillon, W.P., Max, M.D., 2000. Oceanic gas hydrate. In: Max, M.D. (Ed.), Natural Gas Hydrates in Oceanic and Permafrost Environments, Coastal Systems and Continental Margins, 5, Kluwer Academic Press, The Netherlands, pp. 61–76.

Epp, D., Grim, P.J., Langseth, Jr., M.G., 1970. Heat flow in the Caribbean and Gulf of Mexico. J. Geophys. Res. 75 (29), 5655–5668.

Fiduk, J.C., Weimer, P., Trudgill, B.D., Rowan, M.G., Gale, P.E., Phair, R.L., Korn, B.E., Roberts, G.R., Gafford, W.T., Lowe, R.S., Queffelec, T.A., 1999. The Perdido fold belt, northwestern deep Gulf of Mexico, part 2; seismic stratigraphy and petroleum systems. AAPG Bull. 83 (4), 578–612.

Francisca, F., Yun, T.-S., Ruppel, C., Santamarina, J.C., 2005. Geophysical and geotechnical properties of near-surface sediments in the northern Gulf of Mexico gas hydrate province. Earth Planetary Sci. Lett. 237, 924–939.

Frye, M., Shedd, W., Boswell, R., 2012. Gas hydrate resource potential in the Terrebonne basin, northern Gulf of Mexico. Mar. Pet. Geol. 34 (1), 150–168.

Goodwin, R.H., Prior, D.B., 1989. Geometry and depositional sequences of the Mississippi Canyon, Gulf of Mexico. J. Sediment. Petrol. 59 (2), 318–329.

Guan, H.X., Feng, D., Wu, N.Y., Chen, D.F., Roberts, H.H., Wu, D.D., 2010. Fatty-acids and their $\delta^{13}C$ characteristics of seep carbonates from the northern continental slope of Gulf of Mexico. Chin. Sci. Bull. 55 (8), 730–735.

Hart, P.E., Hutchinson, D.R., Gardner, J., Carney, R.S., Fornari, D., 2008. A photographic and acoustic transect across two deep-water seafloor mounds, Mississippi Canyon, northern Gulf of Mexico. Mar. Pet. Geol. 25, 969–976.

Heggland, R., 2004. Definition of geohazards in exploration 3-D seismic data using attributes and neural-network analysis. AAPG Bull. 88 (6), 857–868.

Humphris, Jr., C.C., 1979. Salt movement on continental slope, northern Gulf of Mexico. Am. Assoc. Pet. Geol. Bull. 63 (5), 782–798.

Hutchinson, D., Hart, P., 2003. Gas hydrate in the northern Gulf of Mexico has puzzling characteristics and could pose a hazard to deep drilling. In: Sound Waves, Monthly Newsletter USGS, July.

Hutchinson, D.R., Hart, P.E., Dugan, B., Geresi, E., Sliter, R., Newman, K., 2003. Subsurface gas hydrates in the Northern Gulf of Mexico. In: 2003 Seattle Annual Meeting, November 2–5. Geological Society of America, Boulder, CO.

Hutchinson, D.R., Hart, P.E., Ruppel, C.D., Snyder, F., Dugan, B., 2008a. Seismic and thermal characterization of a bottom simulating reflection in the northern Gulf of Mexico. In: Collett, T.S., Johnson, A., Knapp, C., Boswell, R. (Eds.), Natural Gas Hydrates: Energy Resources, Potential and Associated Geologic Hazards. AAPG Special Publication.

Hutchinson, D.R., Shelander, D., Dai, J., McConnell, D., Shedd, W., Frye, M., Ruppel, C., Boswell, R., Jones, E., Collett, T., Rose, K., Dugan, B., Wood, W., Latham, T., 2008b. Site selection for DOE/JIP gas hydrate drilling in the northern Gulf of Mexico. In: Proceedings of the 6th Conference on Gas Hydrates, Vancouver, British Columbia, Canada, July 6–10. Paper 5506.

Hutchinson, D.R., Hart, P.E., Collett, T.S., Edwards, K.M., Twichell, D.C., Snyder, F., 2008c. Geologic framework of the 2005 Keathley Canyon gas hydrate research well, northern Gulf of Mexico. Mar. Pet. Geol. 25 (9), 906–918.

Johnson, A.H., 2011. Global resource potential of gas hydrate – a new calculation. In: Proceedings of the 7th International Conference on Gas Hydrates (ICGH 2011), Edinburgh, Scotland, UK, July 17–21.

Kastner, M., Claypool, G., Robertson, G., 2008. Geochemical constraints on the origin of the pore fluids and gas hydrate distribution at Atwater Valley and Keathley Canyon, Northern Gulf of Mexico. Mar. Pet. Geol. 25, 860–872.

Kirkland, D.W., Gerhard, J.E., 1971. Jurassic salt, central Gulf of Mexico, and its temporal relation to circum-Gulf evaporates. AAPG Bull. 55, 680–686.

Lapham, L.L., Chanton, J.P., Martens, C.S., Sleeper, K., Woolsey, J.R., 2008. Microbial activity in surficial sediments overlying acoustic wipeout zones at a Gulf of Mexico cold seep. Geochem. Geophys. Geosyst. 9(6), doi: 10.1029/2008GC001944.

Lee, M.W., Collett, T.S., 2011. Three types of gas hydrate reservoirs in the Gulf of Mexico identified in LWD data. In: Proceedings of the 7th International Conference on Gas Hydrates (ICGH 2011), Edinburgh, Scotland, UK, July 17–21. Paper 39.

Lee, J.Y., Santamarina, J.C., Ruppel, C., 2008. Mechanical and electromagnetic properties of northern Gulf of Mexico sediments with and without THF hydrates. Mar. Pet. Geol. 25, 884–895.

Leifer, I., MacDonald, I.R., 2003. Dynamics of the gas flux from shallow gas hydrate deposits: interaction between oily hydrate bubbles and the oceanic environment. Earth Planetary Sci. Lett. 21, 411–421.

Lorensen, T.D., Writers, W.J., Hart, P.E., Paull, C.K., 2002. Gas hydrate occurrence in the Northern Gulf of Mexico studied with giant piston cores. EOS Trans. Am. Geophys. Union 83 (51), 601, 607–608.

Lorenson, T.D., Claypool, G.E., Dougherty, J.A., 2008. Natural gas geochemistry of sediments drilled on the 2005 Gulf of Mexico JIP cruise. Mar. Pet. Geol. 25, 873–883.

Losh, S., Eglinton, L.B., Schoell, M., Wood, J.R., 1999. Vertical and lateral fluid flow related to a large growth fault, South Eugene Island Block 330 Field, offshore Louisiana. AAPG Bull. 83 (22), 244–276.

Lutken, C.B., Macelloni, L., Sleeper, K., McGee, T., Simonetti, A., Knapp, J.H., Knapp, C.C., Caruso, S., Chanton, J., Lapham, L., Lodi, M., Ingrassia, M., Higley, P., Brunner, C., 2011. New discoveries at Woolsey Mound, MC118, Northern Gulf of Mexico. In: Proceedings of the 7th International Conference on Gas Hydrates (ICGH 2011), Edinburgh, Scotland, UK, July 17–21. Paper 626.

MacDonald, I.R., 2009. The HYFLUX Sea-Truth Expedition, 4–19 July 2009. In: Fire in the Ice, Fall 2009, DOE NETL Hydrate Newsletter, vol. 9, issue 4, pp. 12–15.

MacDonald, I.R., Guinasso, Jr., N.L., Ackleson, S.G., Amos, J.F., Duckworth, R., Sassen, R., Brooks, J.M., 1993. Natural oil slicks in the Gulf of Mexico visible from space. J. Geophys. Res. 98 (C9), 16351–16364.

MacDonald, I.R., Guinasso, N.L., Sassen, R., Brooks, J.M., Lee, L., Scott, K.T., 1994. Gas hydrate that breaches the sea floor on the continental slope of the Gulf of Mexico. Geology 22, 699–702.
MacDonald, I.R., Buthman, D., Sager, W.W., Peccini, M.B., Guinasso, Jr., N.L., 2000. Pulsed oil discharge from a mud volcano. Geology 28, 907–910.
MacDonald, I.R., Bender, L.C., Vardaro, M., Bernard, B., Brooks, J.M., 2005. Thermal and visual time-series at a seafloor gas hydrate deposit on the Gulf of Mexico slope. Earth Planetary Sci. Lett. 233, 45–59.
Max, M.D., Johnson, A., Dillon, W.P., 2006. Economic Geology of Natural Gas Hydrate. Springer, Berlin, Dordrecht, 341 pp.
McConnell, D., 2000. Optimizing deepwater well locations to reduce the risk of shallow-water-flow using high-resolution 2D and 3D seismic data. In: Offshore Technology Conference 2000. Paper OTC 11973.
McConnell, D., Kendall, B., 2002. Images of the base of gas hydrate stability, Northwest Walker Ridge, Gulf of Mexico. OTC Paper #14103.
McConnell, D., Zhang, Z., 2011. Detailed seismic amplitude analysis and characterization of a gas hydrate and free gas mixed system in Green Canyon 955. In: Proceedings of the 7th International Conference on Gas Hydrates (ICGH 2011), Edinburgh, Scotland, UK, July 17–21. Paper 683.
McConnell, D., Boswell, R., Collett, T., Frye, M., Shedd, W., Shelander, D., Dai, J., Mrozewski, S., Guerin, G., Cook, A., Dufrene, R., Godfriaux, P., Roy, R., Jones, E., 2009. Initial results of Gulf of Mexico Gas Hydrate Joint Industry Program Leg II logging-while-drilling operations in Green Canyon Block 955. In: American Geophysical Union Conference.
McConnell, D.R., Zhang, Z., Boswell, R., 2012. Review of progress in evaluating gas hydrate drilling hazards. Mar. Pet. Geol. 34, 209–223.
Meckler, A.N., Schubert, C.J., Hochuli, P.A., Plessen, B., Birgel, D., Flower, B.P., Hinrichs, K.-U., Haug, G.H., 2008. Glacial to Holocene terrigenous organic matter input to sediments from Orca Basin, Gulf of Mexico – a combined optical and biomarker approach. Earth Planetary Sci. Lett. 272, 251–263.
Milkov, A.V., Sassen, R., 2000. Thickness of the gas hydrate stability zone, Gulf of Mexico continental slope. Mar. Pet. Geol. 17, 981–991.
Milkov, A.V., Sassen, R., 2001. Estimate of gas hydrate resource, northwestern Gulf of Mexico continental slope. Mar. Geol. 179, 71–83.
Milkov, A.V., Sassen, R., 2002. Resource and economic potential of gas hydrate in the Northwestern Gulf of Mexico. In: Proceedings of the Fourth International Conference on Gas Hydrates, Yokohama, Japan, May 19–23.
Miller, P., Dasgupta, S., Shelander, D., 2012. Seismic imaging of migration pathways by advance attribute analysis, Alaminos Canyon 21, Gulf of Mexico. J. Mar. Pet. Geol. 34 (1), 111–118.
Moridis, G.J., Reagan, M.T., 2007. New simulations of the production potential of marine gas hydrates. In: Fire in the Ice, NETL/DOE Newsletter. Fall Issue, pp. 1–3.
Myshakin, E., Gaddipaati, M., Rose, K., Anderson, B., 2012. Numerical simulation of depressurization-induced gas production from gas hydrate reservoirs at the Walker Ridge 313 site, northern Gulf of Mexico. J. Mar. Pet. Geol. 34, 169–185.
Nagihara, S., Sclater, J.G., Beckley, L.M., Behrens, E.W., Lawver, L.A., 1992. High heat flow anomalies over salt structures on the Texas continental slope, Gulf of Mexico. Geophys. Res. Lett. 19 (16), 1687–1690.
Neurauter, T.W., Bryant, W.R., 1990. Seismic expression of sedimentary volcanism on the continental slope, northern Gulf of Mexico. Geo-Mar. Lett. 10, 225–231.
Paull, C.K., Ussler, III, W., Lorenson, T., Winters, W., Dougherty, J., 2005. Geochemical constraints on the distribution of gas hydrates in the Gulf of Mexico. Geo-Mar. Lett. 25, 273–280.

Peel, F.J., Travis, C.J., Hossack, J.R., 1995. Genetic structural provinces and salt tectonics of the Cenozoic offshore U.S. Gulf of Mexico; a preliminary analysis. In: Jackson, M.P.A., Roberts, D.G., Snelson, S. (Eds.), AAPG Memoir 65, pp. 153–175.
Phirani, J., Mohanty, K.K., 2011. Production strategy for marine hydrate reservoirs. In: Proceedings of the 7th International Conference on Gas Hydrates (ICGH 2011), Edinburgh, Scotland, UK, July 17–21. Paper 21.
Pilcher, R.S., Blumstein, R.D., 2007. Brine volume and salt dissolution rates in Orca Basin, northeast Gulf of Mexico. AAPG Bull. 91 (6), 823–833.
Reagan, M.T., Moridis, G.J., 2008. Dynamic response of oceanic hydrate deposits to ocean temperature change. J. Geophys. Res. 113, C12023–C12044.
Roberts, H.H., Shedd, W., Hunt, Jr., J., 2010a. Dive site geology: DSV ALVIN (2006) and ROV JASON II (2007) dives to the middle-lower continental slope, northern Gulf of Mexico. Deep-Sea Res. II 57, 1837–1858.
Roberts, H.H., Feng, D., Joye, S.B., 2010b. Cold-seep carbonates of the middle and lower continental slope, northern Gulf of Mexico. Deep-Sea Res. II 57, 2040–2054.
Rowan, M.G., Jackson, M.P.A., Trudgill, B.D., 1999. Salt-related fault families and fault welds in the northern Gulf of Mexico. AAPG Bull. 83 (9), 1454–1484.
Ruppel, C., Boswell, R., Jones, E., 2008. Scientific results from Gulf of Mexico gas hydrates Joint Industry Project Leg 1 drilling: introduction and overview. Mar. Pet. Geol. 25, 819–829.
Sassen, R., Joye, S., Sweet, S.T., DeFreitas, D.A., Milkov, A.V., MacDonald, I.R., 1999. Thermogenic gas hydrates and hydrocarbon gases in complex chemosynthetic communities, Gulf of Mexico continental slope. Org. Geochem. 30, 485–497.
Sassen, R., Sweet, T.S., Milkov, A.V., DeFreitas, D.A., Kennicutt, II, M.C., 2001. Thermogenic vent gas and gas hydrate in the GOM slope: is gas hydrate decomposition significant? Geology 29, 107–110.
Sassen, R., Milkov, A.V., de Freitas, D.A., Sweet, S.T., 2002. Molecular and isotopic properties of high-flux gas seeps, northwestern Gulf of Mexico. In: AAPG Search and Discovery Article 90007, 2002 AAPG Annual Meeting, Houston, Texas, March 1–13, AAPG Annual Meeting Extended Abstracts, vol. 155.
Sassen, R., Roberts, H.H., Jung, W., Lutken, C.B., DeFreitas, D.A., Sweet, S.T., Guinasso Jr., N.L., 2006. The Mississippi Canyon 118 gas hydrate site: a complex natural system. In: 2006 Offshore Technology Conference, OTC #18132, Houston, TX, May 1–4.
Savidge, J.L., Adewumi, M.A., 1995/1996. Program targets fundamental understanding for gas hydrate control. GasTIPS 2 (1), 49–50.
Selig, F., Wallick, G.C., 1966. Temperature distribution in salt domes and surrounding sediments. Geophysics 31 (2), 346–361.
Shedd, W., Frye, M., Boswell, R., Hutchinson, D.R., Godfriaux, P., 2009. Variety of seismic expression of the base of gas hydrate stability in the Gulf of Mexico, USA. In: AAPG Annual Convention and Exhibition, Denver, Colorado, June 7–10. AAPG Search and Discovery Article #90090.
Shedd, W., Boswell, R., Frye, M., Godfriaux, P., Kramer, K., 2012. Occurrence and nature of "bottom simulating reflectors" in the northern Gulf of Mexico. Mar. Pet. Geol. 34, 31–40.
Shelander, D., Dai, J., Bunge, G., 2011. Predictions of gas hydrates using pre-stack seismic data, deepwater Gulf of Mexico. In: Proceedings of the 7th International Conference on Gas Hydrates (ICGH 2011), Edinburgh, Scotland, UK, July 17–21. Paper 131.
Sloan, Jr., E.D., 1998. Clathrate Hydrates of Natural Gases. Marcel-Dekker, New York.
Solomon, E.A., Kastner, M., Jannasch, H., Robertson, G., Weinstein, Y., 2008. Dynamic fluid flow and chemical fluxes associated with a seafloor gas hydrate deposit on the northern Gulf of Mexico slope. Earth Planetary Sci. Lett. 270, 95–105.
Trudgill, B.D., Rowan, M.G., Fiduk, J.C., Weimer, P., Gale, P.E., Korn, B.E., Phair, R.L., Gafford, W.T., Roberts, G.R., Dobbs, S.W., 1999. The Perdido fold belt, northwestern deep

Gulf of Mexico, part 1: structural geometry, evolution and regional implications. AAPG Bull. 83 (1), 88–113.
Walker, N.H., Huh, O.K., Rouse, Jr., L.J., 1993. Warm-core eddy discovered in the Gulf of Mexico. EOS Trans. Am. Geophys. Union 74, 337–338.
Wash, D.C., Antonov, J., Levitus, S., Boyer, T.P., Conkright, M., O'Brien, T., Stephens, C., 1998. Temperature of the Atlantic Ocean, World Ocean Atlas 1998. NOAA Atlas NESDIS 27, vol. 1. US Government Printing Office, USA.
Weitemeyer, K., Constable, S., SIO Marine EM Laboratory, 2009. Cruise report: imaging gas hydrate in the Gulf of Mexico using marine electromagnetic methods. In: Fire in the Ice, DOE NETL Newsletter. Winter Issue, pp. 4–6.
Winters, W.J., 2011. Physical and geotechnical properties of gas-hydrate-bearing sediment from offshore India and the Northern Cascadia Margin compared to other hydrate reservoirs. In: Proceedings of the 7th International Conference on Gas Hydrates (ICGH 2011), Edinburgh, Scotland, UK, July 17–21. Paper 408.
Winters, W.J., Dugan, B., Collett, T.S., 2008. Physical properties of sediments from Keathley Canyon and Atwater Valley, JIP Gulf of Mexico gas hydrate drilling program. Mar. Pet. Geol. 25, 896–905.
Wood, W., Gettrust, J., Chapman, N., Spence, G., Hyndman, R., 2002. Decreased stability of methane hydrates in marine sediments due to phase boundary roughness. Nature 420, 656–660.
Wood, W.T., Hart, P.E., Hutchinson, D.R., Dutta, N., Snyder, F., Coffin, R.B., Gettrust, J.F., 2008. Gas and gas hydrate distribution around seafloor seeps in Mississippi Canyon, Northern Gulf of Mexico, using multi-resolution seismic imagery. Mar. Pet. Geol. 25, 952–959.
Zhang, Z., McConnell, D., 2011. Detailed seismic amplitude analysis and characterization of a gas hydrate and free gas mixed system in Green Canyon 955, Gulf of Mexico. In: Proceedings of the 7th International Conference on Gas Hydrates (ICGH 2011), Edinburgh, Scotland, UK, July 17–21. Paper 683.

CHAPTER FOUR

Producing Methane from Offshore Hydrates

After early discoveries and calculations suggested enormous offshore volumes of hydrate gas, investigations into producing the gas have been methodical: (1) drilling and production tests conducted on Mallik gas hydrate reservoir in the Canadian Arctic; (2) offshore gas hydrate reservoirs characterized and organized into classes; (3) producing gas by computer simulation from each of the reservoir classes; (4) drilling, logging, and seismic techniques developed through the Joint Industry Program (JIP); (5) offshore production tests.

In this chapter potential hydrate-gas recovery principles are discussed: (1) depressurization, (2) CO_2 exchange with CH_4 or CO_2 + N_2 exchange with CH_4, (3) thermal stimulation, (4) inhibitor injection, and (5) downhole combustor.

In the series of research and development events marking offshore hydrate-gas utilization, the JIP established basic information for hydrate-gas exploration in deep waters. Inaugural wells were drilled, located in this case on the continental slope of the Gulf of Mexico (GOM), for the express purpose of characterizing hydrate reservoirs beneath deep waters. Discovered within the gas hydrate stability domain near bottom gas hydrate stability (BGHS) were high saturations of hydrates in sands having porosities, permeabilities, and gas-source accesses that historically afford economical production of conventional oil and gas.

Finally, feasibility is discussed of storing and transporting to shore produced hydrate gas after reforming hydrates by an anionic surfactant process. The process would allow delivering hydrate gas from remote offshore locations in the absence of pipeline infrastructures. Details are given of a surfactant process scale-up that demonstrated natural gas storage of 5000 scf. Also discussed is the potential of further developing the hydrate ultrastability phenomenon as an economic benefit for safely storing and transporting reformed hydrates near atmospheric pressure.

Offshore Gas Hydrates. http://dx.doi.org/10.1016/B978-0-12-802319-8.00004-8

4.1 POTENTIAL PRODUCTION METHODS

4.1.1 Getting Started

Offshore hydrates accumulate with dispersed, fracture- and vein-filling, massive seafloor protrusions, nodular, or stratified morphologies. The exceptionally diverse morphology complicates choosing and developing an optimum, hydrate-gas production method. The approach has been to seek near the BGHS hydrate accumulations of high pore saturations dispersed throughout permeable sands but which have sealing borders of low permeability. These hydrate-containing sands must have permeable access to gas sources.

For initial production planning, accumulations such as fracture- and vein-filling, massive seafloor protrusions, nodular, and stratified forms of hydrates are being preempted. Accordingly, the following discussion focuses on production from permeable sands.

4.1.2 Analogies with Conventional Oil and Gas Production

The primary methods to produce hydrate gas derive from established technologies that successfully produce conventional oil and gas. Nevertheless, the uniqueness of gas hydrates makes adaptations of the technology challenging.

Primary procedures initially considered for hydrate production are the following: (1) depressurization, (2) thermal stimulation, (3) thermodynamic inhibition, and (4) carbon dioxide replacement of stored methane. As the technology progresses, some of the conventional primary processes probably will be combined to arrive at a more refined procedure.

For the preceding four prospective hydrate production processes, consider process analogs in current oilfield practice.

4.1.2.1 Depressurization Method

Nonconventional production of methane from coalbeds (cbm) was introduced in the mid-1980s and has become a significant source of methane fuel worldwide (Rogers, 1994). In the case of cbm, according to the Langmuir isotherm, methane desorbs from coal by removing water to reduce the confining hydrostatic head.

Somewhat similarly, in a gas hydrate depressurization method (DM), confining reservoir pressure is reduced to destabilize hydrate structures and release methane. For example, when initial reservoir conditions are well within the hydrate stability envelope, isothermally removing free water can lower pressure into the instability region. In practice, however, endothermic reaction of hydrate decomposition cools the reservoir, causing undesirable trending toward stability. Without external heat input in the presence of a low geothermal gradient, sustained hydrate-gas production solely by progressively reducing pressure becomes limited in rate and extent.

4.1.2.2 Thermal Methods

Historically, thermal stimulations have long helped produce viscous crude oils. Elevating reservoir temperature reduces oil viscosity exponentially. With crude oil, thermal stimulation is accomplished by steam injection, huff "n" puff procedure, hot water flood, fireflood, or wellbore heating. A wealth of data exists from these conventional processes. To produce hydrate gas at a sustained practical rate, an energy addition will usually be needed during a depressurization procedure as compensation for endothermic hydrate decomposition. Injected energy would supplement natural heat flux from the geothermal gradient.

Efficiently delivering warm fluids to the reservoir with minimum heat loss in transit, followed by evenly distributing the heat, is a difficult objective common to both hydrate and crude oil thermal processes.

4.1.2.3 Inhibitor Injection Method

Polymer flooding and surfactant injection for tertiary oil recovery are fitting oilfield analogies of the hydrate inhibition method. Each process has the nemesis of losing polymer, surfactant, or inhibitor by adsorption on reservoir minerals, making good process economics difficult, if not untenable.

Injecting thermodynamic inhibitors would lessen hydrate stability by creating more difficult equilibrium P–T conditions in the reservoir. The inhibition method would shrink the hydrate stability envelope by injecting methanol, saline solutions, or other thermodynamic inhibitors (Kurihara et al., 2011a).

4.1.2.4 Replacement Method

Enhanced production methods evoke hydrate instability by either changing thermodynamic conditions or constricting the hydrate equilibrium envelope within the reservoir. Eventual optimum production of hydrate gas may evolve from combinations of conventional methods to produce oil and gas, or by including new innovative techniques.

For example, an envisioned carbon dioxide replacement method injects CO_2 into a hydrate reservoir to supplant and release methane from the hydrate structure; CO_2 hydrate forms at a less stringent P–T relationship than does CH_4 hydrate. The exchange of carbon dioxide for methane in the hydrate structure is important for two reasons. First, the exothermic reaction forming CO_2 hydrate releases more latent energy than required for the endothermic methane hydrate decomposition. This raises the specter of producing methane while circumventing cost and difficulty of direct heat injection. Second, the process introduces a possible means to sequester captured CO_2 industrial emissions. As an aside, note that the replacement of CH_4 with CO_2 leaves intact *in situ* hydrates, retaining shear strengths of sediments residing on a continental slope.

The venerable CO_2 enhanced oil recovery (EOR) process practiced in aging oilfields of the United States, therefore, has some processing similarities and is somewhat analogous to the envisioned hydrate replacement method. The EOR process injects CO_2 into an oil reservoir to decrease oil viscosity, increase reservoir pressure, and reduce surface tensions of the oil/mineral interaction – recovering oil left after primary production and secondary waterflooding. The hydrate replacement method injects CO_2 into a hydrate reservoir to supplant methane occluded in hydrate and to sequester CO_2.

4.2 CATEGORIZING GAS HYDRATE RESERVOIRS

In these early exploration stages seeking recoverable hydrate gas, it is useful to have organized the many reservoir configurations. Categorization assists computer simulation and the overall development of production techniques. It helps evaluate environmental frailties and seafloor instabilities of recovery processes. Researchers organized hydrate reservoirs according to class and type.

4.2.1 Reservoir Class

The following hydrate reservoir classes were developed (Boswell et al., 2011; Kurihara et al., 2011a,b). In the first three classes, reservoirs are assumed to have coarse grains, high permeability, and pore-filling hydrates.

Class 1: Hydrates occur directly above BGHS with free gas below in sediments of similar intrinsic permeability and porosity. These reservoirs have good communication with gas sources. Reservoirs may be confined or unconfined.

Table 4.1 Hydrate reservoir classifications

Class	Description	Example
1	Free gas underneath gas hydrate zone	Messoyakha
2	Water underneath gas hydrate zone	Alaminos Canyon Block 818, Tiger-shark (Moridis and Reagan, 2007)
3	Gas hydrate zone isolated, not bounded by mobile fluid	a. Eastern Nankai Trough (Kurihara et al., 2009) b. Mount Elbert, Alaska (Moridis and Reagan, 2007) c. Mallik, N.W. Territories (Moridis and Reagan, 2007)
4	Gas hydrates dispersed, not bounded by mobile fluid. Unconfined	Blake Ridge

Class 2: Water underlies Class 2 hydrate reservoirs that are not necessarily at BGHS. Gas supply limits hydrate extent, where inadequate or dispersed free gas within the reservoirs may leave available water starved for gas. These reservoirs may be confined or unconfined.

Class 3: This reservoir class has confining boundaries. Even small permeability decreases of bounding strata improve sealing effectiveness, shale being an effective seal. Hydrates saturate the unit. Unreacted water pockets in the reservoir may be isolated from methane.

Class 4: Sediments exhibit 5–20% S_{gh}. Reservoirs are fine-grained sands or muds, with low intrinsic permeability and no surrounding seals, i.e., unconfined (Moridis and Collett, 2004).

Some examples of the four classes are given in Table 4.1.

4.2.2 Reservoir Type

Type further distinguishes hydrate-bearing sediments, as outlined by Boswell et al. (2011):

Type C: Consolidated sediments; coarse sand grains; S_{gh} 50–85%:
- C_{TB}: Thin, less than 1 m sands interbedded with shales and muds
- C_{MB}: Thicker sands between shales and muds

Type F: Fine grains, silt to clay sizes; unconsolidated; low permeability:
- F_{PF}: Pore-filling mode; low saturation less than 5% hydrates
- F_{GD}: Grain-displacing veins; nodules; fracture fills
  - $F_{GD\text{-}Ch}$: Many fractures approximately vertical in chimney-focused flow

$F_{GD\text{-}S}$: Bound within specific stratigraphic units

Type M: Massive gas hydrates of cold seep; shallow; mounds may outcrop seafloor

Type R: Hydrates accumulated within permafrost setting

Multiple combinations of type/class are encountered. The most common is Type $F_{GD\text{-}Ch}$/Class 4. At the present state of the art for producing gas from hydrates, the most commercially sought combination is Type C_{MB}/Class 1; contrarily, Type R ranks last. Type M mounds protruding from the seafloor have relatively small gas content, burdened with additional difficulties of not disturbing local ecological communities during production. Thus, Type M mounds are least attractive for potential production but intriguing as environmental or climate studies (Boswell et al., 2011).

4.2.3 Examples of Type/Class Reservoirs

Lacking full suites of well logs, seismic data, and core data, some early hydrate reservoir discoveries cannot be fully characterized. However, those reservoirs listed in Table 4.2 illustrate the classification scheme. Note that the Mount Elbert reservoir, Milne Point, Alaska, extends across different

Table 4.2 Hydrate reservoirs, classification system (Boswell et al., 2011)

Reservoir and location	Type/class	References
Walker Ridge	Class 1 and 2 downdip; Class 3 updip	Boswell et al. (2010)
Nankai Trough	C_{TB}/Class 1	Fujii et al. (2008)
Green Canyon 955	C_{TB}/Class 1	Boswell et al. (2010)
Mallik	C_{MB}	Boswell et al. (2010)
Green Canyon 955-Q	Type C/Class 1	Collett et al. (2010)
WR313 downdip	(estimated)	
Nankai Trough (majority)	Type C_{TB}/Class 2	McConnell et al. (2010)
GC955-H		
Nankai Trough (minority)	Class 3	
Alaminos Canyon 818	Type C_{MB}/Class 2u	Boswell et al. (2009)
Mount Elbert downdip	Type C/Class 2	Moridis et al. (2011)
Mount Elbert, crestal	Type C/Class 3	
Mallik (Zone A)	Type C/Class 2	
South China Sea	Type C/Class 3u Transitional Type F_{PF}	Wu et al. (2010)
Andaman Islands	Type C/Class 3	Collett et al. (2008)
Korea		Lee and Ryu (2011)
Cascadia margin		Riedel et al. (2008)
Alaminos Canyon 21		Frye et al. (2010)

Table 4.2 Hydrate reservoirs, classification system (Boswell et al., 2011) *(cont.)*

Reservoir and location	Type/class	References
Blake Ridge	Type F_{PF}/Class 4	Frye et al. (2010)
Mahanadi		Collett et al. (2008)
Keathly Canyon-151	Type F_{GD-S}	Cook et al. (2008)
Walker Ridge-313		Boswell et al. (2010)
Offshore Norway	Type F_{GD-C}	Plaza-Faverola et al. (2010)
Krishna–Godavari Basin		Collett et al. (2008)
Ulleung Basin, Korea		Ryu et al. (2010)
Qilian Mountains, Tibet	Type R/Class 1	Lu et al. (2010)
Messoyakha field		Makogon et al. (1972)
GOM	Type M	Collett et al. (2009)
Cascadia margin		

strata with varying seals and contains both Type C/Class 3 and Type C/Class 2 hydrate reservoirs.

4.3 RESERVOIR CHARACTERISTICS IMPACTING PRODUCTION

4.3.1 General Criteria for Hydrate-Gas Production

Early on Pooladi-Darvish (2004) recognized some random impediments that must be overcome to economically produce hydrate gas: (1) dispersed hydrates at low S_{gh}; (2) locations lacking surface accessibility and infrastructure; (3) low permeability of host media; (4) questionable energy supply to sustain hydrate dissociation.

Other less obvious potential problems are environmental sensitivity, host sediment stability, and susceptibilities of the sediments to mechanical strength alterations from dissociating hydrates. Rutqvist et al. (2009) suggest tests for initial shear stress and sediment elastic properties, especially the Poisson's ratio, preparatory to site selection.

4.3.2 Gas Hydrate Fractional Saturation

At moderate depths below seafloor, hydrates generally have larger gas contents per unit of reservoir volume than other conventional and unconventional gas reservoirs. Collett and Kuuskraa (1998) compared gas volumes at standard temperature and pressure (STP) per unit volume of specific reservoir sources. Those comparisons were for depths less than 1525 m in reservoirs assumed to have porosities of 30%. The maximum energy content of the reservoirs ranges from 50 gas volumes/reservoir rock volumes for hydrates to 5 gas volumes/reservoir rock volumes for Devonian shale. Overall, the comparison sequence ranged as follows: hydrates > conventional gas > coalbed methane > tight sands > Devonian shale.

Under these particular conditions, energy density of hydrate mass compares well with other reservoir types, including superiority to conventional sandstone gas reservoirs.

4.3.3 Heterogeneous Reservoir Permeability

Seol et al. (2010) report that heterogeneous permeability improves methane extraction rates from cores by orders of magnitude. They arrived at this conclusion by analyzing data fed directly into a numerical simulator from X-ray computed tomography (CT) scans of cores during production experiments.

Based on these laboratory core-scale experiments, greater cumulative gas production could be expected from heterogeneous reservoirs. Note that high initial production rates leading to larger cash flows early in production become decisive when determining project profitability based on rates of return or net present value measures of profitability (Thompson and Wright, 1985).

4.3.4 Bounding Strata Seals of Hydrate Reservoirs

Prospects for successful production from a hydrate reservoir are enhanced if bounding strata during hydrate dissociation prevents excessive water entry, gas loss, and heat loss. An outstanding example of desirable hydrate reservoir to boundary permeability contrast is that of Mount Elbert (Winters, 2011). Two sites exhibit seals one to two orders of magnitude less permeable than the encased hydrate reservoir. Contrasts of reservoir to seal permeability for the two sites are 675/49 and 1700/5.7 mD, reservoir sand to bounding strata.

4.4 HISTORICAL HYDRATE-GAS PRODUCTION IN THE ARCTIC

Some of the early, reliable data on feasibility of hydrate-gas production come from wells drilled in Arctic regions. The Mallik and Mount Elbert wells in polar regions of Canada and Alaska, respectively, assume importance for potential marine, hydrate-gas production. In fact, striking similarities exist between these Arctic hydrate reservoirs and those offshore in Nankai Trough (Konno et al., 2010).

4.4.1 Discovery Well L-38

While drilling L-38, a conventional oil and gas exploration well in the Canadian Arctic, a thick zone of gas hydrates was penetrated. Thus, L-38 became the discovery well of the Mallik gas hydrate field. Although no cores from the discovery were analyzed at the time, well logs indicated an unusually large interval of hydrates below permafrost. Based on the logs, gas hydrates were estimated to occupy at least 67% of the media pore spaces (Collett and Dallimore, 1998).

4.4.2 Core-Evaluation Well 2L-38

A consortium of companies followed up the Mallik hydrate discovery, drilling the 2L-38 research well. Based on their analyses of core samples from the 886–952 m interval of 2L-38, some observations proved significant enough to warrant continued hydrate exploration – an important consideration at the time for fledgling universal hydrate programs (Uchida et al., 2002; Winters et al., 2004; Wright et al., 1999):

1. Filling pore spaces of reservoir sands but not cementing the grains, hydrate saturations reached 70–80%.
2. Sands/pebbles interbedded with silts and clays characterized the reservoir media with hydrates occurring in 1.5 m thick sand-dominated intervals of 25–36.3% porosity. No appreciable hydrates accumulated within the silts and clays, although infrequent hydrate veins less than 1 mm thick or hydrate nodules less than 1 cm diameter were observed.
3. Hydrate particles less than 2 mm diameter with the largest reaching 2 cm occurred near 914 m depth.
4. Temperature at 900 m depth was approximately 7°C.

4.4.3 Mallik Drilling Program

4.4.3.1 Objectives

Three gas hydrate research and development wells were then drilled near conventional wells at Richard's Island, Mackenzie Delta. Almost 100 researchers representing 8 national organizations sponsored the drilling program, coordinated by the Geological Survey of Canada. Objectives were to evaluate sand strata having high pore saturations of hydrates. Geophysical, geochemical, and geological evaluations were made as well as abbreviated tests on producing hydrate gas by depressurization (Dallimore and Collett, 2005). Reservoir performance, seismic, and well-log data of the Mallik operation provide basic information for future seafloor gas hydrate evaluations.

4.4.3.2 Predrilling Problems to Overcome

Whether offshore or in Arctic regions, gas hydrates are found in extreme environments, often difficult to access. For example, consider the Mallik site. Permafrost extends 640 m below surface. Dallimore and Collett (2005) relate trucking and barging drill rig and 760 tons of equipment to the site. Harsh weather conditions and environmental standards permit a narrow window each year for work. During the seasonal window, mean air temperatures reach −10°C on the surface. In order to perform analyses, cores were hauled 200 km over constructed ice roads to four laboratories and program offices (Dallimore and Collett, 2005).

Difficulties of producing Arctic wells portend different but difficult problem sets to overcome in extreme conditions of producing deep-sea, hydrate gas.

Two observation wells, 3L-38 and 4L-38, were subsequently drilled on opposite sides of 5L-38 but 40 m distant. In Table 4.3 is a summary of wells drilled at Mallik, production method, and results.

4.4.3.3 Results

Between 890 and 1106 m depths, cores and well logs identified 10 hydrate intervals with cumulative thickness of 110 m at Mallik (Dallimore and Collett, 2005). Gas hydrates in the intervals contain an estimated 0.77 bcm (27 bcf) of in-place gas (Rutqvist et al., 2009).

The Mallik drilling program provided many findings useful for future hydrate projects. Core analyses, density logs, and resistivity logs show increasing hydrate saturations with porosity increases. Hydrate saturations S_{gh} as high as 80–89% exist where porosities reach 40%. High S_{gh} for the predominant porosity range of 0.2–0.4 over 890–1106 m depths are illustrated in Figure 4.1 for well 2L-38 (Jin et al., 2002).

From a best fit of the data over the predominant porosity range, an empirical relationship in Equation 4.1 gives S_{gh} as a function of porosity ϕ, specific to Mallik (Jin et al., 2002):

$$S_{gh} = 4.5(\phi - 0.2) \tag{4.1}$$

Sand gravel, which comprises the porous media, has grain diameters ranging from 11 to 337 μm over the subject intervals (Winters, 2011).

Temperature profiles were taken over a period of 622 days from top to bottom along the 1200 m deep observation well 3L-38. The resident geothermal gradient, representative of the three 2002 Mallik wells, is 23.5–24.1 K/km (Henninges et al., 2005).

Table 4.3 Summary of production research wells at Mallik

Well	Drill date	Hydrate interval	Production method	Results
2L-38*	1998		n/a	1. Log data 2. Core samples
5L-38[†,‡,§,||,¶] 3L-38 Observation well 4L-38	2002	800–1100 m depth	1. Thermal 2. Depressurize	1. Coring, geophysics 2. 468 m^3 gas produced 3. Promising
2L-38** Reentry	Winter 2007	Zone A[#] 1093–1105 m	1.5-Day production test; produced water reinjected to aquifer below	Excessive sand product; depressurize in three stages Approximate cumulative[#] gas production = 830 m^3
2L-38** Reentry	Winter 2008	Zone A[#] 1093–1105 m	6-Day production test. Reduce pressurize in three steps (11 to 7.4 to 4.5 MPa) with downhole electric submersible pump.[††] Recomplete. Add sand screen. Heat bottom 10 h	Gas produced = 2000–3000 m^3/day; total gas = 13,000 m^3 Water produced = 10–20 m^3/day; total water = 70 m^3

[#]Dallimore et al. (2008).
[*] Dallimore et al. (1999).
[†] Hancock et al. (2005a).
[‡] Hancock et al. (2005b).
[§] Kurihara et al. (2005a).
[||] Kurihara et al. (2005b).
[¶] Dallimore and Collett (2005).
[**] Kurihara et al. (2011a).
[††]Yamamoto et al. (2011).

The reservoir parameters and their ranges in the interval 1093–1105 m of well 2L-38 used in computer simulations conducted by Kurihara et al. (2011a) to match water and gas production from the two field tests included 30–35% porosity, 10–20% shale content, 100–1000 mD absolute permeability, and 0.01–1.0 mD initial permeability. Initial pressure and temperature at the bottom of the 1113 m interval were taken as 11.3 MPa and 285.7 K.

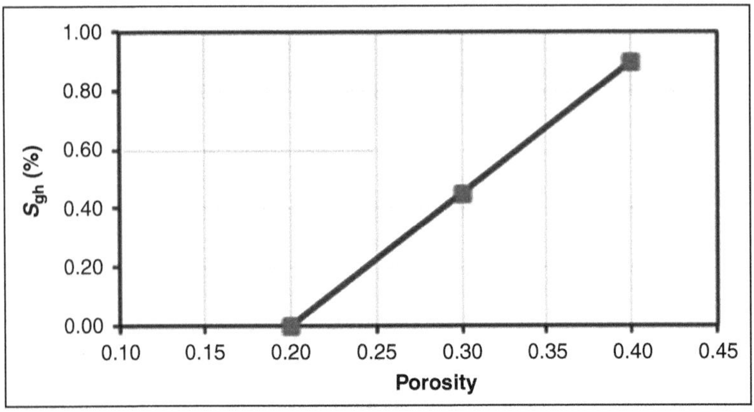

Figure 4.1 ***Hydrate concentrations for Mallik 2L-38 over predominant porosity range.*** *(Modified from Jin et al., 2002.)*

Extensive history matching deduced that production tests extended 7–10 m radially from the wellbore and extended 2 m below and 2 m above perforations. In order to match field observations, the following mechanism was deduced: gas hydrates decomposed, sand eroded as gas flowed to the wellbore, and open channels developed near the wellbore, creating high-permeability ratholes. Large sand amounts produced in the Mallik tests led thereafter to inserting sand screens (Kurihara et al., 2011a).

4.4.4 Prudhoe Bay–Kuparuk River

The Prudhoe Bay–Kuparuk River hydrate field has similarities with the Siberian field of Messoyakha. Similarities include Cretaceous and Tertiary formations and free gas below BGHS. Collett (1992) estimates that the Alaskan field contains more methane than the Siberian analog, although a drill stem test gave 112 m^3/day gas flow rate – a rate similarly obtained at Messoyakha in depressurization tests.

Approximately 1.0–1.2 tcm of methane is stored in the hydrates of Prudhoe Bay and Kuparuk River regions, which is thought to exceed the conventional gas in the Prudhoe Bay field (Lorenson et al., 2008; Collett, 2002).

4.4.5 Mount Elbert, Alaska North Slope

The Mount Elbert gas hydrate field contains 0.99–1.19 tcm (35–42 tcf) of methane within three stratigraphic units of 11.25 m cumulative thickness (Rutqvist et al., 2009). Hydrates within six sands have been identified

Table 4.4 Hydrate test well, Mount Elbert (Lee and Collett, 2011)

	Value
Hydrate Zone D thickness (m)	13.4
Hydrate Zone C thickness (m)	16.5
Zone C, D electrical resistivity (Ω m)	50–100
V_p, compressional wave velocity (km/s)	3.4–4.0
V_S, shear wave velocity (km/s)	1.1–1.8
S_{gh}, hydrate saturation (%)	50–78
Intrinsic permeability (mD)	>2
Porosity, Unit D (%)	38
Porosity, Unit C (%)	40
Mobile water (%)	15
Effective permeability, presence of gas hydrate (mD)	0.12–0.17

(Winters, 2011). Data for the field come from exploratory wells, cores, and well logging.

Hydrates occur at depths of 50–661 m below 590 m thick permafrost. Similar to the Mallik field, the gas hydrates at Mount Elbert are dispersed as high fractional filling of pores in thick sand layers (Kurihara et al., 2011a), where grain diameters usually exhibit a mean of about 210 μm (Winters 2011). Hydrates were found below the base of permafrost in two stratigraphic units termed C and D. See Table 4.4.

Isotopic analyses of core gases reveal a thermogenic gas source $\delta^{13}C$ of −48.6% for predominantly sI methane hydrates. The carbon-13 content was uniform throughout 131 m of cores recovered from the Mount Elbert test well (Lu et al., 2011).

Coring and pressure testing complemented wireline logging results through the reservoir. A mineral-oil-based drilling fluid chilled to 0°C maintained borehole stability, gave hydrate-interval stability, and generally improved data (Lee and Collett, 2011).

Unconsolidated sandstones with grain sizes up to 210 μm characterized cores recovered from Unit C and Unit D; hydrate saturations reached 77–78% (Winters, 2011; Lee and Collett, 2011).

The Mount Elbert hydrate reservoir exhibits important properties of high intrinsic permeability, greater than 50% saturation values, thick sand units, and pore-filling dispersed hydrate morphology. The reservoir has another feature desirable for producing hydrate gas: permeable sands are sealed by low-permeable strata. As Winters (2011) determined with cores from C and D units, the hydrate-laden sands are 14–300 times more permeable than bounding strata.

4.5 PRODUCTION METHODS FOR HYDRATE GAS

Because of the complexity of developing a hydrate-gas production process and the associated field costs of tests in extreme environments, computer simulations attain an importance unprecedented in previous hydrocarbon productions.

Consider some process complexities to be anticipated:

1. During hydrate decomposition and gas release, the endothermic reaction cools the reservoir. Hydrate decomposition solely by progressively decreasing reservoir pressure becomes increasingly difficult – impractical in the limit.
2. Energy must be introduced uniformly and cheaply to supplement a typically inadequate geothermal heat flux.
3. Hydrate reservoir permeability changes with production.
4. Alternating fluid phases affect sediment strengths.
5. Suites of high-quality seismic, well logs, sediment mechanical strength, hydrate-sediment data, thermal conditions, and complete reservoir characterizations become necessary.
6. Operations must be planned for extreme conditions.
7. At hydrate sites, a delicate ecology must be anticipated.

The following sections review most promising methods for economic hydrate-gas production. The methods are based on current state of the art established by reported process simulations, laboratory experiments, and field data.

4.5.1 Depressurization Method

4.5.1.1 Operational Characteristics

The DM for producing hydrate gas has potential for reducing process operating costs when deployed alone or as a complement to other method(s). In the DM a bottomhole pump removes free water to lower reservoir pressure below hydrate stability at prevailing reservoir temperatures. Lowering reservoir pressure near the wellbore imposes a sequence of process events as follows (Kurihara et al., 2011a,b):

1. Hydrates near the wellbore dissociate first.
2. Initial production of dissociated gases and waters increases effective permeability near the wellbore.
3. Localized permeability enhancement allows depressurization to extend further into the reservoir, enlarging the radius of hydrate dissociation.

Table 4.5 Comparison of Mallik/Mount Elbert hydrate deposits (Rutqvist et al., 2009)

Property	Mallik	Mount Elbert
Permafrost zone (m)	606	590
Hydrate layer depth (m)	906–930	650–661
Hydrate layer thickness (m)	24	11.25
Initial pressure at base of hydrate layer (MPa)	9.1	6.6
Initial temperature at base of hydrate layer (°C)	7.2	2.8
Hydrate saturation (%)	75	65
Intrinsic permeability (mD)	1200	1000
Porosity (%)	37	35
Thermal conductivity [W/(m K)]	2.24	3.1

4. Relatively large endothermic latent heats of hydrate dissociation cool the reservoir when dissociation rates release methane rapidly enough to be economically acceptable but faster than can be maintained by the geothermal gradient (Oyama et al., 2009).

Simultaneously, water from hydrate dissociation must be removed along with sediment matrix water to maintain low pressures. Bulk reservoir densities change as solid hydrates decompose and released water and gas phases dissipate. The changes impose stresses on the media matrix. Sediment's hydrate-cementing loss weakens reservoir structures, especially of concern on continental slopes. Subsidence in the case of shallow, unconsolidated sediments becomes another concern (Kimoto et al., 2007; Rutqvist et al., 2009).

4.5.1.2 Simulation: Arctic Productivity

Rutqvist et al. (2009) evaluated by numerical simulation the potential productivity of hydrate gas from two Arctic sites, Mallik and Mount Elbert. Both sites have Class 3 gas hydrate reservoirs (Moridis and Reagan, 2007). See Table 4.5 for a comparison of reservoir conditions at the two sites.

Both Mallik and Mount Elbert have substantial in-place hydrate gas. To help put into perspective the extent of hydrate gas, note that Mount Elbert hydrates contain approximately twice the conventional gas reserves of Prudhoe Bay Field (Lorenson et al., 2008).

In a DM simulation applied individually to Mallik and Mount Elbert, Rutqvist et al. evaluated horizontal drilling effects on production. In terms of fractional hydrate decomposed, Figure 4.2 shows less than 15% of Mallik's reservoir hydrates decomposed solely by depressurization during steady production over 5 years. Of the two reservoirs, Mount Elbert production

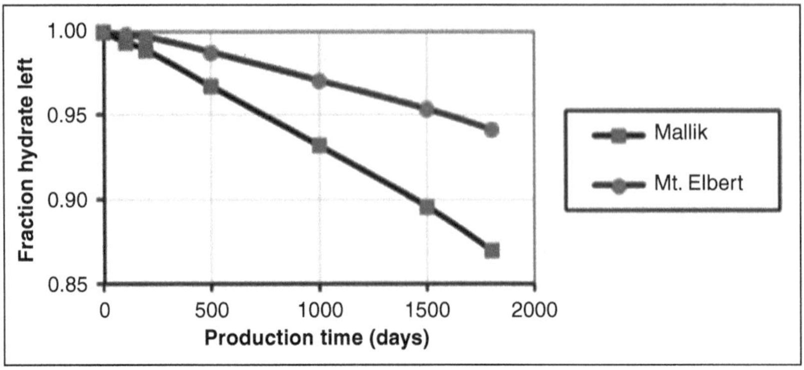

Figure 4.2 ***Productivity comparison: Mallik/Mount Elbert reservoirs (Rutqvist et al., 2009).***

would be much slower, primarily because of its initially lower reservoir temperature (Rutqvist et al., 2009).

Initial temperature at the base of the Mount Elbert hydrate accumulation is 2.8°C versus 7.2°C for Mallik. As a consequence of the temperature difference, lower pressures must be imposed at Mount Elbert to reach hydrate dissociation conditions.

Using DM simulation, Konno et al. (2010) found hydrate reservoirs of high permeability at eastern Nankai Trough, Mallik, and Mount Elbert to exhibit gas productivity roughly proportional to the initial reservoir temperature.

4.5.1.3 Permeability Affects Productivity

In a simulated DM applied to a Class 3 reservoir, Konno et al. (2010) determined gas production rate as a function of absolute permeability when initially reducing flowing bottomhole pressure to 4 MPa. The Konno simulations of Class 3 reservoirs are pertinent to prospective reservoirs in GOM and eastern Nankai Trough having high hydrate saturations of permeable sands. Production rate peaks later with lower value when permeability is reduced to 100 from 500 mD in the Class 3 reservoirs. Generally, higher permeability allows faster propagation of the depressurization front from wellbore to outer reservoir boundaries.

A large early production profoundly improves process profitability. In fact, a threshold initial effective permeability exists for the Class 3 reservoir studied by Konno et al. (2010), below which profitable production becomes unlikely. Initially, effective permeability must exceed 1–10 mD before DM productivity is influenced by reservoir temperature.

During depressurization, a sufficiently high permeability is necessary to transmit pressure reductions and to remove dissociation products. In the simulation of Masuda et al. (2002), effective permeability at any time during the process is related to S_{gh}. By letting k_{do} represent permeability of the host media when no hydrate is present, the permeability k_d at any later time is given by Equation 4.2, where n is a permeability reduction index and S_{gh} is the fractional saturation of gas hydrate:

$$k_d = k_{do}(1 - S_{gh})^n \tag{4.2}$$

In simulating production of gas hydrate reservoirs by depressurization, Masuda determined transmittance to decrease when n increases. Ultimately, high n values reach the point where further lowering of pressure in the wellbore becomes less efficient for decomposing hydrates. That is, at n values between 2 and 11, gas is produced from hydrates, but no production results at an n value of 20.

Pooladi-Darvish (2004) suggests that the most effective depressurization process should be drilling into a free-gas phase trapped below a hydrate permeability cap at BGHS, thereby reducing reservoir pressure by producing the free gas and then coproducing hydrate decomposition gas as a consequence of the lower pressure. Although initial hydrate dissociation occurs rapidly, the technique is self-limiting because the necessary energy must transfer from surrounding media, thus cooling the matrices surrounding hydrates and moving *P–T* of hydrates toward equilibrium. Ice eventually forms as temperature decreases. Therefore, even under conditions of favorable permeability with trapped free gas below the hydrate zone, the ultimate success of a depressurization process depends on rapidly supplying energy to the hydrates to complement sensible heat gained from the reservoir matrix through the geothermal gradient.

4.5.1.4 Thermal Conductivity Affects Productivity

A critical factor for economical methane extraction from offshore hydrates is the efficient delivery of latent heat for hydrate mass dissociations – a task assisted by favorable reservoir thermal conductivities.

Simulations have evaluated impact of reservoir thermal conductivity on DM production (Warzinski et al., 2008). Conditions established for the Warzinski simulations are straightforward: constant boundary temperatures, no free gas or aquifer beneath the reservoir, impermeable overburden, and impermeable underlying boundaries. The simulations employed composite reservoir thermal conductivities of 0.50 and 0.68 W/(m K) in which the conductivities were calculated according to Equation 2.6. Most noteworthy is that conductivity becomes increasingly important to cumulative production as the affected reservoir area expands. Process simulation predicts cumulative hydrate-gas production for a reservoir of 0.68 W/(m K) composite conductivity to exceed production of a similar reservoir of 0.50 W/(m K) conductivity by 14% after 5 years (Warzinski et al., 2008). The higher conductivity allows more efficient heat transfer to an extended reservoir area.

4.5.2 CO_2–CH_4 Exchange Method

4.5.2.1 *Potential Advantages of CO_2–CH_4 Exchange Method*

Injecting carbon dioxide into marine hydrate reservoirs and producing occluded methane as an exchange has intriguing potential. Multiple advantages are envisioned: (1) exothermic heat from forming CO_2 hydrates ($\Delta H_{\text{f-CO}_2}$) supplies energy for endothermic dissociation of methane hydrates ($\Delta H_{\text{dis-CH}_4}$); (2) hydrate-containing sediments maintain uninterrupted structural integrity of the reservoir; (3) CO_2 hydrates are thermodynamically more stable than CH_4 hydrates at relaxed pressure/temperature combinations; (4) added value from sequestering industrial waste CO_2 and supplementary energy from $\Delta H_{\text{f-CO}_2} > \Delta H_{\text{dis-CH}_4}$ improve process economics; (5) reduced production of water accompanies the methane recovery; (6) the exchange offers a possible means for disposing industrial CO_2 waste.

Process innovations are underway in laboratory research and in the field. As an example of the many studies, the German Ministry of Research and Education funds laboratory studies of the CO_2–CH_4 exchange process to produce hydrate-gas from bounded, coarse-sand reservoirs of high S_{gh} (Wallman and Bialas, 2010).

The exothermic heat of CO_2 hydrate formation is compared with the endothermic heat of CH_4 hydrate dissociation in the reactions of Equations 4.3 and 4.4 (Farrell et al., 2010):

$$CO_2\,(g) + nH_2O \rightarrow CO_2(H_2O)_n, \qquad \Delta H_f = -57.98\,\text{kJ/mol} \tag{4.3}$$

$$CH_4(H_2O)_n \rightarrow CH_4\,(g) + nH_2O, \qquad \Delta H_{\text{dis}} = 54.49\,\text{kJ/mol} \tag{4.4}$$

Note from the preceding two equations that the energy liberated in the exothermic reaction to occlude 1 mol of CO_2 in hydrate exceeds the energy absorbed in the endothermic hydrate dissociation to release 1 mol of methane.

4.5.2.2 Phase Diagram of Carbon Dioxide

To inject carbon dioxide into a marine hydrate reservoir at conditions of gas hydrate stability requires close attention to temperature/pressure combinations during transfer and CO_2 injection. Brewer et al. (1999) illustrated CO_2 phase transitions encountered while releasing CO_2 from a submersible into Monterey Bay waters at depths ranging from 349 to 905 m exhibiting temperatures of 8 to 4°C, respectively. At 349 m depth CO_2 exists as gas phase, but at injection depths of 430 and 905 m the liquid form develops. At 2°C and depths greater than 2600 m, the liquid CO_2 becomes denser than seawater. On the seafloor, Brewer et al. (1999) found that CO_2 does not wet seafloor particles.

Representative laboratory tests and numerical simulations of CO_2–CH_4 exchange within hydrates are given in Table 4.6.

Table 4.6 Laboratory tests releasing CH_4 from gas hydrates, replacing with CO_2

References	CO_2 injection	Initial hydrate phase	Final hydrate phase	Transfer time (h)	Final CH_4/ CO_2 hydrate ratio
Schicks et al. (2011a)	1 ml/min flow CO_2 (g) continuous	CH_4 (h) from deionized water	90% CH_4 recovered (Raman spectrometer)	90	10/90
Hirohama et al. (1996)	Replace CH_4 (g) with CO_2 (l)	CH_4 (h)	15% CH_4 (h) + free water to CO_2 (h)	800	85/15
Lee et al. (2003)	Replace CH_4 (g) with CO_2 (l)	CH_4 (h) from powdered ice in silica gel	50% CH_4 (h) recovery	5	50/50
Schicks et al. (2011b)	CO_2 (g)	CH_4 (h) from water or powdered ice	50–90% CH_4 (h) to CO_2 (h)	100	50/50 to 10/90
Park et al. (2006)	N_2 (g) + CO_2 (g) 80 + 20%	CH_4 (h) from powdered ice	85% CH_4 (h) recovery	24	15/85
Kvamme et al. (2007)	CO_2 (l)	CH_4 (h) in sandstone pores	60% CH_4 (h) recovery	300	40/60
Park et al. (2006)	CO_2 (g) CO_2 (g) + N_2 (g)	sII CH_4–C_2H_6	92–95% CH_4 (h) + 93% C_2H_6 (h) recovery	–	5/95 CH_4 + 7/93 C_2H_6

Note in Table 4.6 the necessity of continually removing methane as it is released by hydrate dissociation to prevent its diluting carbon dioxide injection gas. In fact, refreshing those pressurizing fluids with a flow of pure carbon dioxide improves process effectiveness (Schicks et al., 2011a). This emphasizes that CO_2 partial pressure must be maximized during the replacement process – a difficult separation task in laboratory tests as well as seafloor production.

Because exchange rate depends on direct contact of refreshed CO_2 gas phase with methane hydrate, large hydrate surface areas contacting gas phase facilitate exchange. Further, high permeabilities of enclosing sediments to help remove released methane and to accommodate refreshing of the exchange gas therefore become important to CO_2–CH_4 exchange (Hirohama et al., 1996; Schicks et al., 2011a).

4.5.2.3 Difficulties with CO_2 Injection into Reservoir

Bigalke et al. (2011) investigated effects of media permeability and porosity when injecting CO_2 to exchange with CH_4 hydrates. The experiments began by pressurizing ice particles with methane to form methane hydrates, grinding the resulting solid hydrate into 0.3–1.0 mm diameter particles, and then creating alternate layers of hydrate–sand in a test vessel. Carbon dioxide was injected as a 75% saturated saline solution or as a saturated solution in conjunction with liquid CO_2.

An undersaturated saline solution gave no CO_2 exchange with the methane hydrate. That is, CO_2 hydrates did not form after injecting the undersaturated solution into the test cell. The saturated CO_2 solution plus liquid CO_2 injection formed CO_2 hydrates near injection ports, blocking further transport to contact the methane hydrate packing. The experiment of Bigalke et al. (2011) foreshadows practical problems of injecting CO_2 to fully contact hydrate surfaces within reservoir porous media matrices.

4.5.2.4 CO_2 Exchange with CH_4 in sI Hydrates

If the CO_2 exchange method is to become a viable means of recovering methane, the exchange must be kinetically feasible in reservoirs of sI methane hydrates. Methane and carbon dioxide as individual gases or as mixtures form sI hydrates with the customary unit cell having two small cavities and eight large cavities. The CO_2 molecule has approximately the same effective diameter as a small cavity, whereas the CH_4 molecule is about 85% of the small cavity size.

Consider the kinetic data of Lee et al. (2003), where sI methane hydrate rates were determined by first pressurizing powdered ice (270 K, 215 bar) with methane gas while monitoring the reaction by nuclear magnetic resonance (NMR) spectra. The formation rate of methane hydrate was slow. But when the methane hydrate formed from powdered ice is exposed to CO_2 gas at the same temperature and pressure, replacement rate with CO_2 becomes an order of magnitude faster than original methane hydrate formation, where nearly half of the methane is replaced in 5 h (Lee et al., 2003).

Some cell distortion is required for the CO_2 molecule to fit into the sI small cavity (Park et al., 2006), so it is understandable from the NMR data of Lee et al. (2003) that after replacement the CO_2 occupies primarily large cavities of the sI structure, given the molecular diameter to small cavity ratios of CO_2 and CH_4. Therefore, an upper limit of about 75% recovery by CO_2 exchange may be expected because of the sI cell-size frequency of two small cavities per eight large cavities in the sI unit cell.

Simulations give credence to the aforementioned exchange rates and limitations. In the simulation of Qasim et al. (2011), 60% exchange was realized over the allotted time period – diffusion limitations kept the maximum exchange from being fully reached. White and McGrail (2008) performed a field simulation test using a five-spot pattern of wells on a gas hydrate field below permafrost. The center well was the CO_2 injection well; periphery wells were producing wells. The simulated recovery of methane from the hydrates totaled 61.8%.

In the laboratory experiments of Lee et al. (2003), 60% methane recovery was not reached in the exchange of CO_2 for CH_4 from sI methane hydrates generated on powdered ice; CO_2 hydrates began forming from the outside and proceeded toward the center of the particle, reaching a condition where mass transfer was controlled.

4.5.2.5 CO_2 Exchange with CH_4 in sII Hydrates

Some hydrate reservoirs, as occur in the GOM, originate from mixtures of thermogenic and biogenic gases, and sII hydrates prevail.

In the laboratory, Park et al. (2006) investigated recoveries of hydrocarbons from sII hydrates by carbon dioxide exchange. To simplify analyses, their experimental design set an arbitrary ratio of methane/ethane to generate sII hydrates, although each of these components individually would form sI hydrates (Subramanian et al., 2000), and then subjected those sII hydrates to pressurized, pure CO_2. Gas chromatograph, NMR spectrometer, and Fourier transform Raman spectrometer were used in the analyses.

Based on initial gas content in sII hydrates, 92% of the methane was recovered and 99% of the ethane was recovered by CO_2 exchange (Park et al., 2006); under their experimental conditions, most of the ethane and methane was recovered within 20 min. The high yields indicated a carbon dioxide exchange for methane and ethane in all cavity sizes.

Because of the final hydrate composition, a changeover from sII to sI structure apparently occurred during the process. Park et al. (2006) concluded that changeover of hydrate type during the process allowed CO_2 to fit more easily into sI small cavities to displace methane.

4.5.2.6 (CO_2 + N_2) Exchange with CH_4

Injections of CO_2 into methane hydrate reservoirs in an exchange process have the detrimental potential of CO_2 hydrate forming in porous media near injection ports and blocking entry flow of injected gases (Bigalke et al., 2011).

Injections of CO_2–N_2 gas mixtures reduce the blockage problem. *P–T* equilibrium curves for hydrates of methane, carbon dioxide plus nitrogen, and carbon dioxide reveal that more stringent *P–T* conditions are necessary to form the CO_2/N_2 hydrates than to form carbon dioxide hydrates at system temperatures below 10.5°C (Masuda et al., 2011).

Simulations indicate that injecting nitrogen along with carbon dioxide could maintain reservoir pressure while reducing methane partial pressure to help preserve reservoir integrity and dissociate methane, yet help prevent blockage at injection ports (Kurihara et al., 2011b).

Masuda et al. (2011) evaluated the use of CO_2–N_2 gas mixtures to produce methane from sI hydrate in sand cores. Hand-packed cores secured in a Hassler Sleeve having had either 66.5 or 41.4% methane hydrate saturations imposed on them were subjected to gas mixtures containing 60% CO_2/40% N_2. Injection conditions of the exchange gas were 7 or 4°C and 7 MPa. For the lower S_{gh} value, 26.2% of the CH_4 was replaced with CO_2; in the higher S_{gh} case, only 6.5% of the CH_4 was replaced. These productivities were achieved at about five pore volumes of injected gas. Importantly, however, exchange became limited by a thin N_2–CO_2 hydrate film around the methane hydrates that eventually blocked effective mass transfer rates.

Park et al. (2006) found that mixtures of CO_2 + N_2 injection gas improve methane yield through exchange with sI methane hydrate, where the small nitrogen molecule, approximately the size of methane, helps displace methane in the small cavities of sI.

Experiments conducted by Park et al. measured methane recovery from sI hydrate. First, 85% methane recovery resulted from injecting a carbon dioxide/nitrogen mixture; 23% of the recovery was credited to nitrogen displacements in the small cavities, while 62% of the methane was exchanged with carbon dioxide – primarily in the large cavities. Second, 92% recovery of occluded methane from methane hydrates was recovered when injecting pure CO_2 as the displacing gas. While achieving the 92% high recovery, the sI methane hydrate transitioned to sII (Park et al., 2006).

4.5.3 Thermal Stimulation Method

Methane gas hydrates, having a 54.49 kJ/mol endothermic heat of dissociation, decompose according to the general Equation 4.5 (Pooladi-Darvish, 2004; Farrell et al., 2010):

$$(CH_4 \cdot nH_2O) \xrightarrow{\Delta} (CH_4)_{gas} + n(H_2O)_{liquid} \tag{4.5}$$

For sI types, hydrate number n may vary from a minimum of 5.67 to some higher value dependent on gas composition, temperature, and pressure. Any thermal stimulation method of recovering methane from methane hydrate reservoirs must utilize the path of Equation 4.5. Although choice of heat delivery varies, the selection depends on economics, energy availability, and efficiency of transferring to hydrates in the reservoir.

4.5.3.1 Localized Heat Addition

Injecting energy into reservoirs from heating sources within wellbores dissociates hydrates radially from the point source. The effect may extend a disappointingly short distance into the reservoir. Several means to supply the energy are by: (1) downhole heater, (2) hot water circulation downhole, and (3) huff "n" puff injection of hot water in a manner similar to the process used in conventional heavy oil recovery.

These methods have technical merit. However, a downhole heater or hot water circulation within the wellbore relies on heat conduction to extend the heated radius, and conduction delivers heat slowly. The method may conduct heat too slowly to be an economical process in itself but may effectively prevent hydrate reformation near the wellbore while supplementing other methods. Note that any pressure increase near the wellbore, such as would be caused by reforming hydrates, necessitates higher temperatures for hydrate dissociation, further slowing production and raising process costs (Kurihara et al., 2011a,b).

Alternatively, the huff "n" puff method injects hot water directly into the hydrate reservoir from the wellbore. After injection, reservoir temperatures elevate during well shut-in, and dissociated fluids are then produced. Repeated cycles extend radial distances warmed by conduction and advection.

4.5.3.2 Thermal Flood

In the thermal flood technique, injection wells for hot fluids (water or steam) are separate from the producing wells. Convection bolsters conduction heat transfer.

But thermal floods in hydrate reservoirs face potential problems: (1) fluids cool rapidly during delivery. Heat is lost in fluid transit through wellbore and reservoir matrix. Relatively large latent heats must be delivered to dissociate hydrates. (2) Hydrates easily reform if temperatures drop below equilibrium values. (3) Low effective permeability of reservoir media limits rate of fluid injection.

Combining the thermal flood method with the DM alleviates some problems. In a simulation using parameters of eastern Nankai Trough reservoirs, Kurihara et al. applied wellbore heating and hot water injection techniques in conjunction with the DM. As a primary recovery method, depressurization was effective. As a secondary recovery method, hot water injection showed promise (Kurihara et al., 2011b, 2009).

In another simulation, Masuda et al. (2002) employed a reservoir five-spot pattern for heat addition to decompose hydrates and produce liberated methane. Masuda assumed an injection well on each corner of a square while in the center of the square was the production well. Water injection temperatures of 40, 60, and 80°C were simulated. Estimates of methane recoveries indicated almost complete recovery after 30 days.

4.5.3.3 Advantages and Disadvantages of Thermal Stimulation

Disadvantages of producing hydrate gas by thermal injection are similar to those traditionally recognized in thermal stimulation of conventional reservoirs (Moridis and Reagan, 2007):

1. High sensible heat requirements to raise temperatures of inert reservoir media
2. Energy wasted through hydrate layer boundaries
3. Decrease of gas relative permeability with hot water injection
4. Poor heat transfer by conduction, especially with electrical resistance heating

Generating energy at low cost and injecting that energy with minimum losses to heat hydrate zones has familiar technical and economic challenges. For dissociation, relatively high latent heats of 52–54.49 kJ/mol must be transferred to the hydrates (Rydzy et al., 2007; Farrell et al., 2010).

Despite the difficulties, thermal stimulation in some form may be necessary for hydrate-gas production, possibly in conjunction with depressurization – even if only to prevent hydrate and ice reformation near the wellbore (Moridis and Reagan, 2007).

4.5.4 Inhibitor Injection Method

Methanol, glycol, salt solution, or low-dosage synthetic inhibitors may be injected into the reservoir to assist hydrate dissociation. The chemicals constrict the hydrate *P–T* equilibrium envelope, moderating pressure and temperature requirements to create hydrate instability. However, there is a difficult proviso that injected inhibitors must directly contact the hydrate mass (Collett and Kuuskraa, 1998).

One disadvantage of inhibitor injection, deleterious to the degree of being prohibitive, is the threat or perceived threat of environmental damage and toxicity. Other disadvantages are addressed by Moridis et al. (2009): (1) waters from dissociated hydrates dilute the inhibitor solution; (2) synthetic inhibitors are expensive; (3) inhibitor precipitation may damage permeability; (4) similar to chemical injections for EOR, adsorption of inhibitor chemicals on reservoir minerals is costly.

These disadvantages make inhibitor injection a doubtful potential course of independent action. A low, effective permeability of porous media containing hydrates with substantial S_{gh} would make injection of an inhibitor solution difficult as the sole procedure. A more practical application may be the use of inhibitor following one of the hot fluid injection schemes (Kurihara et al., 2011a,b). Then injection contributes to the hydrate decomposition effort as a secondary recovery process. Certainly, thermodynamic inhibitors could help prevent hydrate reformation near the wellbore.

4.5.5 Downhole Combustor Method

Historically, enhanced production of viscous oils with steam flood incurs large energy losses when the hot fluid is injected from a surface generator through the wellbore. An alternative to steam injection, as an attempt to negate generated energy losses in oil recovery, has been the *in situ* ignition of heavy oil in its reservoir – the fireflood. Under favorable conditions the fireflood decreases oil viscosity along the flame front for economical production.

Similar economics, safety, and environmental problems encountered with thermally produced heavy oil reappear in supplying latent heats for thermally decomposing gas hydrates.

Latent heat requirement for endothermic decomposition of hydrate unit mass is miniscule compared with the exothermic chemical energy released in burning methane contained in a unit mass of hydrate (Holder, 1982). Whereas latent heat to dissociate methane hydrate amounts to 52–54.49 kJ/mol (Rydzy et al., 2007; Farrell et al., 2010), burning an equivalent amount of methane releases about 803 kJ/mol of chemical energy (Lee and Trimm, 1995). These considerations support the idea of a downhole combustor to supply required energy for hydrate-gas production.

A downhole, catalytic combustor was designed and built to laboratory scale for hydrate production by Schicks et al. (2011a,b). The inventors report that a 10% (w/w) palladium catalyst on a ZrO_2 support gives flameless methane oxidation below methane's autoignition temperature, a relative low 595°C, and outside an ignition envelope of prevailing CH_4/air concentrations. Once initiated, the combustion self-sustains using released hydrate methane. To make the Schicks combustor more efficient, entering feed is preheated in a countercurrent heat exchanger of a closed system.

4.6 SIMULATIONS OF HYDRATE PRODUCTION ACCORDING TO CLASS, TYPE, METHOD

4.6.1 Class 1 Hydrate Reservoir Simulations

Konno et al. (2010) performed simulations to compare performances of the four reservoir classes subjected to the DM. Important simulator assumptions are listed in Table 4.7.

Class 1 gas hydrate reservoirs are bounded below by the BGHS. Although hydrates are unstable below the lower boundary, free gas may have accumulated there either from hydrate dissociations as the boundary receded or from diffusing gas. Based on cumulative gas production, Class 1 excels over the other classes in the DM simulation results of Konno et al., but much of its production comes from the free gas.

Table 4.7 Assumptions of Konno simulation (Konno et al., 2010)

Parameter	Value
Reservoir temperature, T_R (°C)	14
Permeability, k_{DO} (mD)	500
Flowing bottom-hole pressure (FBHP; MPa)	4
Geothermal gradient (°C/m)	0.03
Seafloor temperature (°C)	3.47
Hydrate saturation, Classes 1–3	0.6
Hydrate saturation, Class 4	0.1
Thickness gas hydrate layer (m)	50
Water depth to seafloor (m)	492–863

4.6.2 Class 2 Hydrate Reservoir Simulations

The simulation of Konno et al. (2010) found Class 2 and Class 3 hydrate reservoirs to perform similarly under depressurization techniques, but the lack of a mobile fluid at the boundary of a Class 3 reservoir slightly improves its productivity over Class 2.

Using the DM, Moridis and Reagan (2007) simulated production from the GOM Class 2 reservoir of Alaminos Canyon, near the Tigershark site. Table 4.8 summarizes some important reservoir properties used in their work.

Noteworthy results of the Moridis and Reagan (2007) simulation are listed as follows:

1. During production, secondary hydrates formed within a radius of less than 20 m from the wellbore, necessitating intermittent thermal stimulation with warm seawater.
2. A 550-day lag occurred initially before significant gas production.

Table 4.8 Properties of the Alaminos Canyon Block 818 hydrate reservoir (Moridis and Reagan, 2007)

Property	Value
Water depth (m)	2750
Hydrate-bearing layer thickness (m)	18.25
Porous media	Sandy
Porosity, ϕ (%)	30
Intrinsic permeability, k	≈Darcy range
Depth, hydrate-bearing layer (m)	3210–3228 (10,530–10,590 ft)
Water zone thickness (beneath HBL) (m)	15

3. During 2015 days of simulated life, the well's gas production rates ranged between 0.95 and 2.66 m^3/s for three designs evaluated. Salinity of produced waters declined less than 10% –implying fewer water disposal problems.

The DM applied to AC-818 in the Moridis and Reagan simulation shows deleterious countereffects from the underlying aquifer during depressurization. In later simulations of AC-818, Boswell et al. (2009) predict that serious production declines by the DM when an infinite hydrate reservoir of high-permeability sand exists in two lateral directions.

4.6.3 Class 3 Hydrate Reservoir Simulations

Nankai Trough offers an example of Class 3 reservoirs.

To help select well sites for the first full-scale drilling of production-test wells in the eastern Nankai Trough, Kurihara et al. (2011b) performed DM simulations on 2D and 3D reservoir models based on seismic, core, and well-log data from two prospect wells, α_1 and β_1, drilled earlier. Distinguishing features of the α_1 reservoir include good layer connectivity with only minor porosity and permeability changes laterally, a 4–6° reservoir dip, and numerous faults contributing to heterogeneity.

The Kurihara simulation by depressurization (pressure reduced to 3 MPa) predicted α_1 sand production over 30 days: (1) 15,000–20,000 m^3/day gas production rate, (2) 150–350 m^3/day water production rate, (3) production greatly benefited due to permeability increase, (4) production drop-off when bottomhole pressures reduced to only 4 MPa instead of 3 MPa, (5) dissociation extended 30–40 m laterally from the wellbore, not reaching a prevailing fault boundary.

The distinguishing features of the β_1 prospect included poor layer connectivity, a 20° reservoir dip, and no detected faults (Kurihara et al., 2011b).

Important productivity results from the Kurihara simulation by depressurization (pressure reduced to 3 MPa) from the β_1 sands over 30 days were the following: (1) 20,000–45,000 m^3/day gas production rate, (2) 250–650 m^3/day water production rate, (3) production greatly benefited due to permeability increase, (4) production drop-off when bottomhole pressures reduced to only 4 MPa instead of 3 MPa, (5) dissociation extended 50–60 m laterally from the wellbore over the 30 production days (Kurihara et al., 2011b).

4.6.4 Class 4 Hydrate Reservoir Simulations

Konno et al. (2010) in their simulations found unbounded and dispersed hydrates of Class 4 reservoirs to show markedly inferior production by the DM in simulations. Poor performance is characteristic of a low S_{gh} of the dispersed hydrates.

Low hydrate saturations are to be expected with Class 4 reservoirs. Moridis and Sloan (2007) studied, by simulating DM, the feasibility of economically producing hydrate gas when dispersed accumulations demonstrate $S_{gh} < 10\%$. They concluded that profitable production from a Class 4 reservoir was improbable because of low gas contents, high water productions, and inherent lack of impermeable reservoir boundaries. With current technology, DM does not appear to be a viable prospect for hydrate-gas production from Class 4 reservoirs.

4.7 ECONOMICS OF HYDRATE-GAS PRODUCTION

Extreme environments in which hydrate gas must be extracted especially affect fixed capital investment, operating costs, and drilling costs. Water depth, remoteness, infrastructure accessibility, duration, and schedule of production impact the economics.

Production simulation results of gas hydrate reservoirs in eastern Nankai Trough suggest profitability to be highly dependent on well spacing and adapting production methods to specific reservoirs (Kurihara et al., 2009). Contemporary drilling costs and prevailing reservoir properties were employed in production simulations of three hydrate wells in eastern Nankai Trough (Kurihara et al., 2011b). Profitability depends on well spacing, gas hydrate saturation, and total gas content of the reservoirs. For the DM of this Nankai simulation, discounted payout occurs 3 years after production commences in the best well evaluated. That is, the discounted profits (cash flows discounted to the beginning of production at 5% per year) will retrieve the capital investments in 3 years. The net present value (NPV) profit (total discounted cash flows in excess of capital investment) amounts to about US$2.5 million for this specific well over a production period of 8 years.

Konno et al. (2010) devised a productivity function based on operating costs that allows preliminary judgment on potential profitability of hydrate-gas production by any devised method. This productivity function E_R is defined in Equation 4.6:

$$E_R = \frac{E_P}{E_i} \tag{4.6}$$

That is, the amount of energy E_p that would be available by combusting the methane product, when compared with process energy input E_i, indicates whether the process can be economical (Masuda et al., 2002).

Applying simulation to three wells drilled in the eastern Nankai Trough, Kurihara et al (2011b) found that the best efficiency as given by Equation 4.6 occurred with the DM when there was a high S_{gh}; the worst efficiency occurred with the thermal method.

However, the energy efficiency of the thermal method improves if injection-water temperatures can be kept low. For example, the energy efficiency decreases from a theoretical maximum of 17.5, when no heat is lost, to an efficiency of 10.8 for 40°C injected water. The efficiency declines even more to 5.0 whenever 80°C water is injected (Masuda et al., 2002).

Substituting expressions for E_p and E_i converts Equation 4.6 to the form given in Equation 4.7, applicable to the thermal method (Masuda et al., 2002):

$$E_R = \frac{890.31 N_{gp}(t)}{\rho_w C_w (T_i - T_f) F_i \Delta t} \tag{4.7}$$

where 890.31, heating value of methane gas (kJ/mol); $N_{gp}(t)$, cumulative moles of gas produced over Δt time period; C_w, specific heat of injected water at average temperature; ρ_w, density of injected water at average temperature; T_i, initial temperature of injected water; T_f, final temperature of injected water; F_i, injection rate of hot water; Δt, time of water injection.

4.8 STORING AND TRANSPORTING PRODUCED METHANE

Wherever offshore hydrate gas is produced, whether in deep waters, remote areas, isolated fields, or anywhere transport infrastructure is nonexistent, the produced gas must be transported to shore.

One conceptual remedy is to reform produced gas into hydrates aboard surface vessels and transport to onshore distribution points (Andersson and Gudmundsson, 2000). The concept has advantages for remote offshore production: (1) approximately 160 volumes (STP) of natural gas per unit volume of hydrate can be practically stored and transported. (2) Handling natural gas in hydrate form is unmatched in safety. Gas releases slowly from hydrate dissociation in the event of storage tank rupture. (3) Relatively low pressures and moderate temperatures are required for storage. (4) Ultrastability of hydrates at atmospheric pressure and −5°C has been demonstrated on laboratory scales; with further R&D, storage and transport at atmospheric pressure might be achieved.

Multiple processes have been envisioned for hydrate-gas storage. Ultimately, the selection must rely on economics.

A most favorable technical and potentially economical process is discussed here – a surfactant-catalyzed process that rapidly forms and self-packs hydrates symmetrically as they form to optimize storage space and to minimize labor, process equipment, and capital costs (Rogers and Zhong, 2002; Rogers and Zhang, 2012). The surfactant-catalyzed process was conceived in a laboratory feasibility study that was followed by a scale-up demonstration with support of the Department of Energy (DOE; Rogers, 1997, 2001). In the scaled-up hydrate process, natural gas in the amount of greater than 142 m^3 (>5000 ft^3) was occluded, stored, and decomposed on command (Rogers et al., 2004, 2005; Rogers and Zhong, 2000).

4.8.1 Criteria for Viable Storage/Transport System

Economic and technical features determine hydrate-process viability to store and transport natural gas on a large scale:

1. Low capital investment. The process must remain simple to minimize capital costs. Simplicity includes rapidly forming and packing hydrates in the field plus dissociation at onshore delivery, efficient self-packing of hydrates as formed, low-pressure storage, and equipment minimization.
2. Minimum operating, maintenance, and labor costs. The goal is minimum equipment in a process of fewest moving parts. To form hydrates, pressurize the chilled solution with incoming gas with no manual or mechanical stirring, pumping, or manual packing of formed hydrates. Decomposition at the onshore point of delivery must be on command with low-grade heat input. Simplicity of the design reduces labor costs.
3. Controlled safety and environment. Reuse water in the cycles of forming and decomposing hydrates. Develop fail-safe rupture of storage vessel. Vessels and tanks accommodate low-pressure storage and transport of the hydrates.

Although an incomplete list, the preceding items are important points to judge a storage/transport process for gas hydrates.

4.8.2 Laboratory Initiation of Surfactant-Catalyzed Hydrate Storage Process

As a basic requirement, hydrates must form rapidly in any economical storage process.

The rate equation for hydrate formation in a stirred laboratory system relates formation rate to pressure, temperature, degree of supercooling, and water–gas interfacial area (Vysniauskas and Bishnoi, 1983). In a nonstirred closed system without surface-active agents, a thin film rapidly forms at the water–gas interface when hydrate equilibrium conditions are reached in the test cell. See Figure 4.3.

Figure 4.3 shows a hydrate barrier film as viewed through the transparent top of a Mississippi State University (MSU) laboratory test cell. Liquid as chilled water solution remains underneath a thin, solid film of hydrates visible on the water's surface below pressurized hydrate-forming gas in the headspace. With the barrier film in place, extremely low hydrate formation rates prevail. In independent experiments, Moudrakovski et al. (2001) estimate the barrier film thickness to be 240–1200 Å, where hydrate reaction rate drastically slows at a film thickness of about 1200 Å. Controlled by slow diffusion of structured water through barrier film capillaries,

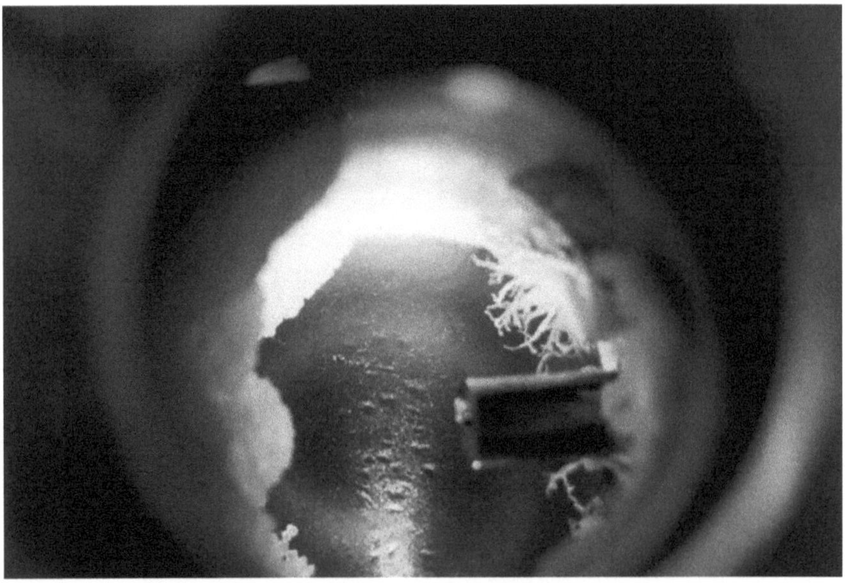

Figure 4.3 ***Hydrate barrier in laboratory quiescent system without surfactant; pressurized gas above film and water below.*** *(Courtesy MSU Gas Hydrate Laboratory.)*

hydrate crystallization occurs in the gas phase above the film, but grows very slowly (Rogers and Zhong, 2000; Servio and Englezos, 2003; Sugaya and Mori, 1996; Kobayashi et al., 2001).

A benign hydrate film may be avoided and rapid hydrate formation achieved in the laboratory by mechanical stirring or, if the test cell is metal, by adding small amounts of anionic surfactant to the active water solution.

Anionic surfactant solutions promote rapid hydrate formation. Surfactant molecules accomplish this without system stirring by nucleating hydrate crystals and adsorbing, with attached developing crystals, onto cold metal surfaces at the water–gas interface. The action prevents a hydrate film from forming across the free-water surface. Microscopic particles of hydrate nuclei attached to surfactant rapidly migrate to metal walls of a test cell in pressurized, chilled, and unstirred systems. Hydrate formation rates of these anionic surfactant systems approach rates achieved in stirred systems (Zhong and Rogers, 2000).

Metal surfaces on which developing hydrate crystals adhere in anionic surfactant systems rapidly remove by conduction the exothermic heats of formation, increasing hydrate generation rates. Metal walls of a test cell become efficient heat exchangers – important for any eventual commercial process to store/transport bulk hydrates. The metal walls of the test cell also conduct electricity. Note in the attached digital film that static charges on hydrate particles cause flocculation as hydrates initially reach the metal surface, possibly slowing agglomeration, but the charges are quickly dissipated by metal conduction.

Intimate contact of adsorbed hydrates on metal surfaces, made to become heat exchangers, facilitates rapid heat removal and makes practical hydrate formation and its symmetrical packing on a large scale. The process functions in reverse: rapid heat addition, hydrate–metal contact, and heat conduction by the metal provide hydrate decomposition in a timely fashion at the delivery site.

An economical hydrate storage process must maximize methane packing density. Whether it is a quiescent or a stirred system, free water becomes trapped in hydrate interstitial spaces as hydrates form to reduce methane content density. Initially, this interstitial water may represent a large percentage of hydrate volume. Englezos (1996) reported 1.4–14% of water initially absorbed in forming hydrates going into the hydrate structure – the remaining absorbed water being interstitially trapped. This would be intolerable in hydrate storage/transport processes. Appreciable volumes of storage tank occupied by interstitial water would decrease methane

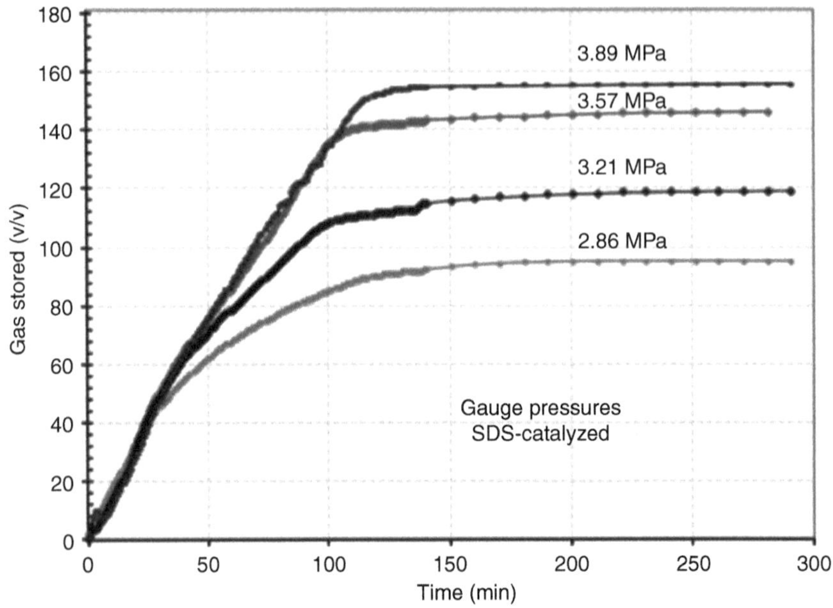

Figure 4.4 ***Pressure effects on hydrate storage capacity and formation rates.*** Hydrates sII formed from natural gas at 275.4 K (Zhong and Rogers, 2000).

storage capacity and increase capital costs. However, anionic surfactant in the system facilitates continuing reaction of interstitial water with diffusing gas, completing water conversion to hydrates. That is, methane diffuses through the porous and permeable hydrate particles packed on the metal walls of the formation vessel, reacting with interstitial water–surfactant solution. Therefore, methane-laden hydrate presents a more dense energy packing.

For a given temperature and gas composition, fractional filling of hydrate cavities with gas increases with pressure. Increased pressures help maximize gas loading and formation rate. But as pressures increase, process economic optimization must consider the increased operating and capital costs. Laboratory experiments were conducted with sodium dodecyl sulfate (SDS), an anionic surfactant, to roughly estimate pressure optimization. See Figure 4.4 (Zhong and Rogers, 2000).

A loading factor of 156 volume gas/(volume hydrate) was achieved at 3.89 MPa (500 psig). This was accomplished with the assistance of 282 ppm SDS in water solution. Hydrates filled the 3900 ml test cell within 2 h and self-packed symmetrically. Significantly, the cell had no moving parts; only constant pressure was maintained. Interstitial water approached full reaction during processing.

Action within the test cell was viewed through a transparent quartz top (3-in. viewing diameter) and by fiber-optic light serving a borescope camera extending into the cell. The photographs in Figures 4.5, 4.6 and 4.7 illustrate action of surfactants in influencing gas hydrate formation.

In Figure 4.5 hydrate accumulations are seen in the test cell after 5–10 days, but hydrates formed without any anionic surfactant present. Then, crystal growths are slow, random, and unevenly distributed throughout the test cell. Note that hydrates grow into the gas phase above the barrier film at the water–gas interface.

When SDS surfactant is added to the water, hydrates grow symmetrically and radially from cylindrical cell walls. Initially, the hydrates form on the cylindrical test-cell walls. After 1 h, the cell is partially filled with hydrates. See Figure 4.6 taken 1 h after process beginning.

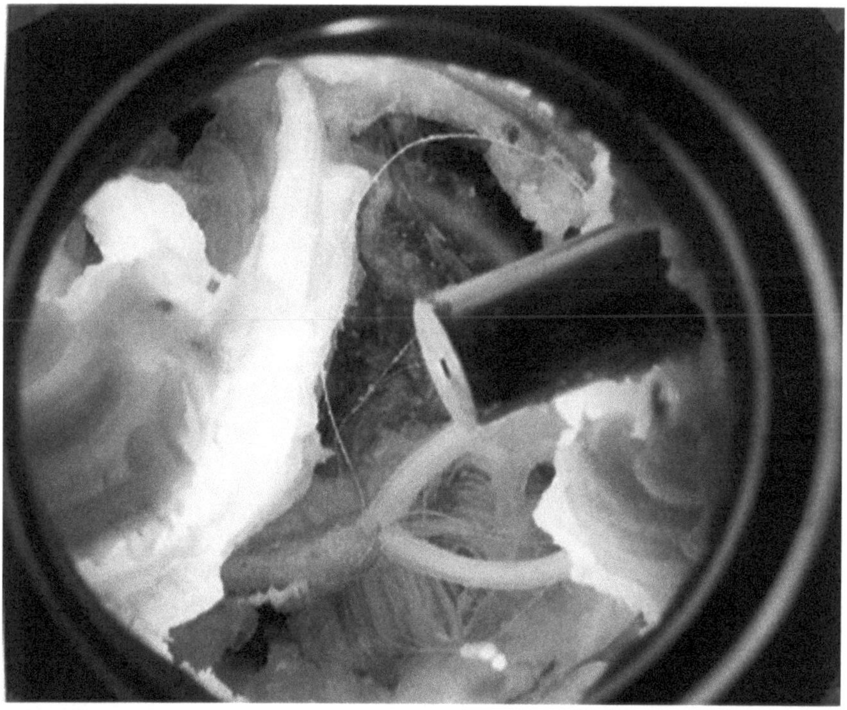

Figure 4.5 ***Nonsymmetrical slow growth of hydrates without SDS surfactant; after 5–10 days, chilled, under gas pressure.*** Hydrates above interfacial film barrier. *(Courtesy MSU Hydrate Laboratory; Zhong and Rogers, 2000.)*

Figure 4.6 ***Symmetry of hydrate self-packing; 282 ppm SDS.*** After 1 h, hydrate partially fills cell; viewed through transparent cell top. *(Courtesy MSU Gas Hydrate Laboratory.)*

After 1 h, hydrate partially fills the test cell as viewed through a transparent cell top (courtesy MSU Gas Hydrate Laboratory).

In Figure 4.7 hydrates fill the test cell after 2 h of growth from the surfactant–water solution. To stress the symmetry of hydrate packing in the cylindrical test cell, note the black hole in the center of the mass that extends to the test-cell bottom. The mass consists of microscopic hydrate particles in a porous and permeable packing.

4.8.3 Proof-of-Concept Scale-Up

The laboratory feasibility study revealed promising attributes of an economical industrial application when using anionic surfactant solutions to form gas hydrates: (1) small concentrations of anionic surfactant solutions increase hydrate formation rates; (2) hydrates self-pack on metal surfaces at gas–surfactant–solution interfaces; (3) hydrates pack symmetrically on metal surfaces under influence of SDS, an anionic synthetic surfactant; (4) porous and permeable hydrate mass adsorbs on metal surfaces, allowing diffusing gases to continue reacting with interstitial water; (5) gas hydrates fill formation vessel rapidly with 156 gas volume per hydrate volume; (6)

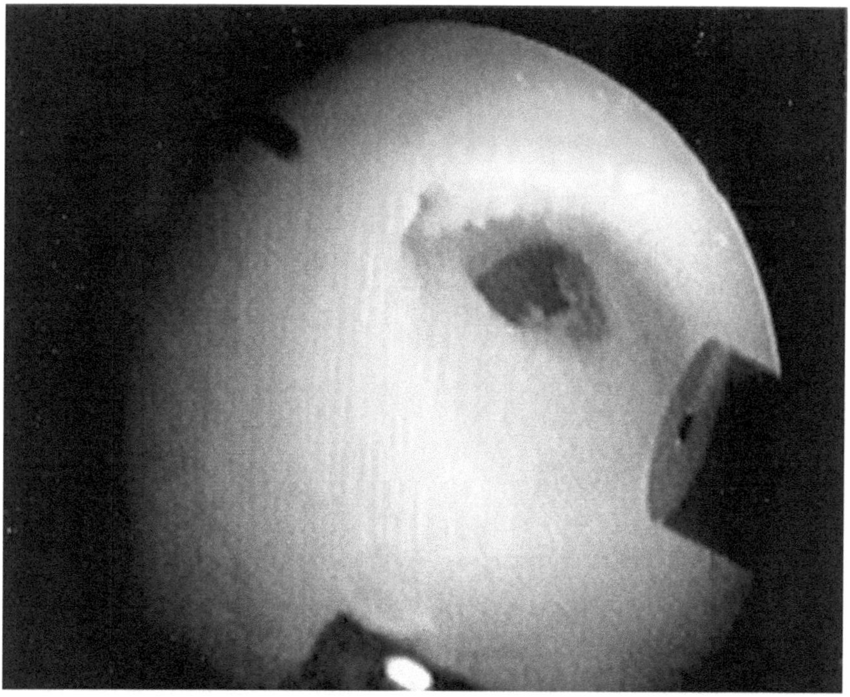

Figure 4.7 ***Symmetrical hydrate self-packing, 282 ppm SDS.*** After 2 h, hydrates fill cell of capacity 156 volume gas/(volume hydrate). *(Courtesy MSU Gas Hydrate Laboratory; Rogers et al., 2007.)*

no mechanical stirring is required with anionic surfactant; (7) after storage, controlled and rapid hydrate-gas release is possible by heat conduction through base metal on which hydrates adsorb; (8) water/surfactant solution remaining after hydrate dissociation retains memory, and the solution may be reused to advantage.

These attributes offer potential for a practical larger-scale process: (1) to lower labor requirements, (2) to lower maintenance costs, (3) to lower capital investment, (4) to form or dissociate gas hydrates rapidly on command.

A US Department of Energy–sponsored proof-of-concept (POC) scale-up was designed by the MSU Hydrate Laboratory with the following features. Initially, an anionic surfactant in water solution partially fills a quiescent formation vessel. On pressurizing the system with natural gas at an operating temperature of 1.7°C (35°F), hydrates form, self-pack, and store until released on command 141.6 m^3 (5000 ft^3 at STP) of occluded gas per cycle. Finned metal tubes provide maximum surface areas on which hydrates collect (Rogers and Zhong, 2002). See Figure 4.8.

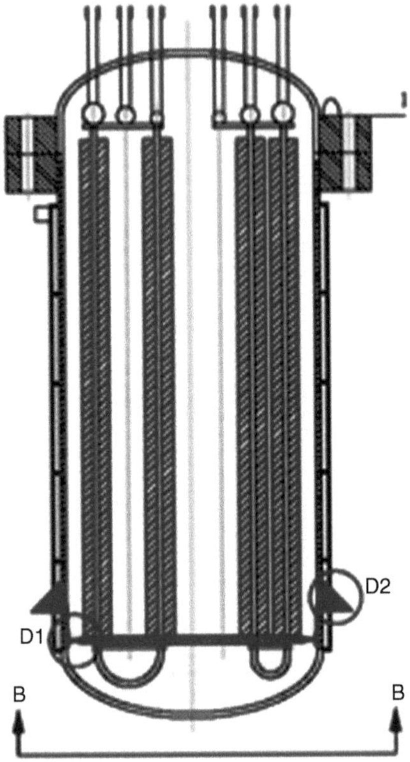

Figure 4.8 ***Sketch showing two representative U tubes with fins.*** Heat-transfer fluid flows through tubes; fins adsorb hydrates and improve heat transfer. With full contingent of tubes, designed to fill tank with hydrates. *(Courtesy MSU Hydrate Laboratory.)*

The process proceeds by controlling natural gas input to maintain constant pressure. As hydrates form and self-pack, levels of water solution recede until the vessel fills with hydrates. Temperature adjustment of heat-transfer fluid flowing through the finned tubes allows selective energy removal or addition during hydrate formation, storage, and decomposition.

A segment of the pilot plant installation is given in Figure 4.9.

In the foreground of Figure 4.9 is the formation/storage tank of 0.91 m (3.0 ft) diameter and 1.83 m (6 ft) length. To the right is a surge tank for discharge gases when the time is chosen to decompose hydrates. In the background is a chiller to circulate heat-transfer fluid through finned tubes dually acting as heat exchanger and hydrate adsorber inside the formation vessel. Natural gas from a bank of gas cylinders pressurizes the formation vessel. Hydrate-occluded gases are liberated at the end of an arbitrary storage time.

Figure 4.9 ***Photograph of hydrate formation vessel POC; chiller background left.*** Combustor, surge tank, bank of source gas not shown. *(Courtesy MSU Hydrate Laboratory.)*

Released gases from hydrate decomposition flow to a surge tank and then to an industrial combustor where the exothermic reaction generates steam. See Figure 4.10. A small portion of the steam is used to provide latent heats of hydrate decomposition.

The POC process successfully demonstrated controlled formation, storage, and dissociation of 150.1 scm (5300 scf) of natural gas, surpassing design criteria.

Two steps remain to arrive at a scaled process large enough to store and transport produced hydrate gases from offshore. First, a R&D step is needed to convert the process into a semicontinuous one. Second, the viability of hydrate ultrastability should be established on a large scale to minimize storage pressure during transport (Rogers and Zhang, 2012).

Figure 4.10 ***Photograph of installed industrial combustor; from liberated methane generates steam, returns small fraction to decompose hydrates.*** *(Courtesy MSU Hydrate Laboratory.)*

4.9 ANOMALOUS STABILITY OF GAS HYDRATES AT 1 ATM

Under 1 atm pressure, methane hydrates reach thermodynamic stability at 193 K (−80.2°C; −112.3°F). However, a pseudostability has been reported near 1 atm over a window of temperatures ranging from about −5 to −20°C (Ershov and Yakushev, 1992; Stern et al., 2001; Circone et al., 2004; Takeya et al., 2001; Zhang and Rogers, 2008). The implications are great for practical storage and transport of natural gas occluded in hydrates if this pseudostability phenomenon could be exploited.

4.9.1 The Window of Hydrate Ultrastability

Most reports of hydrate ultrastability refer to hydrate remnants at 1 atm after the bulk of the hydrate decomposed. For example, Ershov and Yakushev (1992) report a small fractional perseverance for up to 1 year at 1 atm of a hydrate field sample of hydrates within the ultrastability temperature window. Circone et al. (2004) report an anomalous preservation lasting several weeks at atmospheric pressure of remnant laboratory hydrates after dissociation of about 50% original mass.

The reason for a window of ultrastability is not fully understood. Perhaps a sheath of ice from structural water freezing surrounds the remnant hydrate after partial hydrate decomposition (Yakushev and Istomin, 1992; Davidson et al., 1986). But would the ice film have sufficient mechanical strength to stabilize hydrates within? Also, the diffusion coefficient for gas penetrating ice does not check with the rate of gas emanating from ultrastable hydrates (Takeya et al., 2001). Clues to the reason for ultrastability are few. Durham et al. (2003) point out that the mechanical strength of methane hydrates may be as much as 40 times greater than that of hexagonal water ice. Yet hexagonal ice is feasible in the temperature range 160–240 K, a range colder than the ultrastability window (Johari, 1998). Waite et al. (2002) state that ice thermal conductivity is considerably higher than that of hydrate.

The laboratory technique of Stern et al. (2001), as well as Circone et al. (2004), used to prepare hydrates for ultrastability testing was somewhat similar to that of Takeya et al. (2001). They first formed ice particles, melted the particle peripheries, and slowly formed a thickening shell of hydrates with pressurized methane gas. Circone et al. generated hydrate particles of about 250–300 μm diameter, compared with the 20–50 μm diameter particles of Takeya et al.

4.9.2 Micro-CT Scan Distinguishes Ice Tubes in Decomposing Hydrate

Hydrate samples from Black Sea seeps were viewed by micro-CT scan capable of mapping in 3D submicrometer features such as pore sizes and crystal boundaries (Klapp et al., 2011). Within the hydrates was found a natural system of tubes with diameters of 10–25 μm and lengths of greater than 100 μm developing during partial hydrate decomposition. Klapp et al. postulate that tubular ice walls form from released water chilled by endothermic hydrate dissociation. Their images indicate that ice tubes transport exiting gas during hydrate decomposition and those tubes extend throughout massive gas hydrates.

4.9.3 Ultrastable Hydrates Formed from Surfactant Solution

Hydrates carefully nucleated at 3.84–4.52 MPa from surfactant solution onto cylindrical metal surfaces have exhibited ultrastability on reducing confining pressure to 1 atm (Zhang and Rogers, 2008). The procedure resulted in longer stability of a larger mass fraction than previously reported.

Significant procedural changes were used: (1) hydrates were formed from a water–surfactant liquid solution; (2) while forming, hydrates adsorbed onto chilled surfaces of either aluminum or copper cylinders; (3) hydrates grew radially and symmetrically on the cylindrical metal walls. With this procedure, hydrates are thought to solidify without leaving any liquid, ice, or free gas in the otherwise solid cores of hydrate as a result of high freezing rates of small particles and rapid conductive dissipation of ΔH_f (Rogers and Zhang, 2012). In contrast, Circone et al. (2004) relate that cryogenic scanning electron microscope (SEM) analyses of 250–300 μm diameter hydrate particles generated from methane-pressurized ice showed frequent hollow shells of hydrate. Apparently, some ice particles initially had sheaths of hydrate encasing unreacted ice; decomposition possibly microfractured the shell and the exiting melt formed hydrates that annealed the fracture to surrounding particles. Stern et al. (2004) suggest the need for small grain-size ice in order to avoid the center cavity surrounded by a hydrate shell.

In the surfactant procedure, copper as a superior conductor was necessary to achieve the same degree of sII hydrate ultrastability as with methane hydrates. This suggests that more rapid heat removal from the site of hydrate formation helps achieve sII ultrastability. The observations are consistent with the reports of Stern et al. (2004) of incomplete conversion to hydrates in the cores of relatively large ice particles. The importance of the heat-transfer step in ultrastability is also alluded to in the work of Wilder et al. (2002).

Note that hydrates form around metal cylinders symmetrically in the surfactant process. The symmetry would lower stresses in the hydrate mass – fewer miniature cracks, fractures, crevices, and voids that contribute additional surface areas for hydrate decomposition. This seems to be consistent with the work of Klapp et al. (2011) where decomposition is associated with miniscule ice tubes dependent on pore sizes and decomposition boundaries within the hydrate mass.

4.9.4 Sintering Effects

Particles of ice melt where they touch, because molecules at the touching surfaces cross-diffuse. As an example, a common process sinters ceramic powders by pressurizing the contacting particles to form stronger materials of lower porosity. Important net results are decreases of particle surface areas and smaller pore sizes (Callister, 2007).

Sintering of gas hydrate particles in storage and transport would complicate the handling procedure. But hydrate sintering might also contribute to the self-preservation phenomenon of ultrastability at 1 atm.

Uchida et al. (2011) investigated the sintering effect during transport of packed hydrate particles. Their procedure first converted ice particles of 2–3 μm diameters into gas hydrate particles and then by means of an optical microscope observed over time the sintering of two individual particles in a high-pressure test cell. Hydrate sintering rates were determined by means of Equation 4.8 (Uchida et al., 2011):

$$\left(\frac{x}{r}\right)^{q} = Cr^{-p}t \tag{4.8}$$

where t, time (min); C, proportionality constant; p and q, sintering constants; r, radius of particle (mm); x, sintering neck size (mm).

The results of Uchida et al. (2011) included the following: (1) under hydrate P–T equilibrium conditions, CO_2 hydrate particles sinter at a rate approximately equal to that of ice sintering; (2) C_2H_6 hydrate particles do not sinter, suggesting that the ethane–water interactions within the hydrate retard diffusion through the matrix; (3) methane hydrates under equilibrium conditions sintered at a slightly slower rate than did ice or CO_2 hydrates; (4) methane hydrates in the P–T window of self-preservation (263.4 K, 0.1 MPa) sintered as fast as ice.

Their fourth conclusion supports the possibility that ice on the surface of hydrates contributes to the ultrastability effect.

REFERENCES

Andersson, V., Gudmundsson, J.S., 2000. Flow properties of hydrate-in-water slurries. In: Holder, G.D., Bishnoi, P.R. (Eds.), Gas Hydrates: Challenges for the Future (Annals of the New York Academy of Sciences 912). New York Academy of Sciences, New York, NY, pp. 322–329.

Bigalke, N.K., Deusner, C., Kossel, E., Haecked, M., 2011. CO_2 injection into submarine sediments: disturbing CH_4 hydrates. In: Proceedings of the 7th International Conference on Gas Hydrates (ICGH 2011), Edinburgh, Scotland, UK, July 17–21. Paper 591.

Boswell, R., Shelander, D., Latham, T., Lee, M., Collett, T., Guerin, G., Moridis, G., Reagan, M., Goldberg, D., 2009. Occurrence of gas hydrate in Oligocene Frio Sand: Alaminos Canyon Block 818: Northern Gulf of Mexico. Mar. Pet. Geol. 26 (8), 1499–1512.

Boswell, R., Collett, T., McConnell, D., Frye, M., Shedd, B., Mrozewksi, S., Guerin, G., Cook, A., Shelander, D., Dai, J., Godfriaux, P., Dufrene, R., Jones, E., Roy, R., 2010. Gulf of Mexico gas hydrate Joint Industry Project: overview of Leg II LWD results. In: Proceedings of the Offshore Technology Conference, Houston, TX.

Boswell, R., Moridis, G., Reagan, M., Collett, T., 2011. Gas hydrate accumulation types and their application to numerical simulation. In: Proceedings of the 7th International Conference on Gas Hydrates (ICGH 2011), Edinburgh, Scotland, UK, July 17–21. Paper 130.

Brewer, P.G., Friederich, G., Peltzer, E.T., Orr, Jr., F.M., 1999. Direct experiments on the ocean disposal of fossil fuel CO_2. Science 284, 943–945.

Callister, Jr., W.D., 2007. Materials Science and Engineering: An Introduction, seventh ed. John Wiley & Sons, Inc., New York, NY.

Circone, S., Stern, L.A., Kirby, S.H., 2004. The effect of elevated methane pressure on methane hydrate dissociation. Am. Mineral. 89, 1192–1201.

Collett, T.S., 1992. Potential of gas hydrates outlined. Oil Gas J. 90 (25), 84–87.

Collett, T.S., 2002. Energy resource potential of natural gas hydrates; unconventional petroleum systems. AAPG Bull. 86 (11), 1971–1992.

Collett, T.S., Dallimore, S.R., 1998. Quantitative assessment of gas hydrates in the Mallik L-38 well, Mackenzie Delta, N.W.T., Canada. In: PERMAFROST – Seventh International Conference (Proceedings), Yellowknife, Canada, Collection Nordicana No. 55, pp. 189–194.

Collett, T.S., Kuuskraa, V.A., 1998. Hydrates contain vast store of world gas resources. Oil Gas J. 96 (19), 90–95.

Collett, T., Riedel, M., Cochran, J., Boswell, R., Kumar, P., Sathe, A., 2008. Indian continental margin gas hydrate prospects: results of the Indian National Gas Hydrate Program (NGHP) Expedition 01. In: Proceedings of the International Conference on Gas Hydrates, ICGH-6, Vancouver, British Columbia, Canada.

Collett, T., Johnson, A., Knapp, C., Boswell, R., 2009. Natural gas hydrates – a review. AAPG Mem. 1, 146–220.

Collett, T., Boswell, R., Frye, M., Shedd, W., Godfriaux, P., Dufrene, R., McConnell, D., Mrozewski, S., Guerin, G., Cook, A., Jones, E., Roy, R., 2010. Gulf of Mexico gas hydrate Joint Industry Project Leg II: logging-while-drilling operations and challenges. In: Proceedings of the Offshore Technology Conference, OTC-20452.

Cook, A., Goldberg, D., Kleinberg, R., 2008. Fracture-controlled gas hydrate systems in the northern Gulf of Mexico. J. Mar. Pet. Geol. 25 (9), 932–941.

Dallimore, S.R., Collett, T.S., 2005. Scientific results from Mallik 2002 Gas Hydrate Production Research Well Program, Mackenzie Delta, Northwest Territories, Canada. In: Geological Survey of Canada, Bulletin 585. Geological Survey of Canada, Ottawa, Ontario.

Dallimore, S.R., Uchida, T., Collett, T.S., 1999. Scientific results from JAPEX/JNOC/GSC Mallik 2L-38 gas hydrate research well, Mackenzie Delta, Northwest Territories, Canada. In: Geological Survey of Canada, Bulletin 544. Geological Survey of Canada, Ottawa, Ontario.

Dallimore, S.R., Wright, J.F., Nixon, F.M., Kurihara, M., Yamamoto, K., Fujii, T., Fujii, K., Numasawa, M., Yasuda, M., Imasato, Y., 2008. Geologic and porous media factors affecting the 2007 production response characteristics of the JOGMEC/NRCan/Aurora Mallik gas hydrate production research well. In: Proceedings of the 6th International Conference on Gas Hydrates, Vancouver, British Columbia, Canada.

Davidson, D.W., Garg, S.K., Gough, S.R., Handa, Y.P., Ratcliffe, C.I., Ripmeester, J.A., Tse, J.S., 1986. Laboratory analysis of a naturally occurring gas hydrate from sediment of the Gulf of Mexico. Geochim. Cosmochim. Acta 50, 619–623.

Durham, W.B., Kirby, S.H., Stern, L.A., Zhang, W., 2003. The strength and rheology of methane clathrate hydrate. J. Geophys. Res. 108 (B4), 2182.

Englezos, P., 1996. Nucleation and growth of gas hydrate crystals in relation to 'kinetic inhibition'. In: 2nd International Symposium on Gas Hydrates, Toulouse, pp. 147–153.

Ershov, E.D., Yakushev, V.S., 1992. Experimental research on gas hydrate decomposition in frozen rocks. Cold Region Sci. Technol. 20, 147–156.

Farrell, H., Boswell, R., Howard, J., Baker, R., 2010. CO_2–CH_4 exchange in natural gas hydrate reservoirs: potential and changes. In: Fire in the Ice, DOE NETL Methane Hydrate Newsletter, March, pp. 19–21.

Frye, M., Shedd, W., Godfriaux, P., Dufrene, R., Collett, T., Lee, M., Boswell, R., Jones, E., McConnell, D., Mrozewski, S., Guerin, G., Cook, A., 2010. Gulf of Mexico gas hydrate Joint Industry Project Leg II: results from the Alaminos Canyon 21 Site. In: Proceedings of the Offshore Technology Conference 2010, OTC-20560.

Fujii, T., Saeki, T., Kobayashi, T., Inamori, T., Hayashi, M., Takano, O., Takayama, T., Kawasaki, T., Nagakubo, S., Nakamizu, M., Yokoi, K., 2008. Resource assessment of methane hydrate in the eastern Nankai Trough, Japan. In: Proceedings of the OTC, Houston, TX.

Hancock, S.H., Collett, T.S., Dallimore, S.R., Satoh, T., Inoue, T., Huenges, E., Henninges, J., Weatherill, B., 2005a. Overview of thermal-stimulation production-test results for the JAPEX/JNOC/GSC, et al. Mallik 5L-38 gas hydrate production research well. In: Geological Survey of Canada, Bulletin 585. Geological Survey of Canada, Ottawa, Ontario.

Hancock, S.H., Dallimore, S.R., Collett, T.S., Carle, D., Weatherill, B., Satoh, T., Inoue, T., 2005b. Overview of pressure-drawdown production-test results for the JAPEX/JNOC/GSC, et al. Mallik 5L-38 gas hydrate production research well. In: Geological Survey of Canada, Bulletin 585. Geological Survey of Canada, Ottawa, Ontario.

Henninges, J., Huenges, E., Burkhardt, H., 2005. In-situ thermal conductivity of gas-hydrate-bearing sediments of the JAPEX/JNOC/GS, et al. Mallik 5L-38 well. In: Proceedings of the 5th International Gas Hydrate Conference, Trondheim, Norway, June 13–16. Paper 3034.

Hirohama, S., Shimoyama, Y., Wakabayashi, A., Tatsuta, S., Nishida, N.J., 1996. Conversion of CH_4-hydrate in liquid CO_2. J. Chem. Eng. Jpn. 29, 1014–1020.

Holder, G.D., 1982. A thermodynamic evaluation of thermal recovery of gas from hydrates in the earth. J. Pet. Technol. 34, 1127–1132.

Jin, Y.K., Lee, M.W., Collett, T.S., 2002. Relationship of gas hydrate concentration to porosity and reflection amplitude in a research well, Mackenzie Delta, Canada. Mar. Pet. Geol. 19, 407–415.

Johari, G.P., 1998. On the coexistence of cubic and hexagonal ice between 160 and 240 K. Philos. Mag. B 78 (4), 375–383.

Kimoto, S., Oka, F., Fushita, T., Fujiwaki, M., 2007. A chemo-thermo-mechanically coupled numerical simulation of the subsurface ground deformations due to methane hydrate dissociation. Comput. Geotech. 34, 216–228.

Klapp, S.A., Enzmann, F., Walz, P., Huthwelker, T., Tuckermann, J., Schwarz, J.-O, Pape, T., Peltzer, E.T., Hester, K.C., Zhang, X., Mokso, R., Wangner, D., Marone, F., Kersten, M., Bohrmann, G., Kuhs, W.F., Stampanoni, M., Brewer, P.G., 2011. Fluid flow in natural gas hydrates revealed by X-ray tomographic microscopy. In: Proceedings of the 7th International Conference on Gas Hydrates (ICGH 2011), Edinburgh, Scotland, UK, July 17–21. Paper 248.

Kobayashi, I., Ito, Y., Mori, Y.H., 2001. Microscopic observations of clathrate-hydrate films formed at liquid/liquid interfaces. I. Morphology of hydrate films. Chem. Eng. Sci. 56, 4331.

Konno, Y., Masuda, Y., Hariguchi, Y., Kurihara, M., Ouchi, H., 2010. Key factors for depressurization-induced gas production from oceanic methane hydrates. Energy Fuels 24, 1736–1744.

Kurihara, M., Ouchi, H., Inoue, T., Yonezawa, T., Masuda, Y., Dallimore, S.R., Collett, T.S., 2005a. Analysis of the JAPEX/JNOC/GSC, et al. Mallik 5L-38 gas hydrate thermal production test through numerical simulation. In: Geological Survey of Canada, Bulletin 585. Geological Survey of Canada, Ottawa, Ontario.

Kurihara, M., Funatsu, K., Kusaka, K., Yasuda, M., Dallimore, S.R., Collett, T.S., Hancock S.H., 2005b. Well-test analysis for gas hydrate reservoirs: examination of parameters suggested by conventional analysis for the JAPEX/JNOC/GSC, et al. Mallik 5L-38 gas hydrate production research well. In: Geological Survey of Canada, Bulletin 585. Geological Survey of Canada, Ottawa, Ontario.

Kurihara, M., Sato, A., Ouchi, H., Natrita, H., Masuda, Y., Saeki, T., Fujii, T., 2009. Prediction of gas productivity from eastern Nankai Trough methane-hydrate reservoirs. SPE Reservoir Eval. Eng. 12 (3), 477–499.

Kurihara, M., Yasuda, M., Yamamoto, K., Fujii, T., Numasawa, M., Narita, H., Masuda, Y., Dallimore, S.R., Wright, F., 2011a. Analysis of 2007/2008 JOGMEC/NRCAN/AURORA Mallik gas hydrate production test through numerical simulation. In: Proceedings of the 7th International Conference on Gas Hydrates (ICGH 2011), Edinburgh, Scotland, UK, July 17–21. Paper 449.

Kurihara, M., Ouchi, H., Sato, A., Yamamoto, K., Noguchi, S., Narita, H., Nagao, J., Masuda, Y., 2011b. Prediction of performances of methane hydrate production tests in the eastern Nankai Trough. In: Proceedings of the 7th International Conference on Gas Hydrates (ICGH 2011), Edinburgh, Scotland, UK, July 17–21. Paper 450.

Kvamme, B., Graue, A., Buanes, T., Kuznetsova, T., Ersland, G., 2007. Storage of CO_2 in natural gas hydrate reservoirs and the effect of hydrate as an extra sealing in cold aquifers. Int. J. Greenhouse Gas Control 1 (2), 236–246.

Lee, M.W., Collett, T.S., 2011. In-situ gas hydrate saturation estimated from various well logs at the Mount Elbert Gas Hydrate Stratigraphic Test Well, Alaska North Slope. Mar. Pet. Geol. 28 (2), 439–449.

Lee, S., Ryu, B., 2011. 2nd Ulleung Basin gas hydrate expedition: findings and implications. In: Fire in the Ice, NETL/DOE Newsletter, Spring.

Lee, J.H., Trimm, D.L., 1995. Catalytic combustion of methane. Fuel Process. Technol. 42, 339–359.

Lee, H., Seo, Y., Seo, Y.-T., Moudrakovski, I.L., Ripmeester, J.A., 2003. Recovering methane from solid methane hydrate with carbon dioxide. Angew. Chem. Int. Ed. 42, 5048–5051.

Lorenson, T.D., Collett, T.S., Hunter, R., 2008. Preliminary assessment of hydrocarbon gas resources from the Mt. Elbert No. 1 gas hydrate test well, Milne Pt. Alaska. In: Proceedings of the 6th International Conference on Gas Hydrates (ICGH 2008), Vancouver, British Columbia, Canada, July 6–10.

Lu, Z., Zhu, Y., Zhang, Y., Wen, H., Li, Y., Jia, Z., Liu, C., Wang, P., Li, Q., 2010. Gas hydrate features in the Qilian Mountains permafrost, Qinghai Province, China. In: Fire in the Ice, US DOE/NETL Newsletter, March, pp. 6–8.

Lu, H., Lorenson, T.D., Moudrakovski, I.L., Ripmeester, J.A., Collett, T.S., Hunter, R.B., Radcliffe, C.I., 2011. The characteristics of gas hydrate recovered from the Mount Elbert Gas Hydrate Stratigraphic Test Well, Alaska North Slope. J. Mar. Pet. Geol. 28 (2), 411–418.

Makogon, Y., Trebin, F., Trofimuk, A., Tsarev, P., Cherskiy, N., 1972. Detection of a pool of natural gas in a solid (hydrate) gas state. Dokl. Acad. Sci. U. S. S. R. 196, 197–200.

Masuda, Y., Kurihara, M., Ohuchi, H., Sato, T., 2002. A field-scale simulation study on gas productivity of formations containing gas hydrates. In: Fourth International Conference on Gas Hydrates, Yokohama, pp. 40–46.

Masuda, Y., Maruta, H., Naganawa, S., Amikawa, K., Nagao, J., Haneda, H., Konno, Y., 2011. Methane recovery from hydrate-bearing sediments by N_2–CO_2 gas mixture injection: experimental investigation on CO_2–CH_4 exchange ratio. In: Proceedings of the 7th International Conference on Gas Hydrates (ICGH 2011), Edinburgh, Scotland, UK, July 17–21. Paper 498.

McConnell, D., Collett, T., Boswell, R., Frye, M., Shedd, W., Dufrene, R., Godfriaux, P., Mrozewski, S., Guerin, G., Cook, A., Jones, E., 2010. Gulf of Mexico gas hydrate Joint Industry Project Leg II: results from the Green Canyon 955 Site. In: Proceedings of the Offshore Technology Conference, OTC 20801.

Moridis, G., Collett, T., 2004. Strategies for gas production from hydrate accumulations under various geologic conditions. In: Proceedings of the Tough Symposium, LBNL-52568.

Moridis, G.J., Reagan, M.T., 2007. Gas production from oceanic Class 2 hydrate accumulations. In: Proceedings of the Offshore Technology Conference, Houston, TX, April 30–May 3. Paper 18866.

Moridis, G.J., Sloan, E.D., 2007. Gas production potential of disperse low-saturation hydrate accumulations in oceanic sediments. Energy Conversion Manag. 48, 1834–1847.

Moridis, G.J., Collett, T.S., Boswell, R., Kurihara, M., Reagan, M.T., Hoh, C., Sloan, E.D., 2009. Toward production from gas hydrates: current status, assessment of resources, and simulation-based evaluation of technology and potential. SPE Reservoir Eval. Eng. 12 (5), 745–771.

Moridis, G., Silpngarmlert, S., Reagan, M., Collett, T., Zhang, K., 2011. Gas production from a cold, stratigraphically-bounded gas hydrate deposit at the Mount Elbert Gas Hydrate Stratigraphic Test Well, Alaska North Slope: implications of uncertainties. Mar. Pet. Geol. 28 (2), 517–534.

Moudrakovski, I.L., Sanchez, A.A., Ratcliffe, C.I., Ripmeester, J.A., 2001. Nucleation and growth of hydrates on ice surfaces: new insights from ^{129}Xe NMR experiments with hyperpolarized xenon. J. Phys. Chem. B 105, 12338–12347.

Oyama, H., Konno, Y., Masuda, Y., Narita, H., 2009. Dependence of depressurization-induced dissociation of methane hydrate bearing laboratory cores on heat transfer. Energy Fuels 23, 4995–5002.

Park, Y., Kim, D.-Y., Lee, J.-W., Huh, D.-G., Park, K.-P., Lee, J., Lee, H., 2006. Sequestering carbon dioxide into complex structures of naturally occurring gas hydrates. Proc. Natl. Acad. Sci. U. S. A. 103 (34), 12690–12694.

Plaza-Faverola, A., Westbrook, G., Ker, S., Exley, R., Gailler, A., Minshull, T., Broto, K., 2010. Evidence from three-dimensional seismic tomography for a substantial accumulation of gas hydrate in a fluid-escape chimney in the Nyegga pockmark field, offshore Norway. J. Geophys. Res. 115, B08104.

Pooladi-Darvish, M., 2004. Gas production from hydrate reservoirs and its modeling. J. Pet. Technol. 56, 65–71.

Qasim, M., Kvamme, B., Baig, K., 2011. Modeling dissociation and reformation of methane and carbon dioxide hydrate using phase field theory with implicit hydrodynamics. In: Proceedings of the 7th International Conference on Gas Hydrates (ICGH 2011), Edinburgh, Scotland, UK, July 17–21. Paper 241.

Riedel, M., Collett, T., Malone, M., Expedition 311 Scientists, 2008. Cascadia margin gas hydrates. In: Proceedings of the IODP, p. 311.

Rogers, R.E., 1994. Coalbed Methane: Principles and Practices. Prentice-Hall, Inc., Englewood Cliffs, NJ, 345 pp.

Rogers, R.E., 1997. Natural Gas Hydrates Storage Project. Contract no. DE-AC26-97FT33203, Department of Energy, November 1997–October 1999.

Rogers, R.E., 2001. Gas hydrate storage of natural gas. Grant no. DE-FC26-01NT41297, U.S. Department of Energy, October 1, 2001–September 30, 2005.

Rogers, R.E., Zhang, G., 2012. System for stabilizing gas hydrates at low pressures. Patent no. 8,119,078.

Rogers, R.E., Zhong, Y., 2000. Feasibility of storing natural gas in hydrates commercially. In: Holder, G.D., Bishnoi, P.R. (Eds.), Gas Hydrates, Challenges for the Future (Annals of the New York Academy of Sciences 912). New York Academy of Sciences, New York, NY, pp. 843–850.

Rogers, R.E., Zhong, Y., 2002. Surfactant process for promoting gas hydrate formation and application of the same. Patent no. 6,389,820.

Rogers, R.E., Zhong, Y., Etheridge, J.A., Pearson, L.E., 2004. The MSU micellar-solution gas hydrate storage process for natural gas. In: Taylor, C.E., Kwan, J.T. (Eds.), Advances

in the Study of Gas Hydrates. Kluwer Academic, Plenum Publishers, New York, NY, pp. 185–198.
Rogers, R.E., Zhong, Y., Arunkumar, R., Etheridge, J.A., Pearson, L.W., McCown, J., Hogancamp, K., 2005. Gas hydrate storage process for natural gas. GasTIPS 11 (1), 14–18.
Rogers, R., Zhang, G., Dearman, J., Woods, C., 2007. Investigations into surfactant/gas hydrate relationship. J. Pet. Sci. Eng. 56, 82–88.
Rutqvist, J., Moridis, G.J., Grover, T., Collett, T., 2009. Geomechanical response of permafrost-associated hydrate deposits to depressurization-induced gas production. J. Pet. Sci. Eng. 67, 1–12.
Rydzy, M.B., Schicks, J.M., Naumann, R., Erzinger, J., 2007. Dissociation enthalpies of synthesized multicomponent gas hydrates with respect to the guest composition and cage occupancy. J. Phys. Chem. B 111, 9539–9545.
Ryu, B.-J., Riedel, M., Kim, J.-H., Hyndman, R., Lee, Y.-J., Chung, B.-H., 2010. Gas hydrates in the western deep-water Ulleung Basin, East Sea of Korea. J. Mar. Pet. Geol. 26, 1483–1498.
Schicks, J.M., Spangenberg, E., Steinhauer, B., Giese, R., Beeskow-Strauch, B., Klump, J., Erzinger, J., Luzi, M., 2011a. Improving methods for gas production from hydrate bearing sediments: chances and challenges. In: Proceedings of the 7th International Conference on Gas Hydrates (ICGH 2011), Edinburgh, Scotland, UK, July 17–21. Paper 182.
Schicks, J.M., Spangenberg, E., Giese, R., Steinhauer, B., Klump, J., Luzi, M., 2011b. New approaches for the production of hydrocarbons from hydrate bearing sediments. Energies 4, 151–172.
Seol, Y., Myshakin, E., Kneafsey, T.J., 2010. Core-scale heterogeneity of hydrate distribution and its impact on gas production. In: Fire in the Ice, DOE/NETL Newsletter, August, pp. 6–8.
Servio, P., Englezos, P., 2003. Morphology of methane and carbon dioxide hydrates formed from water droplets. AIChE J. 49 (1), 269–276.
Stern, L.A., Circone, S., Kirby, S.H., 2001. Anomalous preservation of pure methane hydrate at 1 atm. J. Phys. Chem. B 105, 1756–1762.
Stern, L.A., Kirby, S.H., Circone, S., Durham, W.B., 2004. Scanning electron microscope investigations of laboratory-grown gas clathrate hydrates formed from melting ice, and comparison to natural hydrates. Am. Mineral. 89 (8–9), 1162–1175.
Subramanian, S., Kini, R.A., Dee, S.F., Sloan, Jr., E.D., 2000. Evidence of structure II hydrate formation from methane + ethane mixtures. Chem. Eng. Sci. 55, 1981–1999.
Sugaya, M., Mori, Y.H., 1996. Behaviour of clathrate hydrate formation at the boundary of liquid water and a fluorocarbon in liquid or vapor state. Chem. Eng. Sci. 51, 3505.
Takeya, S., Shimada, W., Kamata, Y., Ebinuma, T., Uchida, T., Nagao, J., Narita, H., 2001. In situ x-ray diffraction measurements of the self-preservation effect of CH_4 hydrate. J. Phys. Chem. B 105, 9756–9759.
Thompson, R.S., Wright, J.D., 1985. Oil Property Evaluation, second ed. Thompson-Wright Associates, Golden, CO.
Uchida, T., Dallimore, S.R., Lu, H., Collett, T.S., 2002. Occurrences of natural gas hydrates beneath the permafrost in the Mackenzie Delta area with compared to recently obtained natural gas hydrates. In: Proceedings of the 4th International Conference on Gas Hydrates, Yokohama, May 19–23, pp. 228–232.
Uchida, T., Shiga, T., Nagayama, M., Gohara, K., Sakurai, T., 2011. Microscopic observations of sintering processes on gas hydrate particles. In: Proceedings of the 7th International Conference on Gas Hydrates (ICGH 2011), Edinburgh, Scotland, UK, July 17–21. Paper 347.
Vysniauskas, A., Bishnoi, P.R., 1983. A kinetic study of methane hydrate formation. Chem. Eng. Sci. 38, 1061–1072.

Waite, W.F., deMartin, B.J., Kirby, S.H., Pinkston, J., Ruppel, C.D., 2002. Thermal conductivity measurements in porous mixtures of methane hydrate and quartz sand. Geophys. Res. Lett. 29 (24), 82-1–82-4.

Wallman, K., Bialas, J., 2010. SUGAR: marine gas hydrate technology development for environmentally sound energy development and CO2 sequestration. In: Fire in the Ice, NETL/DOE Newsletter, March, pp. 22-23.

Warzinski, R.P., Gamwo, I.K., Rosenbaum, E.J., Myshakin, E.M., Jiang, H., Jordan, K.D., English, N.J., Shaw, D.W., 2008. Thermal properties of methane hydrate by experiment and modeling and impacts upon technology. In: Proceedings of the 6th International Conference on Gas Hydrates (ICGH 2008), Vancouver, British Columbia, Canada, July 6–10.

White, M.D., McGrail, B.P., 2008. Numerical simulation of methane hydrate production from geologic formations via carbon dioxide injection. In: 2008 Offshore Technology Conference, Houston, Texas, May 5–8. OTC Paper #19458.

Wilder, J.W., Seshadri, K., Smith, D.H., 2002. Thermodynamics of the sequestration of carbon dioxide in methane hydrates in porous media. Am. Chem. Soc. Division Fuel Chem. 47 (1), 17–19.

Winters, W., 2011. Physical properties of sediment from the Mount Elbert Gas Hydrate Stratigraphic Test Well, Alaska North Slope. Mar. Pet. Geol. 28 (2), 361–380.

Winters, W.J., Pecher, I.A., Waite, W.F., Mason, D.H., 2004. Physical properties and rock physics models of sediment containing natural and laboratory-formed methane gas hydrate. Am. Mineral. 89, 1221–1227.

Wright, J.F., Taylor, A.E., Dallimore, S.R., Nixon, F.M., 1999. Estimating in situ gas hydrate saturation from core temperature observations, JAPEX/JNOC/GSC Mallik 2L-38 gas hydrate research well. Geol. Surv. Can. Bull. 544, 101–108.

Wu, N., Yang, S., Zhang, H., Liang, J., Wang, H., Lu, J., 2010. Gas hydrate system of Shenhu area, Northern South China Sea: wireline logging, geochemical results, and preliminary resource estimates. In: Proceedings of the Offshore Technology Conference, OTC-20485.

Yakushev, V.S., Istomin, V.A., 1992. Gas-hydrates self-preservation effect. In: Maeno, A.H., Editor, T. (Eds.), Physics and Chemistry of Ice. Hokkaida University Press, Sapporo, pp. 136–139.

Yamamoto, K., Terao, Y., Noguchi, S., Nakatsuka, Y., Inada, N., Matsuzawa, M., Nagakubo, S., Ikawa, T., Ouchi, H., Kanno, T., 2011. The plan of offshore production test of marine methane hydrate and technical challenges. In: Proceedings of the 7th International Conference on Gas Hydrates (ICGH 2011), Edinburgh, Scotland, UK, July 17–21. Paper 606.

Zhang, G., Rogers, R.E., 2008. Ultra-stability of gas hydrates at 1 atm and 268.2 K. Chem. Eng. Sci. 63, 2066–2074.

Zhong, Y., Rogers, R.E., 2000. Surfactant effects on gas hydrate formation. Chem. Eng. Sci. 55 (19), 4175–4187.

CHAPTER FIVE

Hydrate Inhibition During Drilling and Production

In addition to greater needs as conventional oil and gas operations extend into deeper waters, hydrate inhibitors become necessary during production of gas from offshore hydrates: (1) to prevent hydrate re-formations in production lines; (2) to assist initial decomposition of reservoir hydrates; (3) to prevent released hydrate gases and waters from re-forming clathrates and blocking reservoir flow near the wellbore.

It should be kept in mind that memory effects typically linger in decomposed hydrate waters, prompting hydrate re-formations.

Hydrate inhibitors are reviewed in this chapter. The overview examines chemical structures, classifications, mechanisms, and effectiveness of commercially available inhibitors. In addition to commercial synthetic inhibitors, research on applicability as hydrate inhibitors of antifreeze proteins (AFPs) is discussed. The AFPs occur naturally in Arctic fish, plants, insects, and bacteria to prevent cellular ice from destroying the living organism.

Microbial cell walls inhibit gas hydrate formation. Explanation is given on how the phenomenon plays a role in seafloor hydrate nucleation, protecting the living cells from hydrate's heat of formation, yet strategically placing the microorganisms within hydrates to access carbon accumulations. Potential exists to use in low concentrations the cell wall material as an economical hydrate inhibitor for oilfield operations, not only in hydrate-gas production but also in conventional gas production.

5.1 CHEMICAL INHIBITION TO PREVENT HYDRATE FORMATION

5.1.1 Inhibition Needs for Hydrate-Gas Production

Production of hydrate gas from the seafloor may necessitate chemical inhibition to avoid hydrate re-formation at four points: (1) porous media close to the wellbore, (2) lines from subsea wellhead to surface, (3) surface collection and storage facility, and (4) gas transport lines to shore.

Preventing hydrate re-formation near the wellbore is unique to hydrate production, requiring innovations beyond experience with conventional

Offshore Gas Hydrates. http://dx.doi.org/10.1016/B978-0-12-802319-8.00005-X

gas production, while compensating for memory properties retained in the decomposed hydrate waters.

5.1.2 Hydrate Inhibition in Subsea Conventional Gas Production

Industries dealing with hydrocarbon flow lines on land during winter simply use common techniques to prevent hydrate blockage of process lines (Sivaraman, 2002): (1) dehydrate the hydrocarbon gas; (2) elevate operating temperatures; (3) inject inhibitor chemicals. For example, water is removed from the gases at surface wellheads to prevent hydrate occurrence during transport in natural gas pipelines; (4) insulate conduits and place thermal wraps at expansion joints.

But hydrate problems in deep subsea process lines become inaccessible or impractical for most conventional preventive measures. In such cases, annual costs to clear hydrate blockage in lines may exceed \$100 million at a rate near \$1 million per mile of affected lines (Sivaraman, 2002). Inhibitor chemicals must be injected. Three types of hydrate inhibitors are available commercially to alleviate the problem: (1) thermodynamic, (2) kinetic, and (3) antiagglomerate (AA).

5.2 THERMODYNAMIC INHIBITORS

A thermodynamic hydrate inhibitor (THI) retards hydrogen bonding of water molecules forming hydrate structures, altering the hydrate–liquid–vapor equilibrium (HLVE) at given gas composition, water salinity, pressure, and temperature. The THI establishes more stringent P–T values to form hydrates. Furthermore, P–T requirements are eased for dissociation of established hydrates (Zheng et al., 2011).

To evaluate thermodynamic inhibitor effectiveness, degree of subcooling quantitatively answers the following question: how far below normal hydrate equilibrium temperature can the operating temperature be lowered and THI still prevent hydrate stabilization? When present in concentrations of 10 wt%, the amino acid glycine depresses temperature of hydrate formation 1.3–1.6°C (Sa et al., 2011), but the premier THI methanol depresses formation temperatures 3.5–5.0°C at 10 wt% (Sloan and Koh, 2008). Ionic liquids, although inhibiting hydrates, depress the formation temperature only 0.7–1.5°C at the common 10 wt% concentration (Xiao and Adidharma, 2009).

Cost, toxicity, environmental damage, and corrosion are decisive factors in selecting effective thermodynamic inhibitors for conventional oilfield operations. For example, costs of THIs can be expected to easily exceed

$60,000 per day per well in deep water oil and gas production (Sivaraman, 2003).

5.2.1 Methanol

In deep water operations, adding methanol as 10–15% of the weight of the water content of a hydrocarbon production stream may be necessary and indeed is common (Kelland et al., 2006a; Xiao and Adidharma, 2009). Accordingly, methanol requirements increase if fractional water cuts increase during production (Makogon and Sloan, 2002). Likewise, the demand rises for operations in deeper waters. For example, methanol injection operating costs for offshore oil and gas production exceed $500 million per year (Lederhos et al., 1996).

Increasingly large volumes of methanol required for deeper water hydrocarbon product streams raise concerns of toxicity, corrosion, cost, and refinery catalyst poisoning. Those concerns occur in the context of strict federal regulations limiting discharge of methanol-containing waters on offshore platforms (Makogon and Sloan, 2002).

5.2.2 Case Study: Methanol as Hydrate Inhibitor

At one time, production was considered from a discovery in the North Sea of a high-pressured gas field. Production was considered using a subsea wellhead connected by pipeline along the ocean floor to a main platform. Based on prior experience, distance traversed by the pipeline might be as far as 11 km. Before proceeding, the company's primary concern was gas hydrates. During uninterrupted production flow of gas/water in the pipeline, would hydrates form to block the line before hydrocarbons reached the main platform? During shut-in would hydrates form? Data in Table 5.1 and

Table 5.1 Data for North Sea case problem (Blanc and Tournier-Lasserve, 1990)

Reservoir properties	Pressure	91 MPa
	Temperature	155°C
	Depth	4500 mbsf
Gas composition	Methane	84.5 mol%
	Ethane	8.7
	Propane	3.3
	i-Butane	0.5
	Carbon dioxide	3.0
Production rate	3.10 N m^3/day	110 million cfd
Pipeline	Diameter	10 in.
	Maximum length	11 km
Seawater temperature	41–43°F	5–6°C

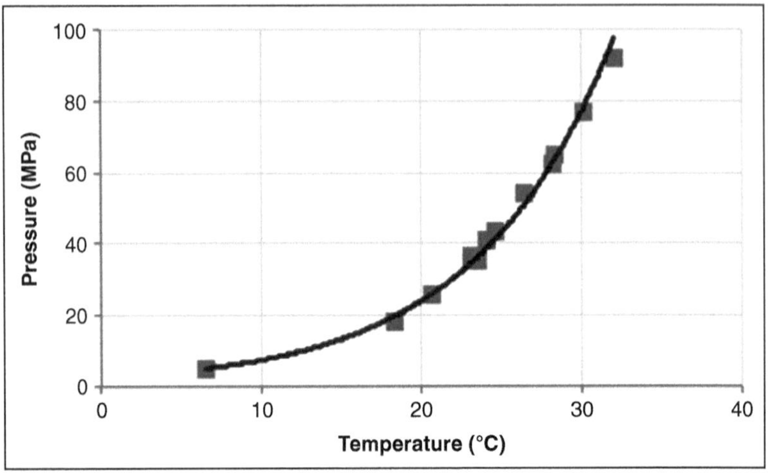

Figure 5.1 ***Hydrate*** **P–T** ***equilibrium of North Sea natural gas (Blanc and Tournier-Lasserve, 1990).***

the *P–T* hydrate equilibrium curve of Figure 5.1 were available (Blanc and Tournier-Lasserve, 1990).

Heat-transfer calculations predicted initial wellhead temperature of the fluids to decrease sufficiently during pipeline flow to induce hydrates after a traverse of 7 km – necessitating methanol injection (Blanc and Tournier-Lasserve, 1990). Further calculations established extent of required methanol injection. See Figure 5.2.

By extrapolating inhibitor effects, Blanc and Tournier-Lasserve determined that 55% methanol, as a weight percentage of water in the line, would prevent hydrate plugging during shut-in.

5.2.3 Amino Acids

Despite their usefulness, common THIs have undesirable properties – non-biodegradability being one common fault. Methanol and ethyl glycol are environmentally unfriendly, non-biodegradable, and toxic; ionic liquids are expensive; electrolytes introduce disposal problems.

Sa et al. (2011) evaluated the amino acids glycine and alanine as potential THIs based on the prospect of these acids being naturally occurring compounds, biodegradable, and nontoxic, and displaying hydrogen bonding with water. Carbon dioxide hydrates were used in their study.

Glycine, whose molecular structure is given in Figure 5.3, was found to inhibit CO_2 hydrate formation.

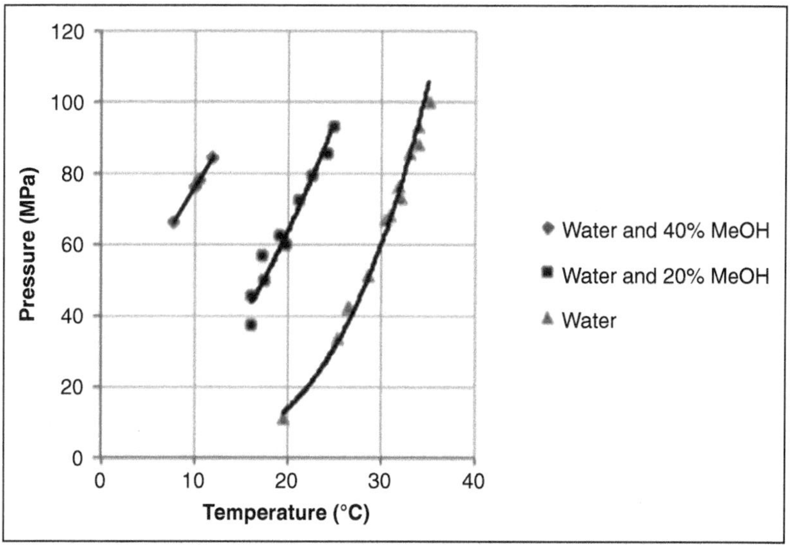

Figure 5.2 ***Methanol requirements to inhibit hydrates at given field conditions.*** Hydrate–liquid–vapor equilibrium when using methanol as a thermodynamic inhibitor at gas composition 84.5 CH_4/8.7 C_2H_6/3.3 C_3H_8/0.5 *i*-C_4H_{10}/3.0 CO_2 (Blanc and Tournier-Lasserve, 1990).

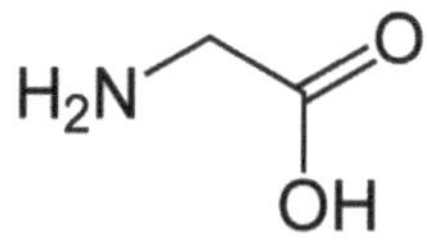

Figure 5.3 ***Glycine molecular structure (Benjah-bmm27, 2007).***

Sa et al. (2011) found more stringent conditions of higher pressures and lower temperatures necessary to form hydrates as glycine concentrations increase. Its hydrate inhibition improves with increasing weight concentration until the amino acid reaches a 20% concentration near its solubility limit. That is, thermodynamic inhibition improves as more glycine molecules go into solution to interfere with the hydrogen bonding (Figure 5.4), but the extent of its inhibiting ability reaches a limit at the point of maximum solubility (Sa et al., 2011).

It is interesting that a glycine of molecular weight 75.07 inhibits CO_2 hydrates better on a weight basis than alanine of molecular weight 89.09, yet on a molar basis the opposite is true. Even though the same numbers of amino acid molecules would be in each molar solution, the extra methyl

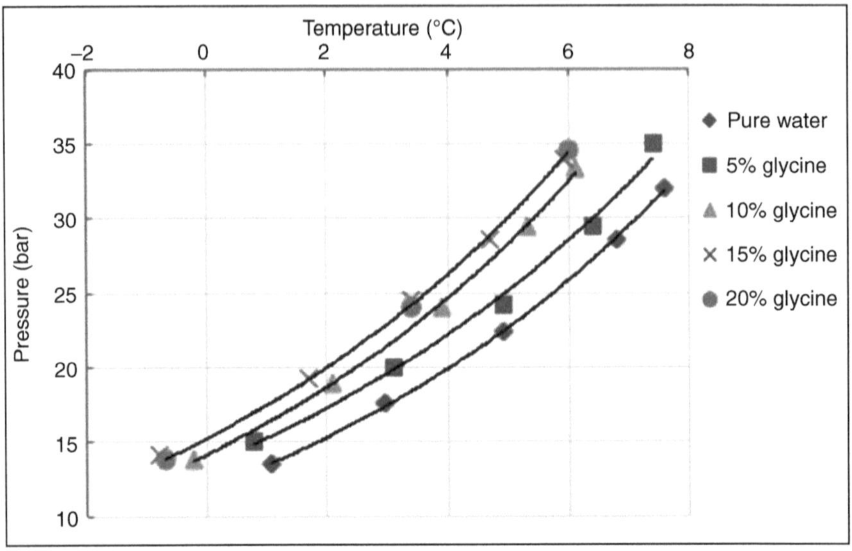

Figure 5.4 ***Thermodynamic inhibition of glycine, amino acid (Sa et al., 2011).*** *(Reprinted with permission from Environmental Science & Technology. Copyright 2011. American Chemical Society.)*

group in the alanine structure translates into a slightly improved hydrate inhibition (Sa et al., 2011).

5.2.4 Electrolytes

Although methanol is the usual THI of choice, sodium chloride might also be selected to prevent hydrate blockage in process lines if the attendant corrosion is tolerable. The addition of 10 wt% NaCl shifts HLVE to require 2.5–4.0°C lower temperatures for hydrate formation, a familiar occurrence in forming natural gas hydrates in saline solutions (Xiao and Adidharma 2009).

Jenson et al. (2011) showed that 3.5 wt% NaCl in water solution lowers by 2–3°C the temperature to initiate hydrates over a pressure range of 20–100 bar.

Salts such as $CaCl_2$, NaBr, K_2CO_3, KBr, KCl, and organic salts provide thermodynamic inhibition in addition to NaCl.

5.2.5 Estimating Thermodynamic Inhibition

Ostergaard et al. (2005) developed the semiempirical Equation 5.1 to estimate the temperature depression of the HLVE from thermodynamic inhibitors – electrolytes as well as organic compounds. Three parameters are needed to use the equation: (1) inhibitor concentration, (2) system pressure, and (3) pressure at which the subject hydrocarbon hydrate, when formed from distilled water, decomposes under a system temperature of 273.15 K;

$$\Delta T = (C_1W + C_2W^2 + C_3W^3)[C_4 \ln(P) + C_5][C_6(P_0 - 1000) + 1] \quad (5.1)$$

where ΔT, suppression of the hydrate dissociation temperature (K or °C); W, concentration inhibitor in liquid water phase (mass%); P_0, dissociation pressure of hydrocarbon hydrate at 273.15 K in distilled water (kPa); C_i, constants (see Tables 5.2 and 5.3); P, system pressure (kPa).

Table 5.2 Constants for Equation 5.1, organic thermodynamic inhibitors (Ostergaard et al., 2005)

Constant	Methanol	Ethylene glycol	Ethanol	Triethylene glycol	Glycerol
C_1	0.478	38.93	2.118	0.1964	0.135
C_2	7.17E−02	−0.522	−4.48E−03	−5.81E−03	8.846E−03
C_3	−1.44E−05	1.77E−02	6.979E−04	1.393E−04	−1.15E−05
C_4	2.95E−02	3.50E−04	5.85E−03	2.855E−02	1.335E−02
C_5	5.96E−01	5.08E−03	0.225	0.854	0.378
C_6	3.10E−05	2.65E−05	3.4E−05	3.24E−05	4.6E−05
Maximum concentration (mass%)	43.3	59.6	31.2	59.5	68.6
Maximum concentration (mol%)	30	30	15	15	30

Table 5.3 Constants for Equation 5.1, inorganic salt thermodynamic inhibitors (Ostergaard et al., 2005)

Constant	NaCl	$CaCl_2$	NaBr	K_2CO_3	KCl
C_1	0.3534	0.194	0.419	0.187	0.305
C_2	1.38E−03	7.58E−03	6.5E−03	−5.7E−03	6.77E−04
C_3	2.43E−04	1.95E−04	1.098E−04	2.551E−04	8.096E−05
C_4	4.06E−02	4.25E−02	2.529E−02	6.917E−02	3.858E−02
C_5	0.7994	1.023	0.303	1.101	0.714
C_6	2.25E−05	2.80E−05	2.46E−05	2.71E−05	2.2E−05
Maximum concentration (mass%)	26.5	40.6	38.8	40.0	31.5
Maximum concentration (mol%)	10	10	10	8	10

Constants for the Ostergaard equation are given in Table 5.2 for thermodynamic inhibitors having organic structures.

For inorganic salts that are thermodynamic inhibitors, the relevant constants to fit Equation 5.1 are given in Table 5.3.

In attempts to overcome high costs, large volumes, and environmental complications of the common thermodynamic inhibitors, particularly methanol, low-dosage hydrate inhibitors (LDHIs) that alter hydrate crystal surfaces effectively at concentrations not exceeding about 0.5 wt% (5000 ppm), while being biodegradable, have been sought by research communities (Lederhos et al., 1996). Thus, kinetic hydrate inhibitor (KHIs), AA inhibitors, and potentially AFPs are emerging to at least supplement THIs.

5.3 KINETIC HYDRATE INHIBITORS

Kinetic hydrate inhibitors (KHIs) retard nucleation of hydrate crystals and then continue to hinder crystal growth after nucleation. But thermodynamic conditions of hydrate formation are typically not significantly affected, as KHIs do not alter the HLVE relationship (Xiao and Adidharma 2009). The main objective of KHI application is the delaying of hydrate crystallization long enough for flowing hydrocarbons and associated water to exit subsea production lines before hydrates have time to plug the line (Huo et al., 2001; Kelland et al., 2006a).

Current state-of-the-art KHIs prolong times of crystal nucleation, followed by extended times of cluster growth. Nucleation periods may be estimated by induction times, representative of retarded nucleation. Practically, growth rate after visual and precipitous hydrate appearance may be approximated by system gas uptake rate or by closed-system pressure decline rate.

Effectiveness is evaluated by how much the KHI counteracts subcooling imparted by prevailing *P*–*T* operating conditions (Natarajan et al., 1994; Ohtake et al., 2005). Greater subcooling conditions, represented by higher ΔT in Equation 5.1, increase hydrate formation rates and decrease hydrate induction times (Villano et al., 2008; Moudrakovski et al., 2001).

Large degrees of subcooling during offshore production operations spur a persistent industry search for KHIs effective in the deeper waters (Makogon and Sloan, 2002; Kelland, 2006). Current state-of-the-art KHIs are effective under subcooling conditions generally less than 10–12°C (Zheng et al., 2011; Villano et al., 2008).

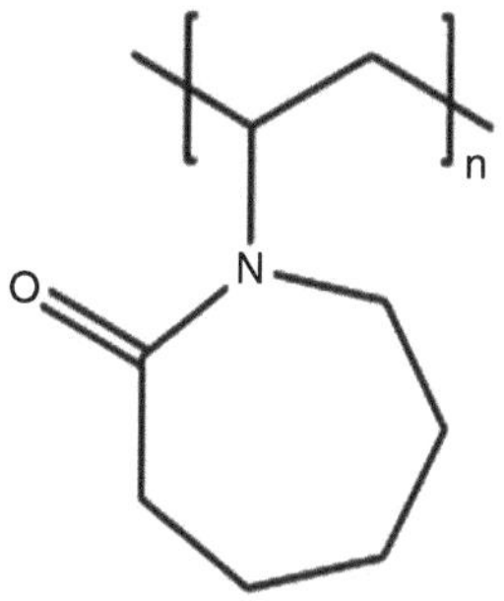

Poly-[*N*-vinylcaprolactam]
PVCAP

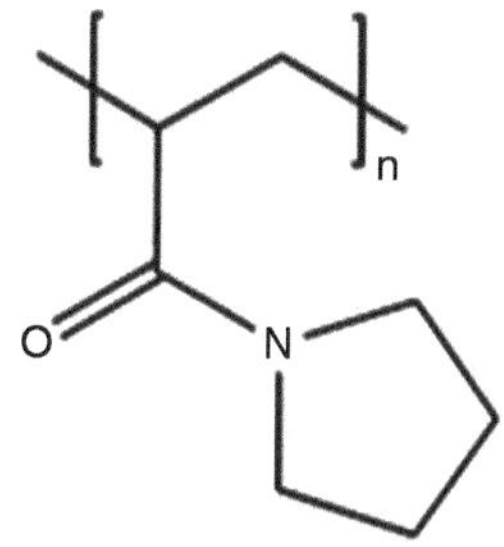

Poly-[acryloylpyrrolidine]
PAPYD

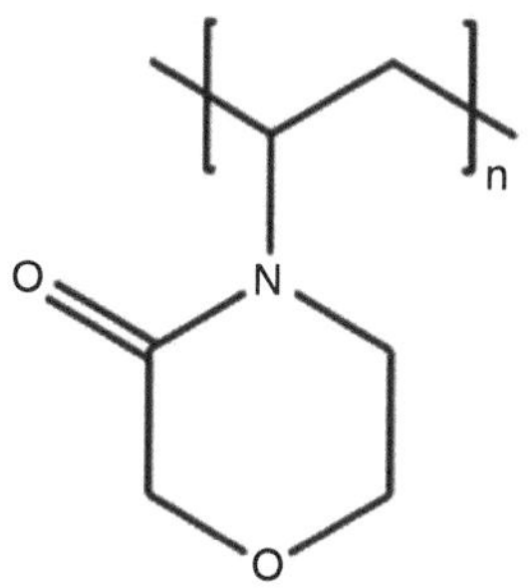

Poly-[acryloylmorpholine]
PAMOR

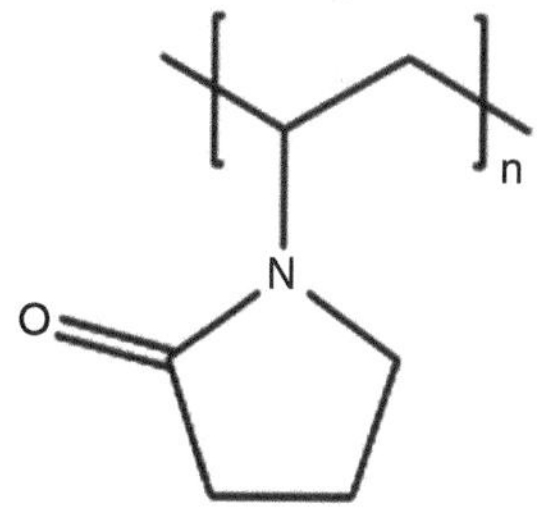

Poly-[*N*-vinylpyrrolidine]
PVP

Figure 5.5 ***Molecular structures of four commercial kinetic hydrate inhibitor (Ohtake et al., 2005).*** *(Reprinted with permission from Journal of Physical Chemistry B. Copyright 2005. American Chemical Society.)*

5.3.1 Molecular Structures of Representative KHIs

KHIs are commonly water-soluble polymers (Zheng et al., 2011). Given in Figure 5.5 are example molecular structures of KHIs used industrially. The effective subcooling limits for poly-[*N*-vinylcaprolactam] (PVCAP), poly-[acryloylpyrrolidine] (PAPYD), poly-[acrylolylmorpholine] (PAMOR), and poly-[*N*-vinylpyrrolidone] (PVP) of Figure 5.5 are 18.0, 17.5, 11.5, and 11.0°C, respectively (Ohtake et al., 2005). Of the four structures, PVCAP most favorably meets the subcooling criteria.

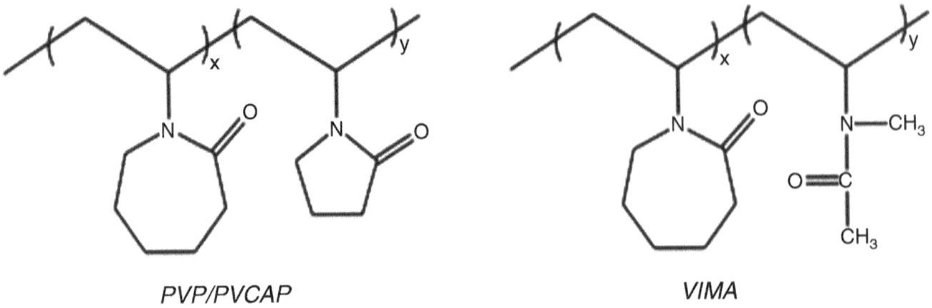

Figure 5.6 ***Molecular structures of two commercial kinetic hydrate inhibitor (Anderson et al., 2005).*** *(Reprinted with permission from Journal of the American Chemical Society. Copyright 2005. American Chemical Society.)*

Other KHIs demonstrate limited degrees of inhibiting hydrate formation: (1) vinyl alcohol adsorbs on developing crystal faces of hydrates to slow continued crystal growth (Makogon and Sloan, 2002); (2) a superior heat capacity and a hydrophilic/lipophilic balance leaning toward molecular hydrophobicity improves inhibition of poly(diethyl acrylamide) (Varma-Nair et al., 2007); (3) quaternary ammonium zwitterions have inhibition properties (Storr et al., 2002).

The two copolymers PVP/PVCAP and (*N*-methyl, *N*-vinylacetamide)/PVCAP are commercially available as inhibitors (Anderson et al., 2005). See Figure 5.6.

5.3.2 PVCAP Performance and Concentration

Based on field performance, laboratory experiments, and molecular dynamics simulations, PVCAP is most often the KHI of choice over PVP and other commercial products available (Kelland and Iversen, 2010; Yang and Tohidi, 2011; Zeng et al., 2008; Zhang et al., 2009; Anderson et al., 2005; Ohtake et al., 2005; Abay et al., 2011b).

Abay et al. (2011a) evaluated PVCAP by matching their laboratory data with the nucleation probability distribution function (Equation 5.2):

$$P(t) = 1 - e^{-J(t_0 - \tau)} \tag{5.2}$$

where $P(t)$, probability distribution function (probability of getting an induction time at time t_0); t_0, induction time between time zero and time t; J, rate nucleation/unit volume; τ, lag time (delay of nucleation by inhibitor; time to attain constant nucleation rate).

The evaluation of sII hydrates by Abay et al. (2011a) indicates a changing nucleation rate dependency on PVCAP concentration when incremented up to 3000 ppm; PVCAP inhibition improved at upper levels of the concentration range but failed to inhibit below 500 ppm. Evaluating sI hydrates similarly, Abay et al. (2011b) found nucleation to be fairly independent of PVCAP at the lower concentrations of their 20–2000 ppm range, but PVCAP retarded sI hydrates at the higher concentrations.

In experiments by Yang and Tohidi (2011), 1.0 wt% of PVCAP retarded sII hydrates during hydrate growth phase, more than lengthening induction time. However, inhibition was significantly improved in both nucleation and growth phases when glycol ether compounds were added to give a mix of 0.50% PVCAP and 0.75% ether, where improvement may be attributed to larger molecular sizes of the glycol ether.

In general for KHI inhibitors, only low dosages of 0.3–1.0% by mass (3000–10,000 ppm) are necessary to be effective (Arjmandi et al., 2002). In independent experiments, Kelland et al., (2006a) find that 0.1–1.0% by mass (1000–10,000 ppm) of the active ingredient of the KHI effectively inhibits hydrate formation.

5.3.3 Proposed Mechanisms of KHI Inhibitions

Anderson et al. (2005) propose a two-step retardation mechanism for KHI hydrate inhibition.

In the first step, molecular structuring of water by hydrogen bonding is disrupted around the developing hydrate crystal. As observed in Figure 5.5, PVCAP has a lactam ring connecting to a polymer backbone through an amide group. Ohtake et al. (2005) showed with the help of mass spectra how this functional group, as contained in *N*-methylcaprolactam, retards hydrate crystal nucleation by decreasing the surrounding free water molecules and nearby water structuring. Both PVP and PVCAP show these reductions during molecular simulations (Storr et al., 2004; Hawtin et al., 2005).

In the second step, KHI molecules adsorb on crystal embryo surfaces and compromise nucleation growth by hindering water or gas diffusion (Kang et al., 2011).

Kang et al. (2011) estimate the degree of surface coverage by KHI. They denote fractional KHI coverage of hydrate crystal surfaces by θ and fractional surfaces remaining for further inhibitor adsorption by $1 - \theta$. Assuming a Langmuir type of monomolecular layer of inhibitor molecules adsorbed, the fractional coverage can be estimated by Equation 5.3 (Kang et al., 2011):

$$\theta = \frac{K(T)^{1/m} C_{\mathrm{KHI}}^{1/m}}{1 + K(T)^{1/m} C_{\mathrm{KHI}}^{1/m}} \tag{5.3}$$

where K, Langmuir constant for specific inhibitor as a function of temperature; C, concentration of inhibitor in solution (wt%); m, number of sites on crystal embryo that could be occupied by a single inhibitor molecule.

Kang et al. anticipate growth rates of embryo hydrate crystals to be proportional to free areas $(1 - \theta)$ and the hydrate induction time inversely proportional to unoccupied areas.

The five-member lactam ring of PVP may cover pentagonal faces of hydrate crystals when the amide associates with the hydrate, whereas the seven-member lactam ring of PVCAP may cover crystal hexagonal faces (Lederhos et al., 1996). Data indicate PVP with its five-member lactam ring has inferior performance to PVCAP with its seven-member lactam ring (Zeng et al., 2008; Zhang et al., 2009; Anderson et al., 2005).

Anderson et al. (2005) hypothesize that KHI adsorbs on embryo crystal faces and physically slows crystal growth. Consequently, when PVCAP binds to a face, crystal growth at that point is repressed and plate-like crystals result (Storr et al., 2004; Sakaguchi et al., 2003). Makogon and Sloan (2002) further found that vinyl alcohol and PVCAP adsorptions create tortuous diffusion paths, slowing hydrocarbon gas movement to that surface; increasing polymer sizes improves effectiveness (Makogon and Sloan, 2002). Furthermore, polymer side groups around the crystal face may block methane gas from some cavities but leave fractional uncovered crystal surfaces accessible, albeit through a tortuous path.

Further, Ohtake et al. (2005) suggest that the size, as determined by polymerization, of the appendage connecting to the functional group of a KHI becomes important to slow crystal growth during the catastrophic growth phase. According to their theory, an optimum size of the lactam–amide functional group better fits size and contour of the developing crystal (Ohtake et al., 2005). Daraboina et al. (2011) formed sII hydrates from a 93/5/2 methane/ethane/propane mixture and then used nuclear magnetic resonance (NMR) spectra to compare uptake differences of hydrocarbon gases between a control and a hydrate with KHI added. The inhibited hydrates took up 7-15% less methane and ethane whose entrance had been blocked by inhibitor.

5.3.4 Molecular Dynamics Simulations

Molecular dynamics simulation extends the kinetic studies of hydrate formations to nanoscale crystal sizes, as it verifies active KHI sites of oxygen double bonds, amide groups, and hydroxyl groups adsorbing on hydrate crystals (Kvamme et al., 2005).

In another dynamics simulation, de Souza and Freitas (2002) studied interactions of cyclic amide functional groups with partially filled $5^{12}6^{2}$ cavities. The cyclic functional groups adsorbed and then seemingly moved within the cavity to equilibrate van der Waals forces, electrostatic forces, and hydrogen bonding.

In simulations of Rodger (2011), hydrate-like clusters of water molecules appear to diffuse as a unit, gaining and losing smaller cluster components in an oscillatory manner until eventually reaching stable critical cluster sizes where thermodynamically favored hydrate particles rapidly form. The Rodger simulation found that PVP adsorbed neither on surfaces of small clusters nor on eventual hydrate particles but nevertheless delayed initial expansion of clusters to a critical cluster size, thus providing a measure of instability while retarding nucleation. In contrast, Rodger found that PVCAP adsorbs on nanoscale crystals before reaching critical cluster size. Thus, PVCAP promotes cluster aggregation rate by stabilizing and increasing clusters of greater than 50 water molecules. On hydrate particles reaching microscale size, the adsorbed PVCAP then hinders water and gas additions essential for fast growth. The apparent paradox of PVCAP promoting as well as inhibiting hydrates can be explained by the scale at the stage of occurrence.

Also by molecular dynamics simulations, Anderson et al. (2005) studied KHI performance by determining KHI bonding strength to the hydrate crystal. Relative bond strengths decrease as follows: VIMA > PVCAP > PVP. The preceding compounds have exothermic binding energies of $-45.8 > -37.5 > -20.6$ kcal/mol, respectively. Anderson et al. report degree of increasing inhibitor effectiveness to be in the order VIMA > PVCAP > PVP. Two primary contributors to strong binding were matching molecular sizes of inhibitor with adsorption site and the matching of charge distribution on inhibitor molecular edge with charge distribution between adjacent water molecules on the crystal surface.

Assuming that degree of KHI-induced subcooling is a good predictor of effectiveness, de Souza and Freitas (2002) found a correlation between KHI's heat of adsorption and its degree of imparted subcooling. Those inhibitors evaluated are presented in Table 5.4.

Again, PVCAP is the most effective and PVP the least effective.

Finally, Varma-Nair et al. (2007) found poly(diethylacrylamide) and PVCAP to perform satisfactorily as KHIs, relating superior inhibition properties of these compounds to higher heat capacities and a hydrophile–lipophile balance (HLB) skewed toward hydrophobicity.

Table 5.4 Inhibitors evaluated for ΔH_{ads} influence on gas hydrate formation (de Souza and Freitas, 2002)

Inhibiting polymer	Chemical formula	$\Delta H_{Adsorption}$ (kcal/mol)	$\Delta T_{subcooling}$ (°C)
PVP	Poly-[vinylpyrrolidone]	−8	11.0
PNEAM	Poly-[*N*-ethylacrylamide]	−16	15.0
PVVAM	Poly-[*N*-vinyl-valerolactam]	−18.5	16.0
PAPYD	Poly-[acryloylpyrrolidine]	−25	17.5
PVCAP	Poly-[*N*-vinyl-2-caprolactam]	−29	18.0

5.3.5 Limitations to Use of KHIs

KHIs in field use are synthetic compounds, limited in application primarily by expense of preparation. A second limitation is poor biodegradability. For example, use in the Norwegian sector of the North Sea requires the KHI and its products to be greater than 20% biodegradable in 28 days (Villano et al., 2008). A third obstacle to greater use occurs whenever wellhead temperatures and water salinities exceed stability or effectiveness limits of the synthetic polymers, especially if injection temperatures must accommodate 70–90°C (Zheng et al., 2011).

Three conditional requirements further limit KHI operational use in oilfield pipelines: (1) a natural gas pipeline or subsea production line must have no liquid hydrocarbon phase (Kelland et al., 2006a). (2) Treated wells or subsea lines must not be shut-in for appreciable periods because KHIs then lose their effectiveness. Since their *raison d'etre* is to delay hydrate formation (Huo et al., 2001), flow curtailment may exceed KHI induction time retardation. (3) At extraordinarily high subcooling conditions of deep water operations, current state-of-the-art KHIs may be unable to sufficiently retard enhanced rates of hydrate crystal growth (Kelland et al., 2006b).

Current KHI applications offshore are limited to a ΔT subcooling maximum of about 10–12°C. Villano et al. (2008) verified this by evaluating in the laboratory a representative, commercial KHI product Luvicap 55W, a 1:1 blend of vinylcaprolactam/vinylpyrrolidone copolymers, i.e., PVCAP/PVP. From a stirred and chilled test cell with natural gas pressurizing 3.6 wt% NaCl–water solution containing 5000 ppm Luvicap 55W, Villano developed the empirical Equation 5.4 relating induction time with degrees of subcooling:

$$\Delta T_{SC} = 28.623\varphi_i^{-0.122} \tag{5.4}$$

where ΔT_{SC}, degrees of subcooling (°C or K); φ_i, induction time (min).

Therefore, hydrate induction time may be estimated from Equation 5.4 under a known subcooling imposed on a 5000 ppm Luvicap 55W–seawater solution. Hydrates form rapidly and natural induction times decrease accordingly in deep waters that may impose 12–18°C subcooling. The representative Luvicap 55W becomes increasingly ineffective to inhibit hydrate formation beyond a subcooling of 10–12°C (Villano et al., 2008).

5.3.6 Drilling Mud Inhibitions with KHIs

Deep water drilling puts formidable constraints on hydrate inhibition in drilling muds as can be surmised from Equation 5.4. Hydrate nucleation accelerates and induction times shorten as degree of subcooling increases with water depth. The residence time for the circulating mud necessarily lengthens in the outlying wells of deeper waters, complicating hydrate formation problems within the mud.

KHI water-soluble polymers in water-based muds encounter other problems: (1) increasing mud salinity reduces KHI polymer solubility, so that thermal hydrate inhibitors may be needed for supplements in muds of high salinity; (2) high mud temperatures reached in deep wells reduce polymer solubility; (3) smectite clays in the cuttings scavenged by the mud enhance hydrate formation and adsorb inhibitors.

Kelland and Iversen (2010) studied *N*-vinyl caprolactam:*N*-vinylpyrrolidone (PVCAP–PVP or Luvicap 55W) copolymers as KHIs in drilling muds. Inhibition was satisfactory in the absence of clays in the mud, but the encroachment of calcium montmorillonite into the mud reduced hydrate induction times by as much as a factor of 20. A quite rapid hydrate formation results when calcium montmorillonite is present in the drilling mud. With no smectite clay present, hydrate initiation is delayed for 18–20 h in the pressurized and chilled system containing PVCAP–PVP, compared with 1 h when calcium montmorillonite is present in a similar system.

5.3.7 KHIs Erase Hydrate Formation Memory Effects

At given operating conditions, gas hydrate re-forms from dissociated hydrate waters with shorter induction time than when formed from fresh waters. This can be seen in Figure 5.7 from data of Seo and Kang (2012). Seo and Kang first formed methane hydrates from deionized water and then re-formed them from remnant water of decomposed hydrates. In each case, hydrate formations were compared during constant-rate cooldown of the system, starting at a temperature above the *P*–*T* equilibrium line (Seo and Kang, 2012).

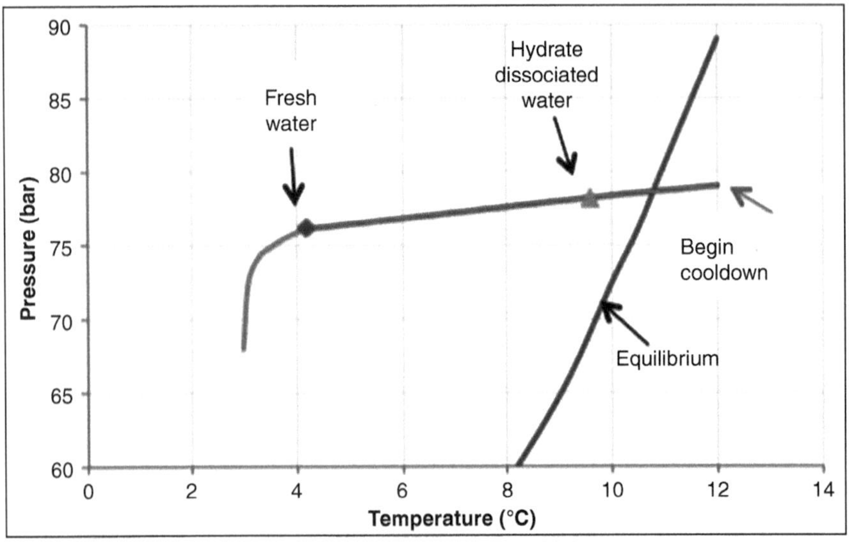

Figure 5.7 ***Demonstration of hydrate memory effect of water; constant-rate cooldown of system (Seo and Kang, 2012).***

Methane-pressurized aqueous solutions were cooled from a starting temperature of 12°C. The water from previously dissociated hydrates re-formed hydrates at a subcooling of 1.2°C, while fresh water required 6.3°C subcooling. Note that subcooling is defined as the temperature difference ΔT_{SC} between the equilibrium value T_{eq} and the temperature of hydrate formation T_{form} as given by Equation 5.5:

$$\Delta T_{SC} = T_{eq} - T_{form} \tag{5.5}$$

Because the memory effect enhances hydrate re-formation, instances may arise in the eventual production of methane from seafloor hydrates where the memory phenomenon becomes significant in hydrate gas recovery processes. For example, production of gas from hydrate reservoirs by the depressurization method may introduce hydrate re-formation along the gas flow path to the wellbore, causing serious permeability restrictions. On pressure drawdown in the reservoir, endothermic hydrate decompositions lower reservoir temperatures. As the waters and gases flow toward the wellbore after initial hydrate dissociation, memory effects from those dissociated waters make hydrates prone to re-form. It is postulated by Seo and Kang (2012) that injection of effective KHIs into the media near the wellbore could prevent hydrate re-formation and the accompanying

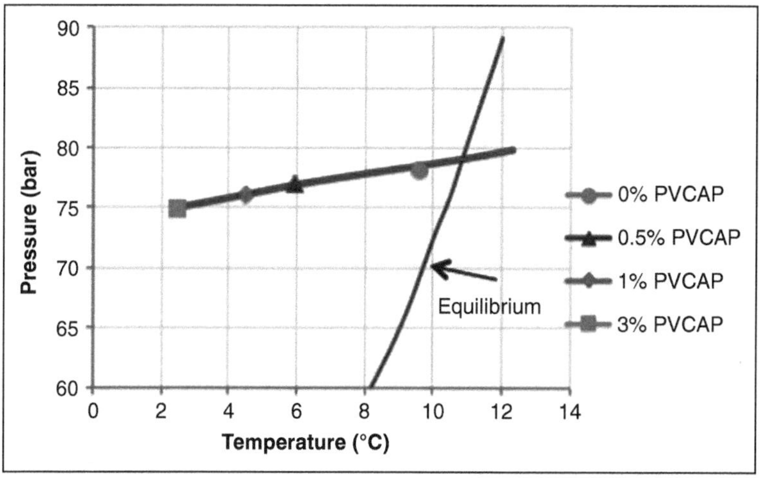

Figure 5.8 ***Dissipation of hydrate memory with PVCAP in water from dissociated hydrates; constant-rate cooldown of system (Seo and Kang, 2012).***

undesirable permeability degradation. Selecting PVCAP as representative of KHIs, they determined KHI concentration effects on induction times of hydrate-dissociated waters containing the memory effect. Their results may be viewed in Figure 5.8.

Hydrate memory of dissociated waters dissipates as more PVCAP is added, as seen in Figure 5.8. PVCAP concentrations elevated from 0.5 to 3.0 wt% progressively improve subcooling ΔT_{SC} at which hydrates form. The hydrate retarding effect at 3.0 wt% PVCAP dictates that subcooling conditions must exceed 7.8°C before hydrates can form, a subcooling which exceeds that of memory-absent fresh water seen in Figure 5.7.

The eventuality of hydrate re-formation near the wellbore during hydrate gas production might technically be circumvented by injecting a soluble KHI and practically be accomplished if the inhibitor is of low cost.

5.4 ANTIFREEZE PROTEINS

AFPs are present in Arctic fish, plants, insects, and bacteria. These proteins prevent destruction of the host from cellular water freezing by adsorbing onto developing ice crystals and limiting ice growth.

One recognizes mechanism similarities of the AFPs functioning in Arctic conditions to KHIs inhibiting hydrate formations in extreme offshore conditions. In fact, interests in AFPs are accentuated because synthetic

LDHI deficiencies become critical in deeper waters of conventional gas and hydrate gas operations. High cost and poor degradation in the environment are serious deterrents to using synthetic hydrate inhibitors. Significantly, AFPs degrade better in the environment than do synthetic LDHIs. For example, a biological half-life of a few weeks is reported for a typical AFP (Fletcher et al., 1982).

Consider three attributes of AFPs contributing to their ice prevention effectiveness: (1) freezing point is lowered below normal melting point of cellular water solution; (2) crystal morphology is altered by adsorbing on specific ice crystal faces; (3) recrystallization is retarded.

5.4.1 Inhibiting Ice Formation with AFPs

AFPs were first discovered in Arctic fish. By inhibiting internal ice formation, these AFPs allow the fish to survive polar waters. They have four classifications, Types I–IV, where proteins of each type associate with specific fish and are distinguished by size and molecular structure (Gordienko et al., 2010; Gordon et al., 1962).

Ramsey (1964) first reported AFPs in insects, where the insects are those ordinarily found in Arctic climes at temperatures regularly dipping significantly below water's normal freezing point. To depress the internal freezing point sufficiently, insect AFPs have become order(s) of magnitude more effective than fish AFPs. Primarily, their extraordinary effectiveness is thought to originate from adsorption on multiple faces of ice crystals (Knight et al., 1991; Graether and Sykes, 2004).

Similarly, AFPs have been identified in Arctic bacteria. For example, an Antarctica lake (1.9% salt) harbors the bacterium *Marinomonas primoryensis* (Gilbert et al., 2005). Isolated from the sea, the lake maintains a 1–2 m thick ice cap on its surface most of the year with a −1 to +1°C water temperature gradient from its ice cap to the seafloor. From *M. primoryensis* samples, Gilbert et al. cultured the species, lysed the cells, and recovered AFP protein in the lysate. Compared with other bacterial AFPs, they found that the protein from *M. primoryensis* exhibits an extraordinarily strong 2°C freezing point depression.

5.4.2 AFP Thermal Hysteresis of Ice

Thermal hysteresis is defined as the number of degrees by which an AFP lowers phase transition of indigenous water solutions to below their normal melting points. For example, up to concentrations of 3.0 mM, a Type I fish

Table 5.5 Freezing point depressions of AFPs

AFP	Freezing depression (°C)	References
Plants: perennial rye grass	0.1–0.5 0.1	Zhang et al. (2004), Sidebottom et al. (2000)
Fish: the four types	0.7–1.5	Barrett (2001), Tyshenko et al. (1997), Jensen et al. (2010)
Bacterium: *Marinomonas primoryensis*	>2	Gilbert et al. (2005)
Insects	5–10	Barrett (2001), Tyshenko et al. (1997)

AFP was found to impose protective thermal hysteresis to an extent given by Equation 5.6 (Chao et al., 1997):

$$\Delta T_{\text{the}} = 0.1286\ln(C_{\text{F1}}) + 0.5128 \tag{5.6}$$

where ΔT, number of degrees lowered (°C); C_{F1}, concentration of Type I fish AFP (mM).

Magnitude of thermal hysteresis varies widely among fish, insect, plant, and bacterium. This can be seen in Table 5.5 where relative freezing point depressions, type of proteins, and references are listed.

Note in Table 5.5 the weak thermal hysteresis of plant AFPs. Although freezing point depression is minor in plants, Sidebottom et al. (2000) found that plant AFPs can restrict size of ice crystals and limit recrystallization more than polar fish or cold-region insect proteins. This important survival characteristic of plant AFPs limits ice crystals to less tissue-damaging sizes (Sidebottom et al., 2000).

Each AFP classification in Table 5.5 has a unique set of proteins. Each set functions differently, suggesting structural variations that may affect the adsorption mechanism of protein to ice (Tyshenko et al., 1997; Barrett, 2001).

Addition of relatively low-molecular-weight solutes such as amino acids, salts, and sugars improves AFP performance, where the degree of enhancement depends on the specific solute, and the effect increases with solute concentration (Li et al., 1998; Kristiansen et al., 2008). Studying ice thermal hysteresis with protein similar to beetle larvae AFP, Li et al. report threefold to sixfold increases when supplementing this protein with sodium salts of five organic acids: succinates, citrates, malates, malonates, and acetates. Independently, Kristiansen et al. demonstrated enhanced thermal hysteresis by using 10 different salts of monovalent cations to complement AFP from the longhorn beetle *Rhagium inquisitor*.

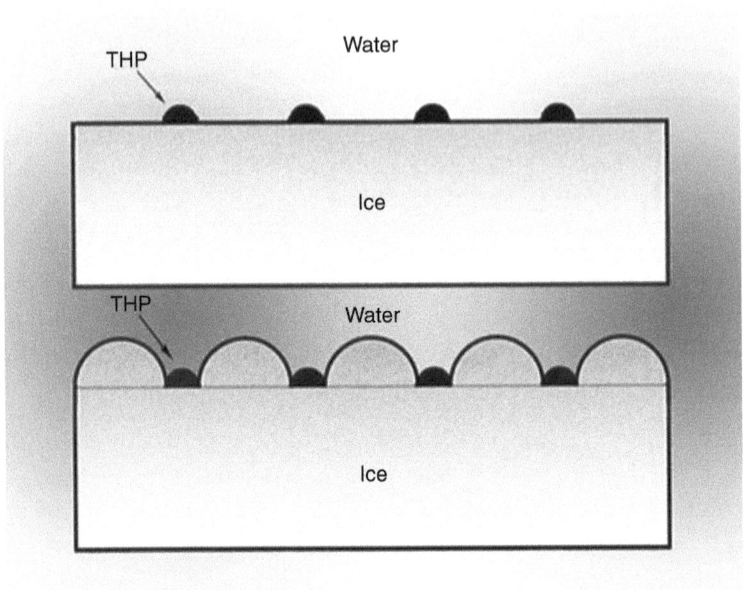

Figure 5.9 ***Protein interference inhibition model of thermal hysteresis in ice (Knight et al., 1991).***

5.4.3 Mechanism: AFP Thermal Hysteresis of Ice

Adsorption of AFPs from fish, bacteria, insects, or plants imposes thermal hysteresis in a proposed mechanism depicted in Figure 5.9 (Knight et al., 1991).

The thermal hysteresis protein, designated THP in Figure 5.9, adsorbs as nodules at the cell–water interface during water's phase transition period. Not being able to form over protein adsorption sites, ice crystals form adjacent to the protein nodules with curvatures less than in the absence of protein. The smaller radius of curvature of the ice crystals leads to greater freezing point depression of those crystals, as presented by Equation 5.7 (Yeh and Feeney, 1996):

$$\Delta T = \frac{2\Omega\gamma T_0}{r_{\min}\,\Delta H_0} \tag{5.7}$$

where ΔT, temperature lowering of freezing point due to small ice crystals; Ω, molal volume ice; γ, isotropic surface energy; T_0, normal freezing point of pure water; $r_{\min}$, radius of ice crystal; ΔH_0, latent heat of fusion.

A universal mechanism to describe thermal hysteresis protein adsorption, in terms of bonding strength or specificity, for the different proteins from plants, fish, bacteria, and insects has not been achieved (Barrett, 2001; Tyshenko et al., 1997; Middleton et al., 2009; Sidebottom et al., 2000).

As solute concentrations increase, so does the thermal hysteresis. Equilibrium exists between the proteins in solution and those adsorbed on the ice surface, so a secondary solute such as salt moves the equilibrium toward adsorption (Kristiansen et al., 2008).

5.4.4 AFPs Change Ice Morphology

Fish AFPs alter ice crystal morphology, causing pyramidal and needle-like columns of ice to form in contrast to sheets from pure water (Barrett, 2001; Tyshenko et al., 1997). Morphology alterations vary with protein source. For example, Tyshenko et al. (1997) report wave-like, smooth-edged crystals forming with the AFP of the spruce budworm. As another example, the AFP from the bacterium *M. primoryensis* develops smooth ice crystals with rounded surfaces (Gilbert et al., 2005).

Molecular binding of the protein may be one reason AFPs alter ice crystal morphology, as binding includes both hydrophilic and hydrophobic affinities (Barrett, 2001). Insect AFPs, such as the spruce budworm, are hydrophilic (Hew et al., 1983), but the fish proteins are hydrophobic (Sönnichsen et al., 1995).

5.4.5 AFPs Retard Gas Hydrate Initiation

AFPs effectively inhibiting ice crystal formation have been evaluated for hydrate inhibition. Jensen et al. (2011) report a Type III protein from the ocean pout performing better, as measured by increased hydrate induction time, than one of the more effective synthetic KHIs, namely polyvinylcaprolactam – that is, PVCAP. The ocean pout protein has three advantages over PVCAP: (1) environmentally neutral, (2) no cloud point, and (3) better solubility in high-salinity water (Jensen et al., 2011).

In another comparison with synthetic inhibitors, Zeng et al. (2006a) report that AFP from winter flounder having Type I AFP with 37 amino acids in its molecular structure inhibits hydrate formation better than the synthetic inhibitor PVP. Effectiveness is indicated by an increase in induction time from 2.3 h without the fish protein to greater than 11.2 h using 0.25 mM of the fish AFP at test conditions of −5°C and 2.2 bar while forming hydrates in tetrahydrofuran–water solutions. Without tetrahydrofuran present, this particular AFP still effectively inhibits sI and sII hydrates at the 0.25 mM concentration. Nucleation rate for the system may be calculated according to Equation 5.8 (Zeng et al., 2006b):

$$\frac{N_t}{N_0} = \exp[-kt] \tag{5.8}$$

where N_t, number of hydrate-free samples at time t; N_0, number of hydrate-free samples at time 0; k, nucleation rate; t, time.

The induction time τ, as the inverse of nucleation rate, is given by Equation 5.9:

$$\tau = k^{-1} \tag{5.9}$$

The Zeng experiments showed that small concentrations of winter flounder Type I AFP slow nucleation rates by a factor of 3 in sII hydrates containing propane and similarly slow methane sI hydrates. In addition to slower nucleation as measured by induction time, hydrate growth is slowed by the AFP, as determined by rates of gas uptake.

It is noteworthy that winter flounder AFP eliminates memory effects on recrystallization of natural gas hydrates from remnant water solution of decomposed hydrates (Zeng et al., 2006b).

5.4.6 AFPs Affect Hydrate Crystal Morphology

Instead of a three-dimensional structural growth, hydrate crystals may grow two-dimensionally with AFP present. Adsorptions of the proteins on hydrate crystal faces redirect crystal growths (Zeng et al., 2006b).

Hydrate crystal morphology may be altered by AFP in a manner somewhat akin to other natural and synthetic polymers. For example, as observed in Figure 8.6, hydrate crystal morphology is altered by the anionic biopolymer emulsan evidenced by its promoting needle-like hydrate crystals on sand surfaces as opposed to nodular crystals on smectite clay surfaces (Rogers et al., 2004). As another example, the synthetic polymer PVCAP adsorbs on developing hydrate crystals to alter morphology (Abay et al., 2011b).

When insect AFP strongly bonds and covers hydrate crystal faces, two phases of crystal growth ensue. In the first phase, crystal size increases according to a second-order polynomial. In the second phase, a linear crystal growth develops (Jensen et al., 2010). Hydrate crystals grow linearly from beginning to end with no AFP present. As the amount of insect AFP progressively increases, phases of nonlinear growth develop and are characterized by second-order polynomials. Jensen et al. (2010) presented mathematical expressions describing hydrate growth rate in the presence of AFPs or other KHIs that adsorb on developing crystals. By letting the moles n of gas be occluded in time t at constant pressure and temperature, they derived Equation 5.10:

$$\left(\frac{dn}{dt}\right)_{P,T} = C_{\mathrm{WO}} K^{*} (x^{\mathrm{i}}_{\mathrm{L\text{-}V}} - x^{\mathrm{i}}_{\mathrm{H\text{-}L}}) \tag{5.10}$$

where dn/dt, moles of gas occluded in time t; C_{WO}, initial concentration of water–AFP solution; K^{*}, overall hydrate crystal growth coefficient; $x^{\mathrm{i}}_{\mathrm{L\text{-}V}}$, gas concentration in liquid water phase at operating conditions; $x^{\mathrm{i}}_{\mathrm{H\text{-}L}}$, gas solubility in liquid water phase at hydrate–liquid equilibrium.

The overall hydrate crystal growth coefficient K^{*} is defined by the equation derived by Jensen (Equation 5.11):

$$\frac{1}{K^{*}} = \frac{1}{A_{\mathrm{p}}}\left(\frac{1}{k_{\mathrm{d}}} + \frac{1}{k_{\mathrm{r}}}\right) + \frac{1}{k_{1} A_{\mathrm{L\text{-}V}}} \tag{5.11}$$

where A_{p}, surface area of hydrate particle free of adsorbed polymer; k_{d}, mass transfer coefficient in diffusion layer around particle; k_{r}, reaction rate constant for particle surface; k_{1}, mass transfer coefficient in liquid film; $A_{\mathrm{L\text{-}V}}$, liquid–vapor interfacial area.

Equations 5.10 and 5.11 predict two regions of hydrate crystal growth. If inhibitor is present in only low concentrations, then the uncovered particle surface area A_{p} becomes disproportionately larger than the liquid–vapor interfacial area $A_{\mathrm{L\text{-}V}}$, and the last term of Equation 5.11 controls.

5.5 MICROBIAL CELL WALLS AS INHIBITORS

5.5.1 Cells Inhibit Hydrate Formation

Tests run at the Mississippi State University (MSU) Hydrate Laboratory found that anionic biosurfactants profoundly promote gas hydrates, especially in conjunction with smectite clays. But what if microbial cells are present in the hydrate-forming media along with biosurfactants?

Surprisingly, when microbial cells are present, hydrate formation may be retarded. That is, polymeric materials comprising the cell walls inhibit hydrate formation. The reason for this property is understandable because the living cells are sensitive to temperature rises, and exothermic reactions of hydrate formation contacting microorganism surfaces endanger cell survival.

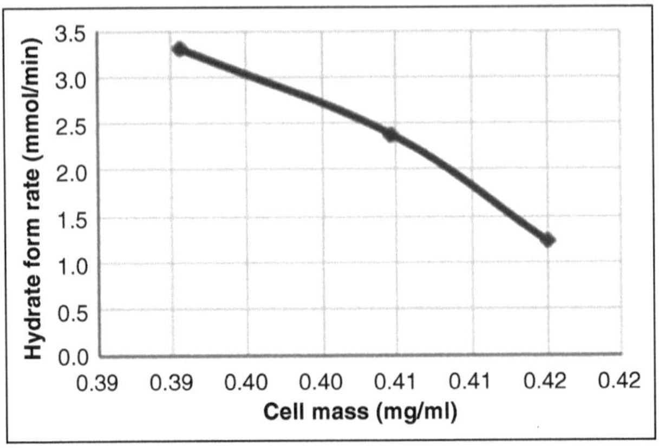

Figure 5.10 **Bacillus subtilis** ***cells retard hydrate crystallization (Radich, 2009).***

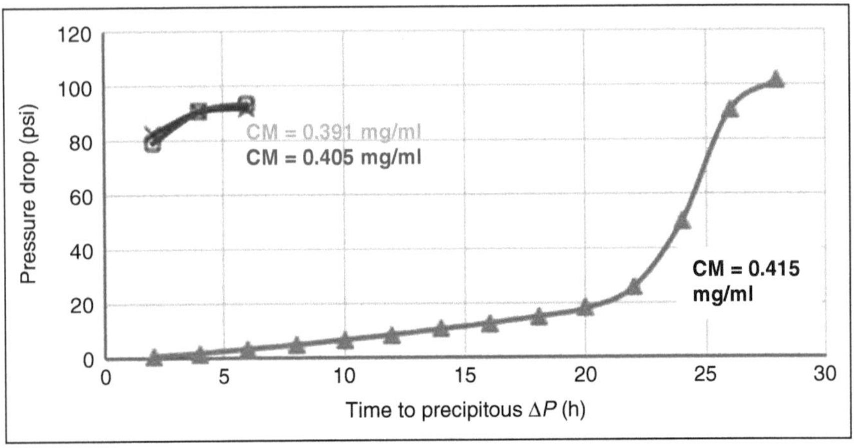

Figure 5.11 **Bacillus subtilis** ***cell mass inhibits gas hydrate formation by increasing induction time (Radich, 2009).***

It is evident in Figure 5.10 that hydrate formation rate declines about threefold when *Bacillus subtilis* cell mass is increased moderately from 0.391 to 0.415 mg/ml in a hydrate-forming system (Radich, 2009).

Hydrate inhibition may be detected if the cell mass lengthens induction time required for hydrate crystal nucleation, where precipitous pressure drop during cooldown of a hydrate-forming closed system marks that time. In Figure 5.11 with *B. subtilis* cell mass concentrations ranging from 0.391 to 0.405 mg/ml in the test medium, a large pressure drop within the first hour indicates hydrates forming extensively with a short induction time. When *B. subtilis* cell mass concentration is increased to 0.415 mg/ml in the medium, onset of rapid hydrate precipitation becomes retarded by about 20 h (Radich, 2009).

To further evaluate inhibition of hydrate formation by cell mass, indigenous microbes from MC-118 sediments were cultured and cell wall materials extracted in the following procedure: (1) cell mass isolated by centrifuging at 1000 relative centrifugal force (RCF); (2) cell mass lysed in 5% NaCl solution for 30 min, (3) lysed sample centrifuged to separate cell fragments, (4) cell fragments washed multiple times in ultrapure water, and (5) washed sample recentrifuged. The procedure gave a much greater surface area of cell wall materials compared with the whole cells. SEM micrographs of the lysed cells are given in Figure 5.12a and b (Radich, 2009).

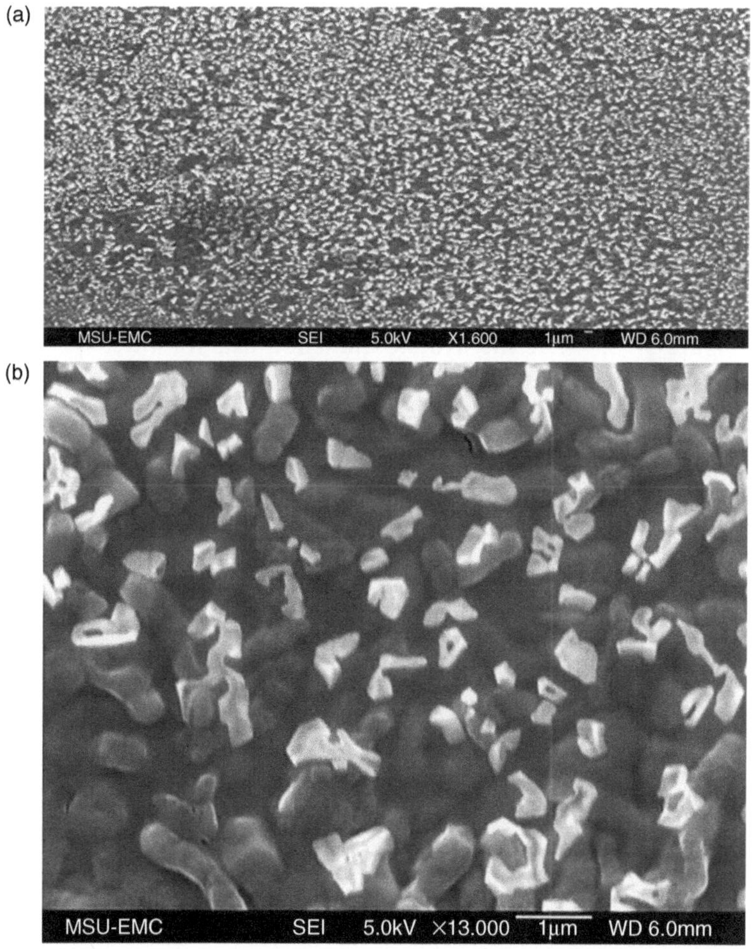

Figure 5.12 ***SEM micrograph of cell fragments from lysing MC-118 indigenous cells (Radich, 2009).*** (a) ×1600 magnification; (b) ×13,000 magnification.

Table 5.6 Lysed cell walls retard hydrate crystallization (Radich, 2009)

Solution content	Cell mass (mg/ml)	Induction time (h)	Maximum formation rate (mmol/min)
Control	0.0	8.3	1.53
MC + S + NM	0.360	18.4	2.15
MC + S + NM	0.368	152.0	1.58
MC + S + NM	0.387	174.8	1.62
Hypertonic-1%	0.350	80.5	1.38
Hypertonic-5%	0.350	>235	≈0

Control, no cell mass present; hypertonic-1%, cell fragments from lysing with 1% NaCl; hypertonic-5%, cell fragments from lysing with 5% NaCl; MC, Mississippi Canyon 118 indigenous cells; NM, nutrient medium; S, surfactant from culture present.

When hydrate-forming conditions are imposed on a system containing lysed and washed fragments of indigenous cells, the augmented cell surface area substantially increases hydrate retardation. Consider the case where MC-118 indigenous cells were lysed by 5% NaCl solution, slurried with original broth, and added to the test system. Hydrates did not form after 235 h at hydrate-forming conditions of 0.5°C and 3.14 MPa under natural gas pressure. The experiment was terminated at the 235-h point without hydrates having formed. A summary of the results is given in Table 5.6.

Another perspective of hydrate retardation by cell wall mass is gained from Figure 5.13, where the hydrate induction data for a test system with lysed cells are superposed on data from a similar test system having only intact cells. As intact cell mass of *B. subtilis* increases from 0.360 to 0.387 mg/ml, the hydrate induction time increases to 18.4 h and then 174.8 h beyond

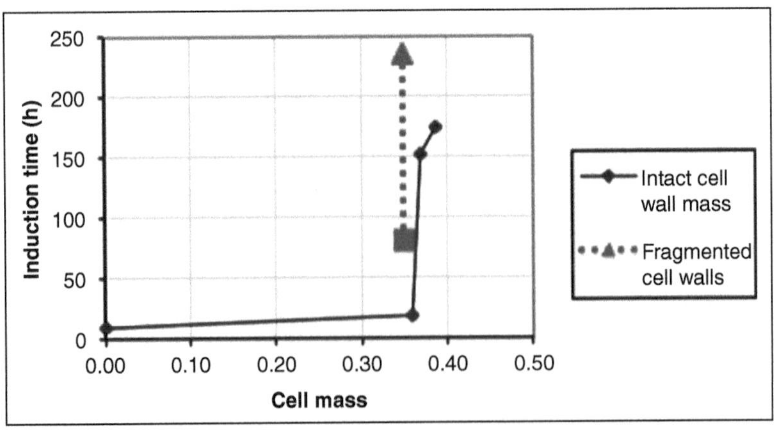

Figure 5.13 *Hydrate retardation dependent on cell wall surface area (Radich, 2009).*

the 8.3 h of the control. However, when a sample contains only fragmented cells (i.e., lysed from 5% NaCl solution), originating from 0.350 mg/ml concentration of whole cells, induction times increase so greatly that hydrates do not form at greater than 235 h.

Cell fragmentation creates greater surface area of the cell's polymeric content to interfere with water's hydrogen bonding in a hydrate-forming system. Radich found that the 5% hypertonic NaCl lysing solution results in a more complete disruption of the cell material and thus gives greater exposed surface areas than a 1% NaCl treatment.

Consider pertinent cell wall properties in the following sections to help explain retardation of gas hydrate formation.

5.5.2 Cell Wall Structure

Cell walls of *B. subtilis*, for example, have average thicknesses of 30 nm and are composed primarily of teichoic acid, peptidoglycan copolymers, and amino acids (Beveridge and Murray, 1980; Birdsell et al., 1975). Strength and rigidity of cell walls of microbial species thriving in the carbon-rich environment surrounding offshore hydrates are due to the association of these three compounds. That is, hydrate retarding effects of the cell fragments viewed in Figure 5.12 must be attributed to the three cell materials, whose structures are derived by the microorganisms from basic building units of amino acids, polysaccharides, and phosphates.

Of the three building blocks of microbial cell walls, first consider teichoic acid, whose α-D-glucopyranosyl glycerol phosphate structure is sketched in Figure 5.14.

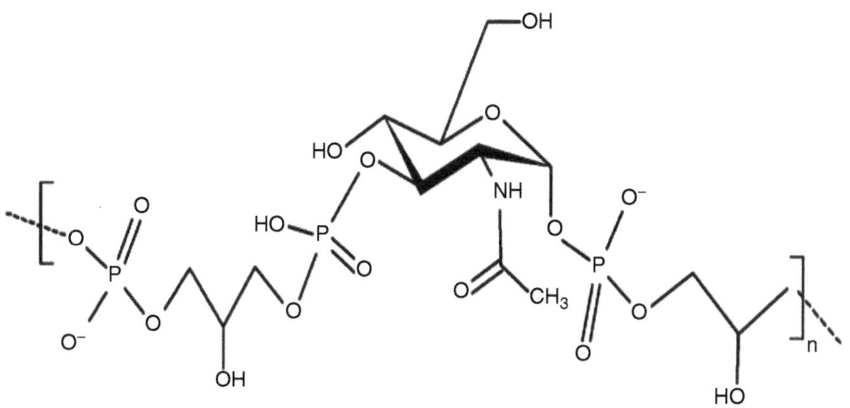

Figure 5.14 ***Structure of teichoic acid (Bolzep, 2014).***

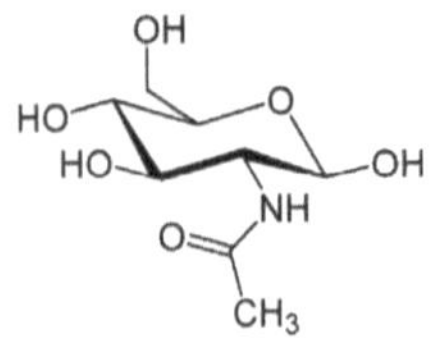

Figure 5.15 ***Structure of* N*-acetylglucosamine, NAGA (Yikrazuul, 2008a).***

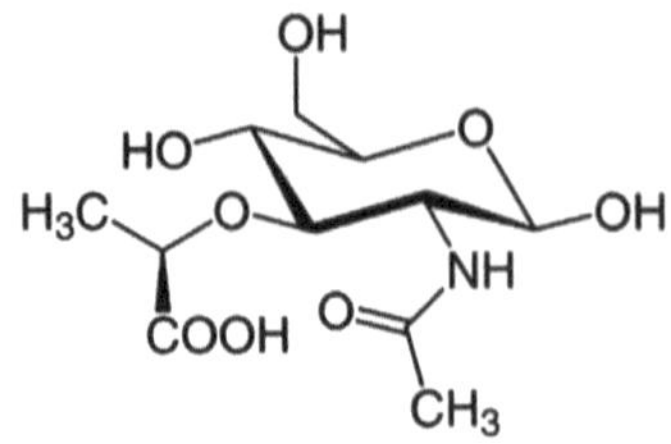

Figure 5.16 ***Structure of* N*-acetylmuramic acid (Yikrazuul, 2008b).***

Teichoic acid represents approximately 54.3% of the dry weight of a *B. subtilis* cell, its polymer being situated perpendicular to the cell's longitudinal axis. It helps cross-link peptidoglycan strands – peptidoglycan strands being composed of two associating copolymers with four-member amino acid appendages (Doyle et al., 1980). In the cell walls of *B. subtilis*, for example, approximately 35% of the glycan strands are cross-linked in an association with teichoic acid as a complex (Beveridge and Murray, 1980).

To delve further into cell wall inhibitive effects that we found during tests in the MSU Hydrate Laboratory, consider next the two copolymers of peptidoglycan and *N*-acetylmuramic acid (NAMA) residues.

Composite peptidoglycan is a high-molecular-weight material, having a foundational structure of copolymers *N*-acetylglucosamine (NAGA) and *N*-acetylmuramic acid residues (Beveridge and Murray, 1980;Vollmer, 2008; Doyle et al., 1980). First, the unit NAGA molecular structure is sketched in Figure 5.15.

Then, in Figure 5.16 the molecular structure of NAMA copolymer *N*-acetylmuramic acid is sketched.

The linkage of the copolymers in a glycan strand is depicted in Figure 5.17. Included in the sketch of Figure 5.17 is the attachment of four amino acids that comprise the third component in a single unit of polymeric cell wall material constituting bacteria.

A better overall view of the composite peptidoglycan can be seen in Figures 5.18 and 5.19.

In Figure 5.18 is sketched how a single strand of four amino acids is attached to *N*-acetylmuramic acid and further illustrates how each strand interlinks with a neighboring polymer unit, which then repeats through *n* units.

Finally, in Figure 5.19 is sketched how single polymer units associate *n* times into a fairly rigid cell wall material.

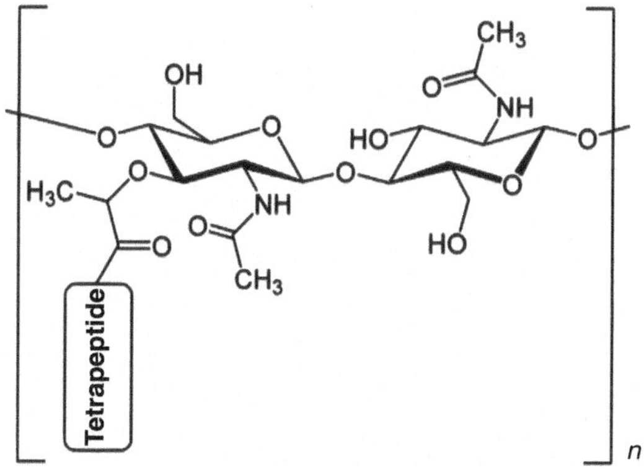

Figure 5.17 ***Linkages of peptidoglycan components (Yikrazuul, 2008c).***

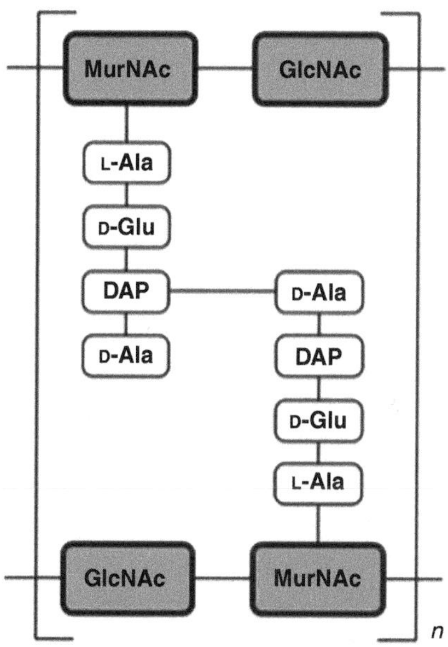

Figure 5.18 ***Linkage of peptidoglycan copolymers with amino acid strands (Yikrazuul, 2008d).***

With visualizations in mind of Figures 5.14–5.19, the following comments and hypothesis are offered to explain possible mechanisms for inhibition of hydrate formation by materials comprising microbial cell walls.

Then, consider the individual inhibitive effect of the three building blocks of microbial cell walls: peptidoglycan copolymers, teichoic acid, and amino acids. Finally, a composite inhibitive effect may be postulated.

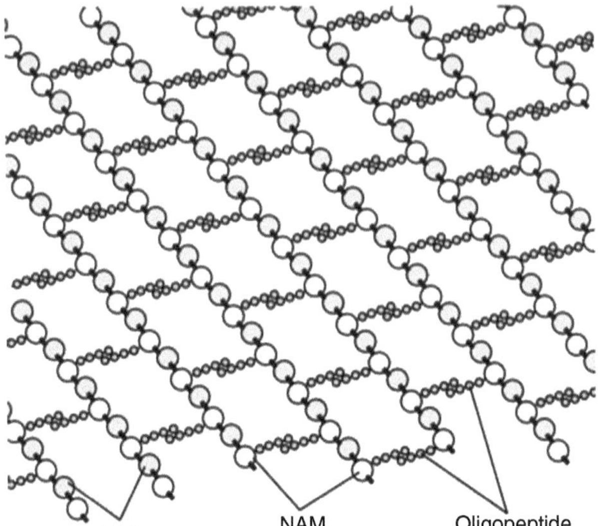

Figure 5.19 ***Composite peptidoglycan structure in cell walls (Mouagip, 2011).***

First, inhibitive effects of the peptidoglycan copolymers were evaluated in MSU research. The water-soluble polymer component NAGA was evaluated as a stand-alone for inhibition of sII hydrate formation. To do so, commercially available NAGA was added before hydrate formation to an aliquot of water-based nutrient broth commonly used for growing microbial cultures. Instead of pure water, nutrient broth was selected as the base because its solution has been found to not only promote microbial growth but also assist hydrate nucleation and thus, as a practicality, reduce laboratory induction times. With a positive influence on hydrate formation, broth use has a secondary value of improving reproducibility of measured hydrate induction times (Radich, 2009). As an aside, Jensen et al. (2011) similarly found that additives such as silica gel help in the reproducibility of the random hydrate nucleation process; Jensen et al. used 0.02 g silica gel/60 cm^3 liquid in the laboratory while comparing hydrate inhibiting agents.

NAGA was added as 0.3 wt% of the water in the nutrient broth solution. Hydrates were formed in a quiescent, closed system in which the test cell was cooled at a constant rate while pressurized with natural gas. In parallel tests, additions of 3.0 wt% NAGA were evaluated in the MSU Hydrate Laboratory for hydrate inhibitive properties. For comparison, 3 wt% of THI methanol was similarly tested. See Figure 5.20.

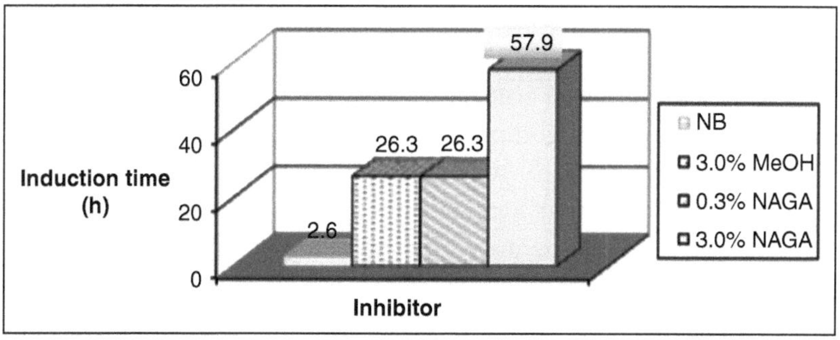

Figure 5.20 ***Gas hydrate inhibition by NAGA as measured by induction time.***

With methanol and NAGA absent, the control exhibited a modest gas hydrate induction time of less than 3 h as shown in Figure 5.20. Also, note in Figure 5.20 that 0.3 wt% NAGA inhibits hydrate nucleation to the same extent as 3 wt% methanol, both having an induction time of greater than 20 h. By increasing NAGA concentration to 3%, hydrate induction time lengthens to greater than 50 h, which is 2.5 times longer than a similar concentration of methanol and 25 times longer than the control solution.

In addition to induction time influence, NAGA effects on hydrate formation rates were evaluated and the results presented in Figure 5.21.

Observe in Figure 5.21 that neither methanol nor NAGA significantly alters hydrate formation rates. Therefore, the main inhibitive effect of NAGA is to lengthen induction time, and the lengthening of induction time is substantially better than that achieved with an equivalent concentration of methanol; that is, an order of magnitude less concentration of NAGA is necessary to achieve the same inhibition as methanol. The

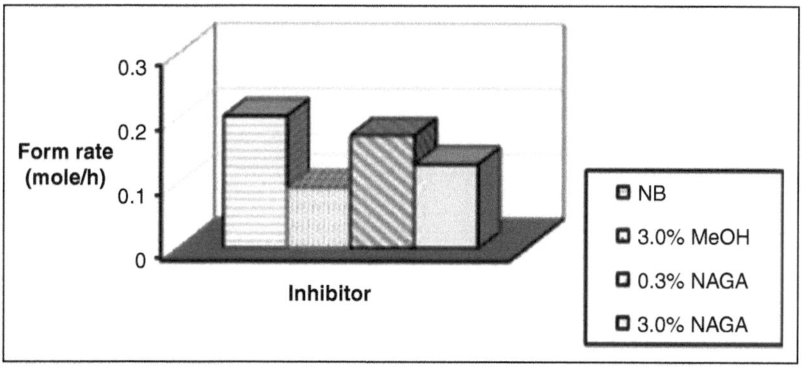

Figure 5.21 ***Hydrate reaction rate inhibition of gas hydrates by*** **N*-acetylglucosamine.***

mechanism is thought to be interference of NAGA with water's hydrogen bonding as well as bonding of inhibitor polymer to hydrate crystal faces to impede growths of embryo crystals.

Therefore, hydrate inhibition of the cell wall material is due, at least partly, to the activity of the NAGA segment, and by itself this water-soluble fragment becomes a potent hydrate inhibitor under conditions evaluated. Based on comparative structures, *N*-acetylmuramic acid probably plays a similar role in hydrate inhibition of microbial cell walls.

Second, inhibitive effects of the amino acid appendages are examined. From Figure 5.18 note specifically the four amino acids involved as an appendage to *N*-acetylmuramic acid. Then recall the previous discussion regarding hydrate inhibition and ice inhibition in winter flounder of AFP, whose protein structure contains 37 amino acids (Zeng et al., 2006a). Also consider the thermodynamic hydrate inhibition of glycine and alanine previously discussed, where Sa et al. (2011) in their work showed the effectiveness of those two amino acids in inhibiting gas hydrate formation.

Third, what contribution to hydrate inhibition would the teichoic acid component of microbial cell walls make? With regard to hydrate inhibition, the outstanding feature of anionic teichoic acid (α-D-glucopyranosyl glycerol phosphate) is its potential for hydrogen-bonding interference.

The net effect of the three wall materials is hydrate inhibitive and, in fact, each individual component seems to exhibit that property. It is interesting to compare the inhibitive structure of microbial cell wall material with the hydrate catalytic effect of anionic biosurfactants. The anionic biosurfactants have hydrophilic heads with many of the groups seen in cell wall materials. For example, the hydrophilic head of surfactin has seven amino acids. But molecular structures of anionic biosurfactants also are balanced by hydrophobic components, usually straight-chain alkanes. Consequently, highly catalytic anionic biosurfactant brings together at a common hydrate nucleation site the hydrocarbon gas and structured water, whereas inhibitive cell wall material appears to reject the hydrocarbon gas.

It is noteworthy that NAGA is biodegradable and occurs naturally in accessible quantities for economy.

5.6 ANTIAGGLOMERATE INHIBITORS

AA inhibitors prevent hydrate crystals from expanding their size, resulting in numerous small particles, which disperse during residence time in a subsea pipeline. Therefore, these AA inhibitors prevent pipeline plugging over a desired transit length. The AA inhibitors are practical at subcooling levels of 15–20°C (Zheng et al., 2011), and may be effective at low concentrations of 0.5–3 wt% of the flow stream's water content (York and Firoozabadi, 2008).

AAs are thought to be effective during particle growth phase, but Lederhos et al. (1996) suggest that effectiveness before crystals reaches critical cluster sizes. That is, hydrate growth is blocked if the crystal surface becomes starved of either water or guest gas. As in the case of KHIs, fractional coverage of the crystal surface by AAs retards gas access. Additionally, AAs may exclude water from hydrate crystal surfaces, limiting crystal size (Makogon and Sloan, 2002). Supporting these conjectures, Varma-Nair et al. (2007) report best performance when surfactant inhibitors bind most of the water.

Two mechanisms are plausible to explain AA effectiveness in preventing hydrate blockages of oilfield pipeline flows when a proper water/oil ratio and a liquid hydrocarbon phase are present (Huo et al., 2001).

First, AA emulsifiers may facilitate water-in-oil emulsions, where the hydrate forms in the small water droplets within the discontinuous phase while surrounding emulsifier polymers segregate droplets to slow agglomeration (Kelland et al., 2006b; Behar et al., 1991; York and Firoozabadi, 2008)). The AA polymers have hydrate-philic head groups attaching to the hydrate-containing water drops and have hydrate-phobic tails extending into the continuous hydrocarbon phase (Huo et al., 2001). As a consequence of relying on a water-in-oil emulsion, AA effectiveness is limited to less than 30% water cut in the produced hydrocarbon stream.

A second mechanism involves polymeric surfactants having hydrate-philic heads of amine or carbonyl groups that attach by electrostatic attraction to the developing hydrate crystal face (York and Firoozabadi, 2008). With adequate agitation, the surfactant's long hydrophobic tail suspends each crystal separately in the liquid hydrocarbon phase to create slurries (Kelland et al., 2006a; Huo et al., 2001). Preferably, the surfactant has limited solubility in both phases, and it may concentrate as a separate layer at the phase interface (Kelland et al., 2009).

Quaternary ammonium surfactants are examples of effective AA inhibitors used in 0.3–1.0% concentrations to substantially increase hydrate induction time and reduce hydrate formation rate during oil and gas production. But they, too, have toxicity and non-biodegradability problems (Kelland et al., 2006b).

Research to improve AAs has been only partially successful, but the mixing of AA and KHIs sometimes increases effectiveness (Arjmandi et al., 2002). In searching for an improved AA, York and Firoozabadi (2008) replaced the quaternary ammonium salt with a biodegradable surfactant that could create emulsions with limited stability.

With the objective to reduce environmental problems, Kelland et al. (2006b) evaluated surfactants with alkylamide groups to replace quaternary ammonium salts. Although some positive results were obtained, the inhibitive properties proved inferior to the quaternary compounds.

In other attempts to improve AA effectiveness, Xiao and Adidharma (2009) evaluated ionic fluids with extremely low vapor pressures based on the premise that strong electrostatic charges might disrupt hydrogen bonding in the water. These organic salts lowered temperature of the HLVE curve for methane hydrate between 0.7 and 1.5°C as compared with about 3.5–5.0°C for methanol.

REFERENCES

Abay, H.K., Hovring, E., Svartas, T.M., 2011a. Does PVCAP promote nucleation of structure II hydrate? In: Proceedings of the 7th International Conference on Gas Hydrates (ICGH 2011), Edinburgh, Scotland, UK, July 17–21. Paper 23.

Abay, H.K., Hovring, E., Svartas, T.M., 2011b. The effect of PVCAP on methane hydrate nucleation and growth. In: Proceedings of the 7th International Conference on Gas Hydrates (ICGH 2011), Edinburgh, Scotland, UK, July 17–21. Paper 32.

Anderson, B.J., Testor, J.W., Borghi, G.P., Trout, B.L., 2005. Properties of inhibitors of methane hydrate formation via molecular dynamics simulations. J. Am. Chem. Soc. 127, 17852–17862.

Arjmandi, M., Ren, S.R., Yang, J., Tohidi, B., 2002. Design and testing of low dosage hydrate inhibitors. In: Proceedings of the Fourth International Conference on Gas Hydrates, Yokohama, Japan, pp. 516–519.

Barrett, J., 2001. Thermal hysteresis proteins. Int. J. Biochem. Cell Biol. 33, 105–117.

Behar, E., Bourmayer, P., Sugier, A., Thomas, M., 1991. Advances in hydrate control. In: Proceedings, Annual Convention – Gas Processors Association, San Antonio, TX, March 11–12, pp. 115–120.

Benjah-bmm27, 2007. By Benjah-bmm27 (own work) (public domain), via Wikimedia Commons. <http://commons.wikimedia.org/wiki/File%3AGlycine-2D-skeletal.png>.

Beveridge, T.J., Murray, R.G.E., 1980. Sites of metal deposition in the cell wall of *Bacillus subtilis*. J. Bacteriol. 141 (2), 876–887.

Birdsell, D.C., Doyle, R.J., Morgenstern, M., 1975. Organization of teichoic acid in the cell wall of *Bacillus subtilis*. J. Gen. Microbiol. 61 (3), 361–369.

Blanc, C., Tournier-Lasserve, J., 1990. Controlling hydrates in high-pressure flowlines. World Oil 211 (5), 63–68.

Bolzep, 2014. By Bolzep (own work) (CC BY-SA 4.0 (http://creativecommons.org/licenses/by-sa/4.0)), via HTML. <http://commons.wikimedia.org/wiki/File%3ATeichons%C3%A4ure.jpg>.

Chao, H.M., Houston, M.E., Hodges, R.S., Kay, C.M., Sykes, B.D., Loewen, M.C., Davies, P.L., Sonnichsen, F.D., 1997. A diminished role for hydrogen bonds in antifreeze protein binding to ice. Biochemistry 36, 14652–14660.

Daraboina, N., Ripmeester, J., Walker, V.K., Englezos, P., 2011. Multi-scale assessment of the performance of kinetic hydrate inhibitors. In: Proceedings of the 7th International

Conference on Gas Hydrates (ICGH 2011), Edinburgh, Scotland, UK, July 17–21. Paper 105.
Doyle, R.J., Matthews, T.H., Streips, U.N., 1980. Chemical basis for selectivity of metal ions by the *Bacillus subtilis* cell wall. J. Bacteriol. 143 (1), 471.
Fletcher, G.L., Addison, R.F., Slaughter, D., Hew, C.L., 1982. Antifreeze proteins in the Arctic shorthorn sculpin. Arctic 35, 302.
Gilbert, J.A., Davies, P.L., Laybourn-Parry, J., 2005. A hyperactive, Ca^{2+}-dependent antifreeze protein in an Antarctic bacterium. FEMS Microbiol. Lett. 245, 67–72.
Gordienko, R., Ohno, H., Singh, V.K., Jia, Z., Ripmeester, J.A., Walker, V.K., 2010. Towards a green hydrate inhibitor: imaging antifreeze proteins on clathrates. PLoS One 5 (2), e8953.
Gordon, M.S., Amdur, B.H., Scholander, P.F., 1962. Freezing resistance in some northern fishes. Biol. Bull. 122, 52–62.
Graether, S.P., Sykes, B.D., 2004. Cold survival in freeze-intolerant insect. The structure and function of β-helical antifreeze proteins. Eur. J. Biochem. 271, 3285.
Hawtin, R.W., Moon, C., Rodger, P.M., 2005. Simulation of hydrate kinetic hydrate inhibitor: the next level. In: The Fifth International Conference on Gas Hydrates, Trondheim, Norway, vol. 1, pp. 317–321.
Hew, C.L., Kao, M.H., So, Y.S., Lim, K.P., 1983. Presence of cystine-containing antifreeze proteins in the spruce budworm, *Choristoneura fumiferana*. Can. J. Zool. 61, 2324–2328.
Huo, Z., Freer, E., Lamar, M., Sannigrahi, B., Knauss, D.M., Sloan, E.D., 2001. Hydrate plug prevention by anti-agglomeration. Chem. Eng. Sci. 56, 4979–4991.
Jensen, L., Ramlov, H., Thomsen, K., von Solms, N., 2010. Inhibition of methane hydrate formation by ice-structuring proteins. Ind. Eng. Chem. Res. 49, 1486–1492.
Jensen, L., Thomsen, K., von Solms, N., 2011. Inhibition of structure I and II gas hydrates using synthetic and biological kinetic hydrate inhibitor. Energy Fuels 25, 17–23.
Kang, S.-P., Lee, H., Potdar, S., Lee, S., 2011. Prediction of the induction of gas hydrate with PVP and/or PVCAP. In: Proceedings of the 7th International Conference on Gas Hydrates (ICGH 2011), Edinburgh, Scotland, UK, July 17–21.
Kelland, M.A., 2006. History of the development of low dosage hydrate inhibitors. Energy Fuels 20 (3), 825–847.
Kelland, M.A., Svartaas, T.M., Ovstus, J., Tomita, T., Chosa, J., 2006a. Studies on some zwitterionic surfactant gas hydrate anti-agglomerants. Chem. Eng. Sci. 61, 4048–4049.
Kelland, M.A., Svartaas, T.M., Ovstus, J., Tomita, T., Chosa, J., 2006b. Studies on some alkylamide surfactant gas hydrate anti-agglomerants. Chem. Eng. Sci. 61, 4290–4298.
Kelland, M.A., Svartas, T.M., Andersen, L.D., 2009. Gas hydrate anti-agglomerant properties of polypropoxylates and some other demulsifiers. J. Pet. Sci. Eng. 64 (1–4), 1–10.
Kelland, M.A., Iversen, J.E., 2010. Kinetic hydrate inhibition at pressures up to 760 bar in deep water drilling fluids. Energy Fuels 24, 3003–3013.
Knight, C.A., Cheng, C.C., DeVries, A.L., 1991. Adsorption of α-helical antifreeze peptides on specific ice crystal surface planes. Biophys. J. 59, 409–418.
Kristiansen, E., Pedersen, S.A., Zachariassen, K.E., 2008. Salt-induced enhancement of antifreeze protein activity: a salting-out effect. Cryobiology 57, 122–129.
Kvamme, B., Kuznetsova, T., Aasoldsen, K., 2005. Molecular dynamics simulations for selection of kinetic hydrate inhibitors. J. Mol. Graph. Model. 23, 524–536.
Lederhos, J.P., Long, J.P., Sum, A., Christiansen, R.L., Sloan, Jr., E.D., 1996. Effective kinetic hydrate inhibitor for natural gas hydrates. Chem. Eng. Sci. 51 (8), 1221–1229.
Li, N., Andorfer, C.A., Duman, J.G., 1998. Enhancement of insect antifreeze protein activity by solutes of low molecular mass. J. Exp. Biol. 201, 2243–2251.
Makogon, T.Y., Sloan Jr., D.D., 2002. Mechanism of kinetic hydrate inhibitors. In: Proceedings of the Fourth International Conference on Gas Hydrates, Yokohama, Japan, pp. 498–503.

Middleton, A.J., Brown, A.M., Davies, P.L., Walker, V.K., 2009. Identification of the ice-binding face of a plant antifreeze protein. FEBS Lett. 583, 815–819.

Mouagip, 2011. By Mouagip. This vector graphics image was created with Adobe Illustrator (own work HTML). <http://commons.wikimedia.org/wiki/File%3AMureine.svg>.

Moudrakovski, I.L., Sanchez, A.A., Ratcliffe, C.I., Ripmeester, J.A., 2001. Nucleation and growth of hydrates on ice surfaces: new insights from ^{129}Xe NMR experiments with hyperpolarized xenon. J. Phys. Chem. B 105, 12338–12347.

Natarajan, V., Bishnoi, P.R., Kalogerakis, N., 1994. Induction phenomena in gas hydrate nucleation. Chem. Eng. Sci. 49, 2075–2087.

Ohtake, M., Yamamoto, Y., Kawamura, T., Wakisaka, A., de Souza, W.F., de Freitas, M.V., 2005. Clustering structure of aqueous solution of kinetic hydrate inhibitor of gas hydrates. J. Phys. Chem. B 109, 16879–16885.

Ostergaard, K., Masouidi, R., Tohidi, B., Danesh, A., Todd, A.C., 2005. A general correlation for predicting the suppression of hydrate dissociation temperature in the presence of thermodynamic inhibitors. J. Pet. Sci. Eng. 48 (1–2), 70–80.

Radich, J.G., 2009. Laboratory and theoretical investigations of direct and indirect microbial influences on seafloor gas hydrates. M.S. Thesis, Dave C. Swalm School of Chemical Engineering, p. 280.

Ramsey, J.A., 1964. The rectal complex of the mealworm *Tenebrio molitor* L. (Coleoptera, Tenebrionidae). Philos. Trans. R. Soc. Lond. B Biol. Sci. 248, 279–314.

Rodger, P.M., 2011. Hydrate nucleation and inhibition: new methods for deeper understanding. In: Proceedings of the 7th International Conference on Gas Hydrates (ICGH 2011), Edinburgh, Scotland, UK, July 17–21. Paper 729.

Rogers, R.E., Zhang, G., Kothapalli, C., French, W.T., 2004. Laboratory evidence of microbial–sediment–gas hydrate synergistic interactions in ocean sediments. ISOPE, Paper 2004-JSC-218. Presented at Toulon, France.

Sa, J.-H., Lee, B.R., Park, D.-H., Lee, K.-H., Han, K., Chun, H.D., 2011. Amino acids as natural inhibitors for CO_2 hydrate formation. In: Proceedings of the 7th International Conference on Gas Hydrates (ICGH 2011), Edinburgh, Scotland, UK, July 17–21. Paper 538.

Sakaguchi, H., Ohmura, R., Mori, Y.H., 2003. Effects of kinetic hydrate inhibitor on the formation and growth of hydrate crystals at a liquid–liquid interface. J. Crystal Growth 247, 631–641.

Seo, Y., Kang, S.-P., 2012. Inhibition of methane hydrate re-formation in offshore pipelines with a kinetic hydrate inhibitor. J. Pet. Sci. Eng. 88-89, 61–66.

Sidebottom, C., Buckley, S., Pudney, P., Twigg, S., Jarman, C., Holt, C., Telford, J., McArthur, J., Worrall, D., Hubbard, R., Lillford, P., 2000. Heat-stable antifreeze protein from grass. Nature 406, 256.

Sivaraman, R., 2002. Flow assurance: understanding and controlling natural gas hydrate. GasTIPS, 18–23.

Sivaraman, R., 2003. Flow assurance: understanding the mechanisms of hydrate nucleation and inhibition. GasTIPS, 28–33.

Sloan, E.D., Koh, C.A., 2008. Clathrate Hydrates of Natural Gases, third ed Taylor & Francis Group, Boca Raton, FL.

Sönnichsen, F.D., Sykes, B.D., Davies, P.L., 1995. Comparative modeling of the three-dimensional structure of type II antifreeze protein. Protein Sci. 4, 460–471.

de Souza, W.F., Freitas, A.M., 2002. Proceedings of the Fourth International Conference on Gas Hydrates, Yokohama, Japan, May 19–23, pp. 478–481.

Storr, M.T., Monfort, J.P., Taylor, P.C., Rodger, P.M., 2002. Natural gas hydrates: modifying stability with low dosage inhibitors. In: Proceedings of the Fourth International Conference on Gas Hydrates, Yokohama, Japan, pp. 504–508.

Storr, M.T., Taylor, P.C., Monfort, J.P., Rodger, P.M., 2004. Kinetic hydrate inhibitor of hydrate crystallization. J. Am. Chem. Soc. 126, 1569–1576.
Tyshenko, M.G., Doucet, D., Davies, P.L., Walker, V.K., 1997. The antifreeze potential of the spruce budworm thermal hysteresis protein. Nat. Biotechnol. 15, 887–890.
Varma-Nair, M., Costello, C.A., Colle, K.S., King, H.E., 2007. Thermal analysis of polymer–water interactions and their relation to gas hydrate inhibition. J. Appl. Polym. Sci. 103 (4), 2642–2653.
Villano, L.D., Kommedal, R., Kelland, M.A., 2008. Class of kinetic hydrate inhibitors with good biodegradability. Energy Fuels 22, 3143–3149.
Vollmer, W., 2008. Structural variation in the glycan strands of bacterial peptidoglycan. FEMS Microbiol. Rev. 32 (2), 287–306.
Xiao, C., Adidharma, H., 2009. Dual function inhibitors for methane hydrate. Chem. Eng. Sci. 64, 1522–1527.
Yang, J., Tohidi, B., 2011. Characterization of inhibition mechanisms of kinetic hydrate inhibitors using ultrasonic test technique. Chem. Eng. Sci. 66, 278–283.
Yeh, Y., Feeney, R.E., 1996. Antifreeze proteins: structures and mechanisms of functions. Chem. Rev. 96 (2), 601–617.
Yikrazuul, 2008a. By Yikrazuul (own work) (public domain), via Wikimedia Commons. < http://commons.wikimedia.org/wiki/File%3AN-Acetylglucosamine.svg>.
Yikrazuul, 2008b. By Yikrazuul (own work) (public domain), via Wikimedia Commons. <http://commons.wikimedia.org/wiki/File%3AN-Acetylmuramic_acid.svg>.
Yikrazuul, 2008c. By Yikrazuul (own work) (public domain), via Wikimedia Commons. <http://commons.wikimedia.org/wiki/File%3APeptidoglycan.svg>.
Yikrazuul, 2008d. By Yikrazuul (own work) [public domain], via Wikimedia Commons. <http://commons.wikimedia.org/wiki/File%3APeptidoglycan_E_coli.png>.
York, J.D., Firoozabadi, A., 2008. Comparing effectiveness of rhamnolipid biosurfactant with a quaternary ammonium salt surfactant for hydrate anti-agglomeration. J. Phys. Chem. B 112 (3), 845–851.
Zeng, H., Wilson, L.D., Walker, V.K., Ripmeester, J.A., 2006a. Effect of antifreeze proteins on the nucleation, growth and the memory effect during tetrahydrofuran clathrate hydrate formation. J. Am. Chem. Soc. 128, 2844–2850.
Zeng, H., Wilson, L.D., Walker, V.K., Ripmeester, J.A., 2006b. Effect of antifreeze protein on nucleation, growth and memory of gas hydrates. AIChE J. 52 (9), 3304–3309.
Zeng, H., Lu, H., Huva, E., Walker, V.K., Ripmeester, J.A., 2008. Differences in nuclear adsorption may explain distinct inhibition activities of two gas hydrate kinetic hydrate inhibitor. Chem. Eng. Sci. 63, 4026–4029.
Zhang, D.Q., Liu, B., Feng, D.R., He, Y.M., Wang, S.Q., Wang, H.H., Wang, J.F., 2004. Significance of conservative asparagines residues in the thermal hysteresis activity of carrot antifreeze protein. Biochem. J. 377, 589–595.
Zhang, J.S., Lo, C., Couzis, A., Somasundaran, P., Wu, J., Lee, J.W., 2009. Adsorption of kinetic hydrate inhibitor on clathrate hydrates. J. Phys. Chem. C 113, 17418–17420.
Zheng, J., Musa, O.M., Alexandre, M., Edris, S., 2011. Development of innovative polymers as kinetic hydrate inhibitors. In: Proceedings of the 7th International Conference on Gas Hydrates (ICGH 2011), Edinburgh, Scotland, UK, July 17–21. Paper 106.

CHAPTER SIX

Hydrate-Associated Seafloor Instabilities

Hydrates in ocean floors represent potential hazards as a consequence of one property – free gases compress within structured water by phase change into solid-solution hydrates where, if the hydrates become confined within impermeable sediments, subsequent decomposition creates extremely high pressures.

Diverse hazards of offshore hydrates have been manifested in extraordinary ways: (1) at the bottom of gas hydrate stability (BGHS) plane on continental slopes, evidence indicates that hydrate dissociation resulting from sea-level decline or sea warming has caused submarine mud slides resulting in tsunamis and methane releases of climate-changing magnitudes. (2) Hydrate masses confined by impermeable sediments become highly pressurized pockets of gas subject to catastrophic release by any disrupting phenomena. (3) Heat from drill bit friction, drilling fluids, and exothermic cementing reactions may portend blowouts or drill string damage. (4) Buoyancy losses from hydrate-released gases may destroy surface vessels or seafloor pipelines. (5) Lethal amounts of hydrogen sulfide may be released in hydrate decomposition gases.

Large seafloor pockmarks, gas wipeout zones, gas chimneys, submarine mud volcanoes (SMVs), gas plumes from seismic events, gas plumes from salt-body movements, and gas eruptions from plate tectonics are naturally occurring instability events widely observed in hydrate zones.

The aforementioned hydrate hazards are discussed in this chapter, as well as a case study describing the Lake Nyos disaster in Cameroon. The case illustrates potential, but difficult to foresee, marine dangers of the hydrate overpressure phenomenon.

6.1 HYDRATE DISSOCIATION HAZARDS IN OILFIELD OPERATIONS

6.1.1 Drilling and Completions

Oilfield operations in the Gulf of Mexico (GOM) move inexorably into deeper waters where slide dangers on continental-slope inclines greater than

Offshore Gas Hydrates. http://dx.doi.org/10.1016/B978-0-12-802319-8.00006-1

2–3° increase as do other perceived hydrate hazards of drilling and production (Ruppel et al., 2008; Kvalstad et al., 2011). Water depths of 2800 m are encountered at Alaminos Canyon, which Joint Industry Program (JIP) now studies for feasible hydrate-gas production. Generally, prospects located to the seaward limits of the Sigsbee Escarpment involve similar depths (McConnell et al., 2012). To help put into perspective the omnipresence of hazards while drilling under prevailing ultradeep conditions, blowout of the Deepwater Horizon well occurred in 1500 m water depths.

Formation rates and gas contents of hydrates increase with the lower temperatures and higher pressures of ultradeep waters, assuming obvious stipulations of similar water salinity and gas access. Gases in seafloor hydrates occupy 150–180 standard temperature and pressure (STP) gas volumes per 1.0 hydrate volume, so the potential hazard is evident with inadvertent heat addition to an enclosed hydrate mass – that is, drilling into impermeable shale enveloping hydrates.

A Schlumberger simulation showed that if heat transfer from drilling fluid to hydrate-bearing formations adjacent to the wellbore is kept below a critical level, then hydrate decomposition problems remain manageable (Birchwood et al., 2007). That is, slower mud flow rates through the overlying water column give longer mud residence times in those cooler surroundings, accommodating heat loss from the mud to lower its temperature. The threshold temperature of hydrate stability as the mud flows through the hydrate zone may then not be exceeded, circumventing decomposition hazards.

When spotted in predrill preparations, dispersed hydrates, massive hydrates, and hydrates in shallow muds of the hydrate zone may be avoided during conventional oil and gas drilling. McConnell et al. (2012) report redirecting wells around seismic bright spots, seismic anomalies inferring gas trapped by hydrate impermeable barriers, as an effective but expensive procedure. On the other hand, a 2005 JIP Leg I well was drilled through truncated bottom-simulating reflectors (BSRs) at KC-151, penetrating hydrate-filled vertical fractures in fine-particle sediment intervals. Drilling through these low hydrate levels in muds was accomplished with only moderate modifications to drilling procedures (Birchwood et al., 2008; Cook et al., 2008; Boswell et al., 2012; Winters, 2011).

Another potential drilling problem involves penetrating the bottom of the gas hydrate stability zone where overpressures may exist below the impermeable gas hydrate interface. Again, the danger is not all encompassing, for in a 25-year period, greater than 1000 wells safely penetrated the hydrate zone of the GOM (Smith et al., 2005). In some cases, overpressures below BGHS may have been relieved through natural strata weaknesses (mud volcanoes, gas chimneys, fractures/faults) prior to drill bit penetration. In other cases fine-grained sediments containing less than 4% gas below BGHS, although distinctly producing BSRs, do not necessarily represent blowout dangers (Lane, 2005; Hyndman and Spence, 1992).

An ongoing concern is the presence at BGHS of permeable sands extending below the hydrate stability where free gas accumulates beneath the hydrate mass; high-pressured gases then may be masked from seismic detection by the hydrates (McConnell et al., 2012).

Finally, during well completions in a gas hydrate zone, exothermic cement reactions raise possibilities of transferring heat to decompose contacting hydrates.

6.1.2 Production

Potential hydrate disruptions of infrastructure are routinely avoided by placements of surface structures or seafloor lines. For example, the early discovery of seafloor gas hydrates near Bush Hill at GC-185 redirected locating the first tension-leg platform to a nearby hydrate-free site – a move that preceded safe development of the Jolliet field (McConnell et al., 2012).

Potential hydrate-gas production in the eastern Nankai Trough is in an area tectonically active, where process design must foresee earthquakes of seismic 8.1–8.4 magnitude. For example, hydrates are located on seabed inclines already exhibiting erosions and submarine slides from past seismic events (Kvalstad et al., 2011).

The most promising Nankai hydrate reservoirs are composed of sands highly saturated with hydrates and interbedded with clay layers. To accomplish hydrate-gas production, a depressurization method could feasibly reduce reservoir pressures from 13 to 4 MPa radially at the wellbore. Kvalstad et al. (2011) use these process conditions to investigate by simulation possible reservoir compaction that might result during production.

6.2 MARINE SLIDES ON CONTINENTAL SLOPES

When massive slides occur on continental slopes in hydrate zones, hydrates in the sediments decompose to emit large volumes of methane. Such slides may be extensive. Paull et al. (2002) propose slide mechanisms and consequences. As sediments build over time in the gas hydrate zone, BGHS temperatures increase, and when temperature eventually exceeds hydrate equilibrium values along the interface, hydrate decomposition occurs there and weakens sediment shear strengths. Hydrate decomposition along the lower plane also causes gases to collect below the impermeable sediments still containing hydrates. Between the original BGHS position and the newly elevated BGHS, an unconsolidated, unstable, and fluid base becomes susceptible to the slope's shear forces. Driven by gravity forces bolstered by the upward thrust of underlying pressurized gases, massive sediment slides ultimately result (Kennett and Fackler-Adams, 2000).

During the slide, turbulence of the cascading mass shears particle-to-particle cementing. The violent slide creates a gradation of sediments according to particle size – smaller particles concentrate on top. With hydrate specific gravities less than or equal to 0.91, solid hydrates buoy through the water column to be accelerated by dissociating gases at the upper boundary of the gas hydrate zone (Paull et al., 2002).

As determined by ^{14}C dating, the Storegga Slide is thought to have occurred about 7300 years ago off the Norwegian coast. Violence of the submarine slide resulted in a tsunami strike of Scotland and Norway (Winters et al., 2004; Buenz et al., 2003; Berndt et al., 2002). Noteworthy is that the slide's rubble now rests along the same plane where a contemporary BSR exists in adjacent undisturbed areas.

Another submarine slide, apparently triggered by hydrate decomposition, lies on the Mediterranean seafloor between France and Algeria in 2800 m deep waters. Dated by ^{14}C, the slide occurred 22,000 years ago as a consequence of the last ice age; ocean levels were 20 m below current levels. Slide debris approaches the area of Manhattan Island and the height of Mt. Everest (Monastersky, 1998).

6.3 POCKMARKS, GAS CHIMNEYS, SEAFLOOR INSTABILITY

Seafloor pockmarks are often prominent above marine gas hydrate zones. The sizes of the elliptical to circular craters vary. Hovland and Judd (1988) observed the following: (1) crater depths range from 1 to 80 m with a mean of less than 10 m; (2) crater diameters range from 1 m to greater than 300 m; (3) small craters may merge to form larger ones. Typically, periodic gas venting continues after the causative violent surface event forms the crater. Reduced gas volumes in emissions directly after the causative event indicate a cyclic nature and may foretell similar cratering events in the future, although precise times are not yet predictable.

The lower gas hydrate stability boundary is prone to trap free gas below. An underlying gas column may originate from hydrate dissociation during a warmer climate that caused BGHS to recede. Lower permeabilities of hydrate-containing sediments increase resistance to fluid flow through the boundary interface, prompting a dynamic gas accumulation below. As the gas column height increases, its attendant buoyant forces build. The column may expand from gas migrations having hydrocarbon reservoir or microbial emission sources. Likewise, rising fluids warmer or more saline than interfacial fluids may disrupt equilibrium conditions of hydrates along the lower stability boundary.

As the gas column heightens, the hydrostatic head of seawater at the bottom of the gas column exceeds the hydrostatic pressure at the impermeable interface. Gas column buoyancy increases upward forces on the confining interface. At sufficient net upward force, a breakthrough at the interface allows dissipation of the confined gas column.

Under constraints of fine-grained overburdens, Cathles et al. (2010) theorize that capillary flow dissipates underlying gas columns and capillary seals focus flows to create gas chimneys with consequential seafloor craters and pockmarks.

In theory, capillary bundles partially filled with flow-restricting gas hydrate crystals transmit buoyant and pressurized fluids. Single-phase gas flow exceeds two-phase gas–water flux in the fine-grained, sediment capillaries. At the gas column apex, the water-wetting phase displaces from the larger-diameter capillaries into those of smaller diameter with gas moving into the larger capillaries. This is the beginning of a chimney. Displaced water seals sides and top of the embryo chimney. Perpetuating its upward extension requires a progressively increasing driving force, which comes from net hydrostatic force between chimney top and free-gas base plus buoyant forces deriving from the free-gas column.

Steps in the Cathles Theory are outlined as follows: (1) underlying gas columns grow to critical heights; (2) chimney height extends to near midway between BGHS and seafloor; (3) at midway, catastrophic gas–sediment–water expulsion creates seafloor craters; (4) the approaching chimney causes domed protrusions from the seafloor; (5) as overburden weight on the chimney declines, fluid flows accelerate; (6) gas/mud/water as a fluidized bed bursts catastrophically from the seafloor to form craters (Cathles et al., 2010).

The equation determining time required for chimney extension through fine sediments from BGHS to the seafloor by capillary sealing is given by

Cathles et al. (2010) as Equation 6.1. The calculated time $t_{chimney}$ might reach hundreds of years, depending on prevalent site parameters:

$$t_{chimney} = \frac{\varphi \mu_w R}{k(\rho_w - \rho_g)g} \ln\left(\frac{h}{d} + 1\right) \tag{6.1}$$

where $t_{chimney}$, time for gas chimney to reach seafloor; φ, porosity of sediment; μ_w, viscosity of water; R, radius of gas chimney; k, permeability; ρ_w, water density; ρ_g, gas density; g, acceleration of gravity; h, thickness hydrate stability zone; d, thickness of gas column below BGHS.

By rule of thumb, then, mud ejection may begin when the gas chimney reaches approximately halfway from BGHS to seafloor, that is, near h/2 where fluidized flow initiates. Capillary diffusion rate determines time necessary to get there.

Fluidized flow of mud is common in mud volcanoes. For fluidization, upward fluid velocity exceeds settling velocities of sediment particles; mud loses particle-to-particle contact and adhesion, becoming suspended and entrained by exiting water–gas. Laboratory experiments show fluidization to be enhanced by low concentrations of anionic biosurfactants. It is noteworthy that these bioproducts come from microbes active in the methane-rich areas of hydrate zones, gas vents, and mud volcanoes. Anionic biosurfactants are prolific foaming agents, more so at the low temperatures prevalent in gas hydrate zones (Zhang et al., 2007).

Foaming properties of anionic biosurfactants promote sediment fluidization on hydrate decomposition. See Figure 8.5 where hydrate decomposition fluidizes a packed porous media of sand/clay in the laboratory. The sII hydrate formation had been catalyzed by rhamnolipid (Rogers et al., 2007), where rhamnolipid is the anionic biosurfactant from the parent microbe *Pseudomonas aeruginosa*. Lanoil et al. (2001) identified this microbial species around seafloor gas hydrates. (As an aside, such fluidized flow may damage a drill string when drilling into high-pressurized gas pockets where frictional heat may further decompose hydrates to dramatically increase pressures.) Foaming action is evident in other laboratory experiments to form hydrates in porous media saturated with seawater and containing biosurfactants. It is therefore feasible that bioproducts in fine seabed sediments may promote fluidizations on hydrate decompositions.

Finally, the theory of Cathles et al. predicts chimney extensions (dome development, fluidized bed flow, accelerated fluid flow, rapid gas venting, mass ejection) culminating in seafloor pockmarks. However, the material expulsions reduce driving pressures of the underlying gas column and prompt initiation of another cycle: capillary seals re-form, water/gas columns rebuild, and gas chimneys develop near prior locations. Minor gas volumes vent periodically during the rebuilding process. Equation 6.1 indicates parameters controlling cycle frequency and timing of catastrophic event to depend on gas column height *d* and fine-sediment permeability *k*. If both parameters are small, full chimney growth could take many years (Cathles et al., 2010).

6.4 MUD VOLCANOES, GAS HYDRATES, SEAFLOOR INSTABILITY

Fluids are expelled from SMVs at the seafloor from deep overpressured sediment traps. Mass ejects as a fluidized bed, containing methane and to a lesser extent other hydrocarbon gases, fine-grained sediments, water, and solid-phase gas hydrates (Johnson et al., 2003). Faults, fractures from seismic or salt movement, and weak areas along the BGHS interface may also account for fluid releases at the seafloor as manifestations of seafloor instability.

The large methane volumes emitted by SMVs raise environmental concerns (Tinivella et al., 2008). There is also interest in their effect on seafloor chemosynthetic communities (Werne et al., 2004). Fortunately, these interests and concerns stimulate further studies into the phenomena.

6.4.1 Quantification, Location, Size of Submarine Mud Volcanoes

Probably only a fraction of existing SMVs have been discovered. Sauter et al. (2006) estimate about 1000 mud volcanoes to be onshore, and although offshore numbers are less certain, they estimate that 10,000–100,000 probably exist on continental slopes and abyssal plains. Although approximately 1800 SMVs are documented in 40 basins (Dimitrov, 2002), lack of data characterizes them. For example, very few of these SMVs can be characterized by fluid flux measurements (Bohrmann et al., 2003). Representative SMVs with limited characterizations are presented in Table 6.1.

SMV relief structures for overpressured, deep sediments exist in all oceans and some inland seas (Milkov, 2000). Bohrmann et al. (2003) suggest that overpressured sediments adjacent to fault conduits generate SMVs under the influence of compression from active plate tectonics. For example,

Table 6.1 Representative mud volcano features

Mud volcano	Location	Height (m)	Diameter (km)	Water depth (m)	Shape	References
Bonjardim	Gulf of Cadiz	100	1.0	3059	Conical	Pinheiro et al. (2003)
Carlos Ribeiro	Gulf of Cadiz	80	1.5	2200	Conical	Pinheiro et al. (2003)
Håkon Mosby	Norwegian-Barents-Svalbard margin	<10	1.5	1250	Rounded and flat	Sauter et al. (2006)
Kazan	Eastern Mediterranean	50	1.0	1700	Small dome, elliptical summit	Werne et al. (2004)

in the Gulf of Cadiz, convergence of Africa–Eurasia plates contributes to deepwater SMVs (Pinheiro et al., 2003).

Morphology and location of representative SMVs are given in Table 6.1. It is seen that these SMVs have in common base diameters of 1.0–1.5 km, conical shapes, and cone heights reaching less than 10–100 m above seafloor. The SMVs selected for Table 6.1 reside in water depths that insure pressures and temperatures well within the gas hydrate stability regime (Egorov and Rozhkov, 2010).

6.4.2 Mechanism of SMV Formations

Milkov (2000) proposes three steps for the mechanism of SMV formation. First, the site has a history of rapid sediment depositions. Second, a weak area of clay exists in the strata through which fluidized, high-pressured mud extrudes. Third, fluid flow through fracture or fault conduits relieves pressurized sediments.

Structural forms SMVs may take are those sketched in Figure 6.1 (Milkov, 2000).

Figure 6.1 depicts confined sediments, gas-saturated and overpressured, where shales represent weaknesses in the bounding strata through which pressures may be relieved. In part A of the sketch, diapirs form – one of which pierces the seafloor – but there is insufficient fluid flow to generate a mud volcano. Part B of the sketch depicts fluidized mud flowing through the shale to form a mud volcano at the seafloor.

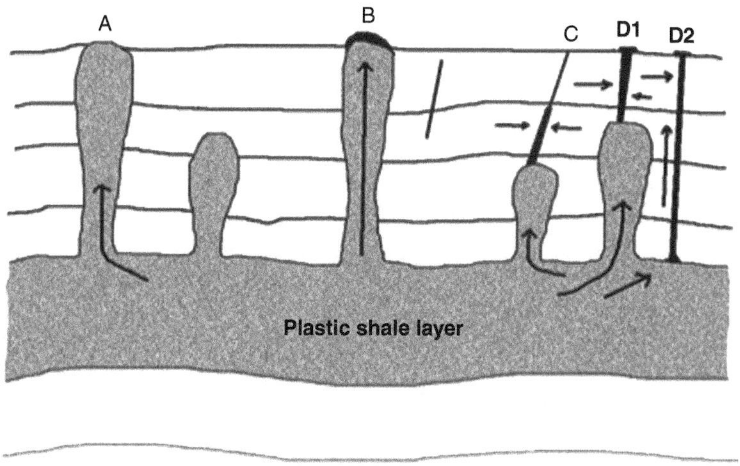

Figure 6.1 ***SMV forms illustrate mechanism (Milkov, 2000).***

In parts C and D1 of Figure 6.1, diapirs are shown to generate fractures and faults, creating outlet flow paths to relieve pressures. Gas may seep only through the C occurrence, but fluidized sediments erupt through D1 to form a mud volcano at the seafloor. Also, fluidized sediments vent through D2 faulting, where enclosing strata is weakest, to form a mud volcano (Milkov, 2000).

6.4.2.1 Håkon Mosby Mud Volcano

The Håkon Mosby mud volcano (HMMV), one of the most thoroughly studied SMVs associated with gas hydrates, is located on the Norwegian shelf in 1250 m water depth (Sauter et al., 2006). Its diameter is 1500 m with rim rising less than 10 m above the seafloor.

Thermal profiles and mass emissions of HMMV were reported by de Beer et al. (2011). As measured 750 m from the center outwardly, the HMMV has three concentric regions: (1) high-temperature center of 50 m diameter exhibiting periodic mass expulsions, (2) annulus around a high-temperature center in which white *Beggiatoa* mats cover the surface, and (3) elevated outer ring populated by tube worms and gas hydrates.

6.4.2.2 Thermal Gradient, Håkon Mosby

Major fluid flow of an SMV occurs in its cone-shaped center, and those fluids originate far below the seafloor – 2 km to greater than 3 km in the case of the HMMV (Hjelstuen et al., 1999). High-temperature profiles at the cone center decrease radially. Heat emanating from central fluid flow creates a cylindrical zone of about 200 m radius in which the temperature is too high for gas hydrate formation.

Ginsburg et al. (1999) found the HMMV thermal gradient to be a very high 3000°C/100 m near the center. An approximate temperature profile of the cone-shaped HMMV is given by Equation 6.2:

$$G = 2 \times 10^6 R^{-1.755} \tag{6.2}$$

where G, thermal gradient (°C/100 m); R, radius of HMMV cone (m).

6.4.2.3 Mud Breccia Compositions

Although some SMVs occur in waters too shallow to meet the temperature–pressure requirement for hydrate formation, many occur in deep ocean sediments well into the temperature–pressure regime of gas hydrates. It is therefore interesting to examine presence of clays, especially smectites, in fluidized sediments expelled by mud volcanoes. In doing so, keep in mind that smectite clays together with anionic biosurfactants catalytically promote gas hydrate formation; the effective anionic biosurfactants include those indigenous to seafloor sediments of gas hydrate zones (Zhang et al., 2007; Rogers et al., 2003; Radich, 2009).

Werne et al. (2004) report compositions of Kazan mud volcano clays in the Eastern Mediterranean to be dominated by smectites, the makeup being 85% smectite, 10% kaolinite, 3% palygorskite, and 2% illite. As one explanation, clay content may be influenced in the following manner, based on laboratory experiments of Dearman et al. (2009). Smectite platelets associate with indigenous anionic biosurfactants, becoming nucleating sites for gas hydrate crystals. Embryo hydrate crystals diffuse to SMV methane-rich sites, entraining attached clay platelets in transit. Hydrates forming throughout the methane-containing sediment enrich that sediment with smectites.

Data from other SMVs show their mud breccia to contain high clay percentages. For example, Bohrmann et al. (2003) report SMV extrusions of the eastern Black Sea, Sorokin Trough, to be enriched in clays. Likewise, Ginsburg et al. (1999) report that fine sediments associated with Håkon Mosby contain 70–80% clays and silts.

Therefore, it is hypothesized that smectite clays, dominant in SMV breccia and associating with bioproducts in the carbon-rich muds, kinetically promote gas hydrate formation and are indirectly instrumental in the fluidization of SMV effluents.

6.4.3 Gas in Submarine Mud Volcanoes

Part of the intensified interest in SMVs centers on greenhouse gases they emit – namely whether the magnitude of methane significantly impacts environment. SMV greenhouse gas dangers, as well as slope instabilities, depend on gas volumes, emission rates, and compositions. Depending on SMV size, rate of methane release may range from hundreds to tens of tons per year (Etiope et al., 2004, 2011).

Estimates of total hydrate gas within SMVs are limited, primarily because of incomplete accounting of SMV numbers and locations. Milkov (2000), however, estimates 10^{10}–10^{12} m^3 of gas at STP associated with worldwide SMVs. In a later estimate, Milkov et al. (2003) puts the total worldwide rate of methane release by SMVs at 27 MT/year.

Ginsburg et al. (1999) estimate methane storage of the HMMV to be 3×10^8 m^3 STP in hydrates, where they assume gas volumes to hydrate volumes to be 160 (v/v). Although the amount of occluded methane is huge, it may represent only 10% of methane associated with Håkon Mosby since inception. In such case, the difference represents methane ejected into waters during the interim period (Ginsburg et al., 1999). For example, one HMMV vent discharges 200 mmol/s of methane (Sauter et al., 2006). This translates into 108 ton/year methane ejecting into overlying waters from one vent.

During a year-long on-site study of HMMV by de Beer et al. (2011), mass eruptions near the center of the volcano expelled about 200,000 m^3 methane. Most of the methane escaped to the atmosphere since the volcano's high temperature near its center negates methane dissipation by anaerobic oxidation or hydrate formation.

6.4.4 Gas Hydrates in Submarine Mud Volcanoes

Gas hydrate content of seafloor mud-volcano sediments is reported as a percentage of sediment volume or as a percentage of pore volume occupied by the hydrates. Conversion of pore hydrate volume percentage H_p to sediment hydrate volume percentage H_v is accomplished by multiplying H_p by porosity ϕ as in Equation 6.3 (Mazurenko et al., 2003; Ginsburg et al., 1999):

$$H_v = H_p \phi \tag{6.3}$$

Water salinities of retrieved SMV cores depend on mud-volcano water salinity before forming hydrates and desalinated water comprising hydrate structure, which is a residue of hydrate decomposition. Water salinities of 19 cores retrieved from HMMV were studied by Ginsburg et al. (1999) to arrive at Equation 6.4 for estimating percentage of pore volumes occupied by hydrates:

$$H_p = \left[\frac{\mathrm{Cl_{mv}} - \mathrm{Cl_s}}{\mathrm{Cl_{mv}}}\right] \times 100\% \tag{6.4}$$

Substituting Equation 6.4 into Equation 6.3 gives the hydrate volume of the sediments, as presented in Equation 6.5:

$$H_v = \left[\frac{\mathrm{Cl_{mv}} - \mathrm{Cl_s}}{\mathrm{Cl_{mv}}}\right] \times \phi \times 100\% \tag{6.5}$$

where H_v, percent gas hydrate based on volume of sediment; H_p, percent gas hydrate based on pore space; $\mathrm{Cl_{mv}}$, chloride concentrations in mud volcano fluid originally (mM); $\mathrm{Cl_s}$, chloride concentrations in mud volcano fluid sample, cores taken from SMV (mM); ϕ, porosity.

Average hydrate saturation, based on sediment volume, was found to be 1.2% in the cores retrieved from HMMV, although a wide range of H_v exists (Ginsburg et al., 1999).

Mazurenko et al. (2003) applied the equations to Yuma and Ginsburg's SMVs in the Morocco margin where original seawater chloride content was 563.7 mM. In the Yuma structure, Mazurenko's calculations indicated that 2–6% of the sediment volumes were occupied by hydrates, that is, 4–11% of the pore volumes. For the Ginsburg structure, 4–11% by sediment volume and 8–19% by pore volume were calculated to be gas hydrates.

Gas hydrates appear in plumes of gases emitted from deep-sea mud volcanoes. Sauter et al. (2006) observed this interesting phenomenon at the HMMV where 99.9% methane gas was emitted at 1250 mbsf into −0.9°C seawater. The conditions, well within hydrate equilibrium *P*–*T* limits, formed hydrate shells around the emitted 5.2 mm diameter methane bubbles, stabilizing the gas as a plume until buoyed to about 500 mbsf – the upper limit of hydrate stability. Similar plume dimensions were observed over a several-year span (Sauter et al., 2006). The protective hydrate coating accommodates additional gas reaching the atmosphere.

SMVs are linked to prolific microbial activities. Bacterial mats are one visual confirmation. But there are secondary contributions of the microbial activity. First, bioproducts kinetically promote hydrate formation within the mud. Second, it is hypothesized that thin biosurfactant films form around emitted gas bubbles, kinetically promoting hydrate shells and preventing gas dissipation as the plumes ascend through the water column. In support of the hypothesis, consider the following: (1) microbes are exceptionally active around seafloor gas vents and seafloor gas hydrate accumulations (Cragg et al., 1996); (2) anionic biosurfactants are produced by microbes around seafloor hydrates and vents (Radich, 2009); (3) anionic biosurfactants are foaming agents at the temperature of subsea hydrate formation (Zhang et al., 2007); (4) when laboratory biosurfactant-promoted hydrates decompose, the liberated gas bubbles become coated with the hydrate-affiliated biosurfactant (Zhong and Rogers, 2000).

Generally, Ginsburg et al. (1999) describe plate-like particles reaching lengths of 4 cm dispersing within the HMMV muds. Clay contents are high and sand contents relatively low in the fine-particle mud.

6.4.5 Chemosynthetic Communities and Microbial Links to SMV

Intensive bacterial activities and chemosynthetic communities commonly mark the sites of SMVs, although Bohrmann et al. (2003) report as an exception that no chemosynthetic communities are to be found at SMV sites below 100 m water depth in the anoxic waters of the Black Sea.

Archaea-oxidizing methane and bacteria-reducing sulfate in consort anaerobically oxidize methane at SMVs. The oxidation is described by the chemical reaction given in Equation 6.6 (Boetius et al., 2000; Werne et al., 2004; Reeburgh, 1976):

$$SO_4^{2-} + CH_4 \rightarrow HS^- + HCO_3^- + H_2O \tag{6.6}$$

As anticipated from Equation 6.6, methane along with sulfate and calcium ions in seawater converts to hydrogen sulfide and carbonate precipitates; the pungent smell of hydrogen sulfide and carbonate rubble characterize SMV sites.

Anaerobic oxidation of methane does not occur in the 50 m diameter center of the HMMV conical volcano (de Beer et al., 2011). First, although fluid flow contains 100–250 mM methane, a relatively high upward velocity prevents sulfate-containing waters to adequately diffuse downward for the reaction. Second, the high volumetric flow rate of reduced mass mixes and replaces overburden, so the slow microbial process is left short of time

Table 6.2 Evidence of anaerobic oxidation of methane (AOM) at submarine mud volcanoes

SMV	Carbonates	Chemosynthetic	Microbial	References	H_2S
Kazan, Eastern Mediterranean	Authigenic carbonates: crusts, nodules, cemented sediments	Pogonophoran tube worms, bivalves	*Beggiatoa* mats	Werne et al. (2004)	–
Black Sea	Carbonate cement slabs; carbonate cement sediments	No	Mats associated with carbonates	Mazzini et al. (2004), Bohrmann et al. (2003)	Strong smell
Yuma, Ginsburg	Authigenic carbonates	Pogonophoran tube worms	–	Mazurenko et al. (2003), Pinheiro et al. (2003)	Strong smell

(Knittel and Boetius, 2009). It should be noted that radially beyond about 50 m, upward velocities lessen and AOM increases.

In Table 6.2 are markers indicative of the level of microbial activity at selected SMVs.

6.4.6 Mud Volcanism in the Freshwater of Lake Baikal

Lake Baikal is a 1640 m deep rift lake of freshwater, salinity only 0.76%, in southwestern Russia near the China–Mongolia border. Bottom sediments 7.5 km thick accommodate a deep gas hydrate zone, and the lake's 3.2–3.5°C surface water temperatures maintain that hydrate zone throughout the year (Egorov et al., 2011;Vanneste et al., 2003).

Although geophysical data have long predicted hydrates to be in the lake sediments, verification finally came from core retrievals. Cores show biogenic gas hydrates with gas compositions approaching 99% methane (de Batist et al., 2002; Kuzmin et al., 2000), while seismic data show that BGHS extends 300–350 m below the floor of the lake.

However, the BGHS extends only 150 m below the lake floor near vertical faults because of heat emanating from warm gases flowing through the faults. Seismic data show that those venting fluids originate from deep beneath the lake and are directed upward along fault planes. BGHS varies widely with depth because of highly uneven heat flow through the sediments. For example, heat flux ranges between 40 and 195 mW/m^2 around a mean of 71 mW/m^2 (Vanneste et al., 2003). Fifty or more hot springs are reported onshore.

As the gas chimney and faults are approached in the gas hydrate zone, BSRs become infrequent. Measurements reveal elevated heat flows near the gas conduits with temperatures exceeding background values and temperatures too high for hydrate stability (de Batist et al., 2002). Larger gas flux occurs in chimneys near the faults as a consequence of gas hydrate absences and the attendant higher sediment permeability.

Vanneste et al. (2001) report free gas below BSRs in Lake Baikal, suggesting pressurized gases. Consequently, mud volcanoes are active along the lake floor, parallel to faults. Indicative of mud volcano activities are elevated methane concentrations in the water columns above gas chimneys (de Batist et al., 2002). For example, a 10 m deep crater residing above one gas chimney gives evidence of high-rate gas/mud expulsion in the past, probably as a sudden fluidized mass ejection from a mud volcano.

Supported by the Russian Academy of Sciences, a manned submersible explored three primary areas in Lake Baikal. Interestingly, all three areas of the lake are within the hydrate stability envelope during all seasons but have wide ranges of water depths (Egorov et al., 2011). Note corresponding features of these freshwater hydrate sites with offshore ocean hydrate zones: (1) St. Petersburg at 1400 m water depth features a mud volcano; (2) Gorevoy Utes at 860 m water features a gas seep; and (3) Goloustnoye at 405 m water depth features a gas seep. At a large hydrate mound 25 m diameter and 6 m high outcropping the St. Petersburg site, bacterial mats partially top the mound. Decomposition gas from the mound exhibits a primarily biogenic origin with a −63.6‰ methane isotopic content (Egorov et al., 2011).

6.5 GAS PLUMES

6.5.1 Background

Gas plumes, synonymously described as flares, are large gas fluxes streaming from seafloor vents ejecting cascades of concentrated hydrocarbon gas bubbles of 1–5 mm diameter high into the water column.

These plumes originate from seafloor instabilities. Sharp physical property contrasts between gas cascade and background water help detection by acoustic means. That is, sound wave velocity contrasts define the boundaries in an acoustically generated image of the plume. For example, flare boundaries may be defined, say, by interior gas concentrations of 1000 nl/l contrasting with gas concentrations exterior to a plume of 40 nl/l.

Plumes continue to be discovered on ocean floors, and some generated by SMVs may even be discovered inadvertently. For example, a ship mapping in 1830 m of water off the Northern California coast by happenstance recorded a plume extending 1400 m above the seafloor (Gardner et al., 2009).

If plumes occur in ocean waters greater than 500 m deep, gas hydrates are likely involved.

6.5.2 Gas Plumes Initiated in Gas Hydrate Zone

6.5.2.1 Theory

Using an ROV near the seafloor of Monterey Bay, a group led by Klapp et al. (2011) conducted experiments releasing compressed methane at 1 pulse/s into 3.5°C waters of 1028 m depth. A hydrate skin immediately formed around each spherical gas bubble, which computed tomography (CT) scan revealed to be hydrate approximately 10–40 μm thick. The hydrate shell helped perpetuate bubble shape, size, and longevity. These experiments have implications within any offshore gas hydrate zone (GHZ) for gas eruptions from mud volcanoes, seismic events, cold seeps in acoustic wipeout zones, or blowouts during oil/gas drilling: hydrate encases methane bubbles, stabilizing plumes and allowing extended residence times during drift within the overlying water's *P–T* envelope of hydrate stability.

In the absence of hydrate shells, large methane concentration differences between vented gas bubbles and surrounding seawater rapidly dissipate the gas – a classic case of concentrated methane gas diffusing into an infinite water reservoir (Maksimov and Sosedko, 2009). Rehder et al. (2009) find methane bubble radii typically decrease near 7.5 μm/s rates.

An anticipated gas dissipation rate is given by Equation 6.7 (Clift et al., 1978):

$$F_{\mathrm{M}} = A k_{\mathrm{B}}\, \Delta C \tag{6.7}$$

where F_{M}, methane flux from bubble; A, surface area of bubble; k_{B}, individual bubble gas transfer rate; ΔC, concentration difference, bubble – surrounding water.

Generally, methane forms hydrate below water depths of about 500 m, a threshold depth varying with gas composition and water salinity/temperature. Although hydrate shells can feasibly form around methane bubbles

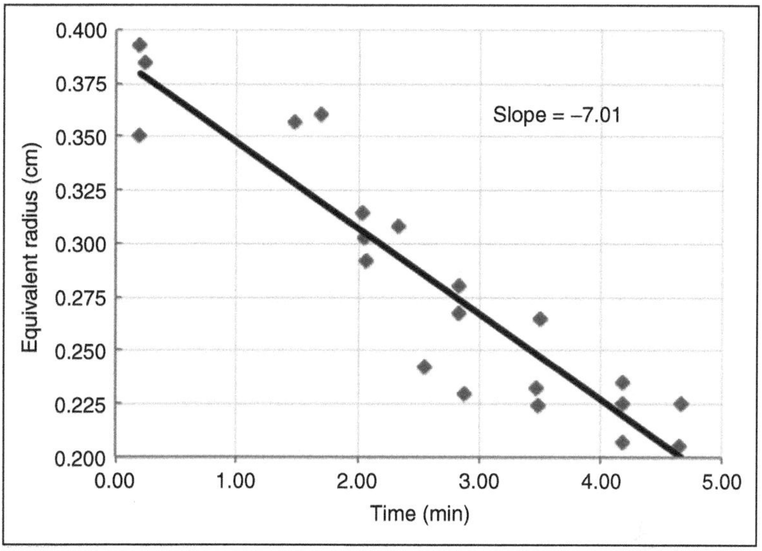

Figure 6.2 ***Bubble dissipation rates at injection depths, 394–471 m (Rehder et al., 2009).***

when released into the water column below the threshold depth, their formation rates increase with depth below the threshold.

As methane bubbles encased in hydrate rise in the water column, Rehder et al. (2009) find a rate of decrease in bubble radii on the order of 1 μm/s, compared with a rate of 7.5 μm/s in the absence of a hydrate shell. Further, Rehder et al. report a time lag accompanying hydrate formation when methane bubbles enter water at depths between 500 and 1200 m. During this hydrate formation lag, an induction time, gas diffuses from the bubble into surrounding water at a 7.5 μm/s rate. Because hydrate induction time decreases with greater gas entry depths, hydrates form almost instantaneously without any induction time delay at 1200 m depth or greater.

Note in Figure 6.2 the lack of hydrates forming in the early stages at 394–471 m water depths of gas injection. With the hydrostatic pressure of these shallow depths, induction times would be longer than the time period represented by Figure 6.2. Methane dissolution results in a −7.0 μm/s relative rapid decrease of bubble radii.

Figure 6.3 represents the methane bubble dissipation rates found by the study by Rehder et al. at intermediate water depths of 691–802 m. Note in Figure 6.3 that hydrate induction occurs within 3.7 min at the intermediate water depths and rate of methane dissipation, as measured by bubble radius size decrease, slows substantially after a hydrate shell encases the methane bubble.

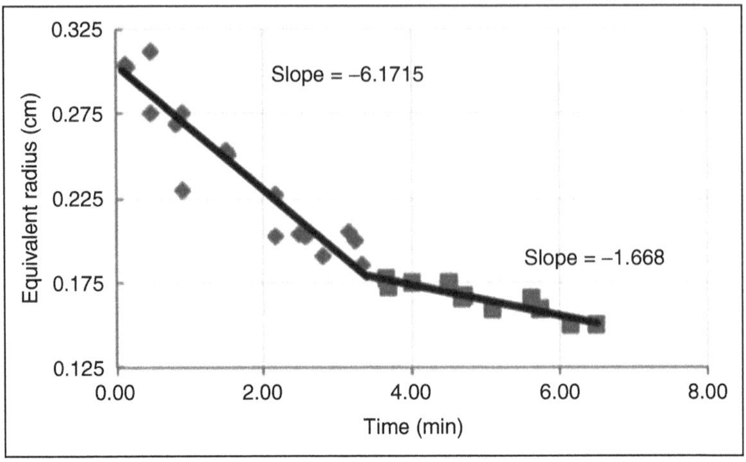

Figure 6.3 ***Bubble dissipation rates at injection depths, 691–802 m (Rehder et al., 2009).***

As hydrostatic pressures increase at 1092–1495 m water depths, hydrate shells form very quickly and further slow methane dissolution. See Figure 6.4.

At 1500 mbsl depths, hydrate formation becomes instantaneous on methane injection into the seawater.

6.5.2.2 Plumes from Håkon Mosby Mud Volcano

Bottom water temperature of HMMV is an exceptionally low −0.9°C. Further, the water column divides into two distinct layers: Atlantic Ocean

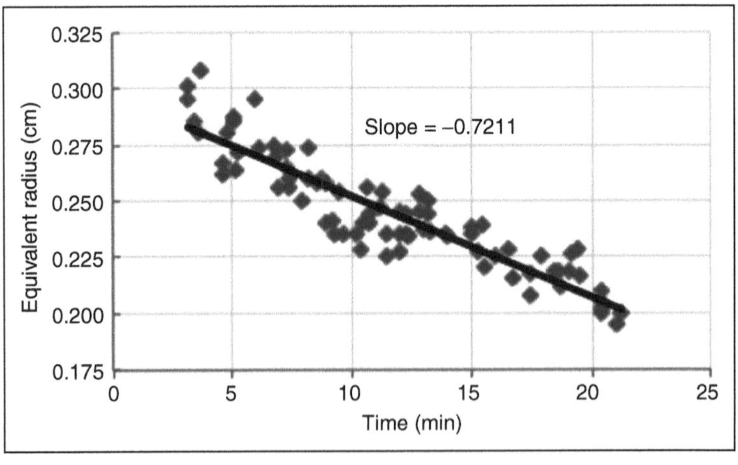

Figure 6.4 ***Bubble dissipation rates at injection depths, 1092–1495 m (Rehder et al., 2009).***

waters for the top 600 m and colder Arctic waters for the bottom 650 m. Under these conditions, hydrates must instantaneously coat bubbles of an estimated 44–200 ton/year of methane entering the waters. In the water column above HMMV, a plume develops whose momentum and buoyancy carry it 750 m above the seafloor. Rising rapidly (31 cm/s), the plume develops momentum that propels it a short distance beyond the hydrate stability boundary (Sauter et al., 2006).

6.5.2.3 Plumes from Sea of Okhotsk Seismic Events

BSRs suggest a 200 m thick gas hydrate zone lying below the seafloor in the Sea of Okhotsk. In further verification of gas hydrates, acoustic images show numerous plumes from gas ejections after seismic events activated fissures. Maksimov and Sosedko (2009) captured seismic images of gas plumes from Sakhalin slope seepages and superposed temperature profiles as a function of seawater depth. The flares are seen to originate at 600–700 m water depth from seafloor faults and dissipate near 300 m of water depth at the upper boundary of gas hydrate stability.

Frequent seismic activity occurs in the Sea of Okhotsk. Methane background concentrations as well as gas plumes in the water column correlate with those seismic events. During a 14-year period, approximately 700 flares emanated from seafloor fractures in the western Okhotsk Sea (Shakirov and Obzhirov, 2011; Obshirov et al., 2004). Earthquakes initiate new faults and increase methane venting into the Sea of Okhotsk. Plots of methane gas concentration in surrounding waters, as well as number of subsea gas flares as a function of time, show spikes following seismic events (Obshirov, 2009). Recorded dates of earthquakes near Sakhalin Island correlate with methane emissions.

6.5.2.4 Limited Seismic Activity, Gulf of Mexico

An infrequent earthquake (6.0 on the seismic scale) in the GOM, 260 miles southwest of Tampa, increased seafloor gas venting at the MC-118 Gas Hydrate Observatory. Two days after the event during manned submersible reconnaissance, it was evident that seafloor venting locations changed at the site. On later expeditions, some of the gas vents had become inactive and others had been displaced. Fracture networks generated by the minor earthquake had conclusively altered gas flow.

Another interesting confirmation of random changes in hydrocarbon gas vent locations comes from analyses of indigenous microbes in the MC-118 sediments. Laboratory cultures grown from *in situ* microbes of cores taken near currently active vents respond with faster biosurfactant production than cultures of microbes from cores extracted near inactive vents. With methane reimposition in the laboratory test cell, a time lag was necessary to reactivate the microbial spores taken from inactive sites (Radich, 2009). The results intimate that venting sites vary historically across the areal extent of the MC-118 wipeout zone, again indicating a fluid flow network of faulted and fractured vents changing locations with time from seismic events, hydrate and carbonate filling of flow conduits, or salt tectonics.

6.5.2.5 Plumes from the Deepwater Horizon Blowout Well

During completion, the Deepwater Horizon well in MC-252 blew out, ejecting gas and light crude with a gas-to-oil ratio of 3000 ft^3 of gas (STP) per barrel of oil. The well flowed continually for 59 days, dumping an estimated 1.5×10^{10} mol of gas into the ocean (Valentine et al., 2010) and 4×10^6 barrels of oil (Camilli et al., 2010).

In a manner similar to large gas eruptions from mud volcanoes or seismic events, gas plumes formed. Data following the GOM blowout provide insight into extremely large gas emissions in deep water within hydrate zones (Kujawinski et al., 2011). Gas entered the water column at 1544 m depth. A hydrate shell immediately formed around each gas bubble without any due process of induction time. Immediate hydrate formation in 1544 m deep water is predicted theoretically as well as demonstrated experimentally by Rehder et al. (2009).

During blowout eruptions, 2.9×10^6 l of liquid dispersants was injected into the seawater at the subsea wellhead (Kujawinski et al., 2011). A major component of the dispersants was dioctyl sodium sulfosuccinate (DOSS). Expectation of instantaneous hydrate formation at this depth without a characteristic induction time is further bolstered by the presence of DOSS in the dispersant, for it is a member of the anionic surfactant group that kinetically increases hydrate formation rates.

During uncontrolled flow of the well, a gas plume was detected 1099–1219 m below the water surface in water temperatures of 4.7°C. The plume drift extended an estimated 10–35 km in length from its origin (Camilli et al., 2010; Hazen et al., 2010). Lutken et al. (2011) documented a plume from the blowout 16.1 km away at the MC-118 Gas Hydrate Observatory. With directions controlled by ocean currents, other plumes were located at similar depths (Valentine et al., 2010). Analyses of samples taken from within a plume show that DOSS content correlates with methane concentration, as would happen with anionic surfactants having hydrophobic and hydrophilic moieties associating with hydrate particles. Equation 6.8 shows the correlation (Kujawinski et al., 2011):

$$C_{\mathrm{D}} = 0.0419C_{\mathrm{M}} + 0.0221 \tag{6.8}$$

where C_{D}, DOSS concentration (μg/l); C_{M}, methane concentration (mM × 1000).

Therefore, most of the injected DOSS surface active agent accumulated in the plume (Kujawinski et al., 2011). DOSS surfactant remains associated with the plumes after hydrate formation at the 1544 m injection depth. The association occurs because DOSS attaches by adsorption to hydrate shells surrounding methane bubbles, consigning dispersant to plumes. The upper limit of plume travel, which is the *P–T* limit of hydrate equilibrium, becomes the limit of upward travel of the surfactant while the plume exists.

Methane, ethane, and propane in approximate ratios 45/4/2, respectively, composed gas initially emitted from near the wellhead. Microbes remove these gas components in the preferential order propane > ethane > methane (Valentine et al., 2010). Microbial cell concentrations within the plume were twice as plentiful as outside the plume (Hazen et al., 2010). In contrast, DOSS was not significantly altered microbially in the deep plume over the first few weeks. The chemical persisted at least 64 days after injection, during which time its associated plume traveled 300 km (Kujawinski et al., 2011). Gas bubble longevity in the plume substantiates the theory by Rehder et al. (2009) of hydrate shells retarding gas dissipation into surrounding seawaters. Also, prior studies verified that at 1544 m water depths hydrates rapidly form around gas bubbles without significant induction times.

Egorov et al. (2011) closely scrutinized methane bubbles ejected from Lake Baikal cold seeps and SMV sites at water depths of 405, 860, and 1400 m. Their manned submersible at these depths allowed close-up observations of gas hydrate formation/dissociation, including dissociation characteristics of ascending methane gas bubbles.

Methane bubbles at 405 m depths of the Goloustnoye seep did not readily form hydrates. That is, hydrate induction times were slow in the shallow freshwater. However, at 680 m depth of the Gorevoy Utes site, powdery hydrates formed within minutes of gas entry. At the 1400 m water depth of the St. Petersburg mud volcano, hydrates rapidly formed; hydrate shells immediately surrounded methane bubbles on gas entering the water column. When the submersible with hydrate mass secured in its external mechanical arm rose to depths shallower than the 380 m stability limit, hydrate rapidly dissociated (Egorov et al., 2011).

6.6 CASE STUDY: GAS HYDRATE INVOLVEMENT IN LAKE NYOS DISASTER

6.6.1 Disaster Events

During the night of August 21, 1986, the quiescent waters of Lake Nyos in Cameroon erupted to send a wave washing 75 m up the western cliff walls surrounding the freshwater crater lake. Within minutes 40–60 million m^3 of high-specific-gravity gases (Freeth, 1992) erupted, spilled over the natural weir in the northwest, and rolled 15 km through two valleys and the village of Nyos below the lake. The massive cloud of dense gases, 30–61 m high, uprooted trees in its path (Freeth, 1992; Barberi et al., 1989; Freeth and Kay, 1987). In the cloud's path, 1700 people died suddenly; 3500 cattle were killed where they stood (Stager, 1987).

See Figure 6.5 for a photograph of the pristine lake before the disaster.

Note the natural weir at the end of the lake and the high cliff-like walls of the crater.

6.6.2 Description of Lake Nyos

Lake Nyos is a crater lake along a chain of extinct volcanoes in Cameroon. It has 208–229 m deep waters and a surface area of 1.48 km^2 (Barberi et al., 1989; Kanari, 1989). The lake has high crater walls, but water height

Figure 6.5 ***Lake Nyos before eruption (Lockwood, 1986).***

is set by a natural weir in the northwest. Verdant hills above the lake supply large masses of organic matter that decay in the lake bottom. Piboule et al. (1990) evaluated cores rich in organic matter that confirmed abundant clays and decaying plant life. The village of Nyos lies in a valley 250 m below the lake surface.

Lake temperatures are 22.9°C near the surface and increase to 23.9°C at the bottom (Fletcher, 1992; Nojiri et al., 1990; Freeth, 1992). The pressure is approximately 2.13 MPa at the lake bottom. Entering the lake through fissures in base rock are spring waters laden with CO_2. Lake Nyos carbon dioxide concentrations were measured by scientific expeditions sent by multiple countries: data taken in October 1988 (Kusakabe et al., 1989); data taken by the German expedition during November 1986 (Tietze, 1987); data taken during January, March, and May 1987 (Kling et al., 1987); and data taken during December 1988 (Nojiri et al., 1990). Results showed carbon dioxide concentrations near the lake bottom again nearing saturation, and the data indicated that CO_2 concentrations would have been closer to lake saturation values at the time of disaster.

Superposition of CO_2 equilibrium concentrations at subject pressures and temperatures onto the measured data made it clear that if bottom waters were uplifted, those waters would emit CO_2 gas on entering shallower waters of lower solubility. This was confirmed during a postdisaster expedition when samples retrieved from deep waters effervesced vigorously nearer the surface (Freeth and Kay, 1987). Unpressurized cores taken after the event revealed thick bottom sediments, close to saturation with CO_2 gas, and showed mixing of sediment's upper layers (Freeth, 1992; Piboule et al., 1990).

6.6.3 Consensus Explanations of Lake Nyos Disaster

After the event, teams of volcanologists representing multiple countries traveled to the remote area. They concluded that seismic activity had been absent and could not have triggered the incident. The postevent explanation describes an upward burst of bottom waters carrying entrained sediments. Bottom waters surged upward to a depth shallower than their CO_2 solubility, freeing excess CO_2 that increased in magnitude as the upward surge progressed. Carbon dioxide liberated from the water then became a gas lift (Kling et al., 1987; Sano et al., 1990). Bottom waters and sediments accelerated, causing waves to reach 75 m up surrounding cliffs. A 40–60 million m^3 gas cloud, denser than air, spilled over the natural weir. The waters hold 300 million m^3 of CO_2 in the near-saturation state. The

Figure 6.6 ***Lake Nyos, 3 weeks after eruption.*** US Geological Survey.

difference between a CO_2 saturation content before the event and the amount of measured CO_2 dissolved in the waters after the catastrophe gives approximately 40–60 million m^3 of gas in the killer cloud (Fletcher, 1992; Kanari, 1989).

Reduced sediments and debris littered the lake surface a few weeks after the event. See Figure 6.6.

The consensus explanation of expeditionary groups leaves four unresolved anomalies, including unexplained details contributed by survivors in the lower valleys at the edge of the cloud path:

1. A phenomenon that would have triggered the massive and sudden circulation of bottom waters to the surface could not be determined (Kling et al., 1989; Kanari, 1989).
2. Survivors reported a smell in the killer cloud similar to rotten eggs (Freeth and Kay, 1987; Freeth, 1992; Kling et al., 1987). However, analyses showed very low concentrations of hydrogen sulfide or sulfur compounds in the water (Freeth et al., 1990a; Freeth, 1990b,c). *Note*: Human death occurs in a greater than 40% CO_2 atmosphere, but sudden death occurs when H_2S concentration is only 800 ppm. Interview statements of survivors were discounted as hallucinations or as language translation problems because later data showed a lack of H_2S in waters and sediments (Freeth, 1993; Kling et al., 1987; Freeth et al., 1990a; Freeth, 1990b,c).
3. Skin lesions (Kusakabe et al., 1989), chemical burns (Barberi et al., 1989), and skin blisters (Freeth, 1992) of victims implied acidic skin injuries (Kling et al., 1987). Respiratory problems characterized H_2S damage. Autopsies of lung, kidney, and muscle tissues 11 days after the incident indicated sulfate ions three to five times in excess of the norm (Kusakabe et al., 1989).
4. Legend related by the natives indicates similar event(s) in the distant past.

6.6.4 Gas Hydrate Theory of Trigger Mechanism

6.6.4.1 Bases

Calculations (Rogers and Yevi, 1996) made with CSMHYD software (Sloan, 1990), using known lake-bottom temperature and hydrostatic pressure, determined a wide range of carbon dioxide, methane, and hydrogen sulfide gas compositions that would form stable solid hydrates, where hydrogen sulfide is the essential ingredient for hydrates to form at those lake conditions (Rogers and Yevi, 1996). Of these gases, carbon dioxide entered the lake from carbonated springs, but methane and hydrogen sulfide would have been generated by microbial activity from abundant organic matter in the oxygen-depleted sediments. Hydrogen sulfide readily forms gas hydrates, so much so that sulfides on generation are scavenged and sequestered by hydrates to isolate them from sediments/water. Calculation results in Table 6.3 show ranges of free gas compositions that would form gas hydrates at the pressure and temperature prevalent on the floor of Lake Nyos. Note the concentrated H_2S in occluded gases.

Free gas concentrations of CH_4, CO_2, and H_2S in sediment pores calculated to form hydrates at Lake Nyos conditions of temperature and pressure given in Table 6.3 are shown on a triangular graph in Figure 6.7. Hydrates were found to form from any of the gas compositions within the cross-hatched area.

Solid hydrate blocks could trigger the deadly gas expulsion from Nyos waters in either of the two mechanisms, which will be discussed in the following section.

6.6.4.2 Hydrate Formation in Nyos Sediments

Hydrate densities are 0.87–0.92 g/cm^3 for all hydrate compositions in the ranges presented in Table 6.3. Densities within the range depend on molecular weights of occluded gases and the fraction of hydrate cavities filled by each gas. Dislodgement of these hydrates from the lake floor would result in vigorous buoyance toward the lake surface. Hydrate densities were

Table 6.3 Composition of gases to form hydrates

Composition	CH_4 (mol%)	CO_2 (mol%)	H_2S (mol%)
Free gas	0–40	0–45	55–100
Occluded gas in hydrates	0–4.5	0–9.1	90.4–100

From Rogers and Yevi (1996).

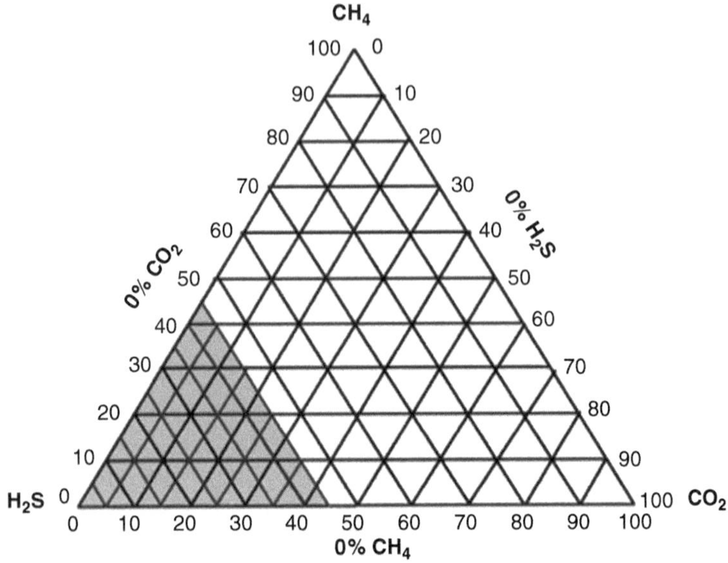

Figure 6.7 ***Feasibility of hydrate formation in Lake Nyos sediments (Rogers and Yevi, 1996).***

calculated as a function of occluded gas content based on the method of Makogon (1981).

For example, if the three gases 55% H_2S, 40% CO_2, and 5% CH_4 are present under a column of water at 23.9°C and a hydrostatic pressure of 2.04 MPa, sI hydrates form; the density of the hydrate and its potential buoyancy were calculated to be 0.87 g/cm^3 by the method of Makogon. Then, fractional fillings of the cavities with each gas were calculated with Sloan's CSMHYD program, finding that sI hydrate forms with an occluded gas composition of 90.4% H_2S, 9.1% CO_2, and 0.5% CH_4 (Rogers and Yevi, 1996). If released at the bottom of the water column, the hydrate block would be buoyed to the surface. The buoyant dislodgement of hydrates is commonly observed in ocean sediments (MacDonald et al., 2005).

Therefore, two possible mechanisms for the disastrous event are suggested that involve gas hydrates.

First, hydrates composed of carbon dioxide, methane, and predominantly hydrogen sulfide form and collect in bottom sediments, possibly over lengthy time periods, to create an impermeable cap of sediments in the deepest part of the lake. As pressure below the impermeable hydrate cap builds from *in situ* biogenic gas generation and gas flow through base-rock fissures, a solid block of low-specific-gravity hydrates/cemented sediments breaks loose. The buoyant hydrate block pulls CO_2-saturated bottom waters and sediments to shallower depths, during which exsolved CO_2 accelerates the rising block by gas lift. At the lake surface, hydrates decompose and release occluded H_2S, CH_4, and CO_2 gases to mix with exsolved CO_2 gases. Hydrogen sulfide and carbon dioxide gases of high specific gravity compose the cloud that sweeps valleys below Lake Nyos. Hydrogen sulfide provides a more deadly component of the cloud that would be instantly fatal and burn living tissues.

Second, the hydrate zone at the bottom of Lake Nyos might have developed gas hydrates as do fine-grained ocean sediments in hydrate zones. Essentially, hydrate particles form in sediment pores and impede capillary pathways to fluid flow. According to the Cathles Theory, when gas columns below BGHS reach critical heights, gas chimneys begin to form. On reaching roughly half the distance from BGHS to sediment surface, fluidized mud/gas/water bursts free to crater the lake-bottom sediments. For example, Cathles et al. (2010) estimate a probable time of 1000 years for a capillary seal mechanism to develop gas chimney, catastrophic mass expulsion, and craters at Blake Ridge. The process is cyclic.

REFERENCES

Barberi, F., Chelini, W., Marinelli, G., Martini, M., 1989. The gas cloud of Lake Nyos (Cameroon, 1986): results of the Italian technical mission. J. Volcanol. Geotherm. Resour. 39, 125–134.

Berndt, C., Mienert, J., Vanneste, M., Bunz, Z., Bryn, P., 2002. Submarine slope-failure offshore Norway triggers rapid gas hydrate decomposition. In: Proceedings of the 4th International Conference on Gas Hydrates, Yokohama, Japan, May 19–23.

Birchwood, R.A., Noeth, S., Tjengdrawira, M.A., Kisra, S.M., Elisabeth, F.L., Sayers, C.M., Singh, R., Hooyman, P.J., Plumb, R.A., Jones, E., Bloys, J.B., 2007. Modeling the mechanical and phase change stability of wellbores drilled in gas hydrates by the Joint Industry Participation Program (JIP) Gas Hydrates Project Phase II. In: SPE Annual Technical Conference, SPE 110796.

Birchwood, R., Noeth, S., Jones, E., 2008. Safe drilling in gas-hydrate prone sediments: findings from the 2005 drilling campaign of the Gulf of Mexico Gas Hydrates Joint Industry Project (JIP). In: Fire in the Ice, US DOE-NETL Newsletter. Winter Issue, pp. 1–4.

Boetius, A., Ravenschiag, K., Schubert, C.J., Rickert, D., Widdei, F., Gleseke, A., Amman, R., Jergensen, B.B., Witte, U., Pfannkuche, O., 2000. A marine microbial consortium apparently mediating anaerobic oxidation of methane. Nature 407, 623–626.

Bohrmann, G., Ivanov, M., Foucher, J.-P., Spiess, V., Bialas, J., Greinert, J., Weinrebe, W., Abegg, F., Aloisi, G., Artemov, Y., Blinova, V., Drews, M., Heidersdorf, F., Krabbenhoft, A., Klaucke, I., Krastel, S., Leder, T., Polikarpov, I., Saburova, M., Schmale, O., Seifert, R., Volkonskaya, A., Zillmer, M., 2003. Mud volcanoes and gas hydrates in the Black Sea: new data from Dvurechenskii and Odessa mud volcanoes. Geo-Mar. Lett. 23, 239–249.

Boswell, R., Collett, T., Frye, M., Shedd, B., McConnell, D., Shelander, D., 2012. Subsurface gas hydrates in the northern Gulf of Mexico. J. Mar. Pet. Geol. 34, 4–30.

Buenz, S., Mienert, J., Berndt, C., 2003. Geological controls on the Storegga gas-hydrate system of the mid-Norwegian continental margin. Earth Planetary Sci. Lett. 209, 291–307.

Camilli, R., Reddy, C.M., Yoerger, D.R., Van Mooy, B.A.S., Jakuba, M.V., Kinsey, J.C., McIntypre, C.P., Sylva, S.P., Maloney, J.V., 2010. Tracking hydrocarbon plume transport and biodegradation at Deepwater Horizon. Science 330, 201–204.

Cathles, L.M., Zheng, S., Chen, D., 2010. The physics of gas chimney and pockmark formation, with implications for assessment of seafloor hazards and gas sequestration. Mar. Pet. Geol. 27 (1), 82–91.

Clift, R., Grace, J.R., Weber, M.E., 1978. Bubbles, Drops, and Particles. Academic Press, New York, 380 pp.

Cook, A., Goldberg, D., Kleinberg, R., 2008. Fracture-controlled gas hydrate systems in the northern Gulf of Mexico. J. Mar. Pet. Geol. 25 (9), 932–941.

Cragg, B.A., Parkes, R.J., Fry, J.C., Weightman, A.J., Rochelle, P.A., Maxwell, J.R., 1996. Bacterial populations and processes in sediments containing gas hydrates (ODP Leg 146: Cascadia margin). Earth Planetary Sci. Lett. 139, 497–507.

de Batist, M., Klerkx, J., van Rensbergen, P., Vanneste, M., Poort, J., Golmshtok, A.Y., Kremlev, A.A., Khlystov, O.M., Krinitsky, P., 2002. Active hydrate destabilization in Lake Baikal, Siberia? Terra Nova 14 (6), 436–442.

de Beer, D., Boetius, A., Camilli, R., Feseker, T., Foucher, J.-P., German, C., Kinsey, J., Mienert, J., Schluter, M., Waldmann, C., Wenzhofer, F., Yoerger, D., 2011. One year direct observations on volcanic activity of a cold seep, the Hakon Mosby mud volcano (HMMV). In: Proceedings of the 7th International Conference on Gas Hydrates (ICGH 2011), Edinburgh, Scotland, July 17–21. Paper 188.

Dearman, J.L., Wilson, W.W., Rogers, R.E., Zhang, G., 2009. Gas-hydrate promotion by smectite–bioproduct interactions. Mar. Chem. 115 (1–2), 21–30.

Dimitrov, L.I., 2002. Mud volcanoes: the most important pathway for degasing deeply buried sediments. Earth Sci. Rev. 59, 49–76.

Egorov, A.V., Rozhkov, A.N., 2010. Formation of gas hydrate reservoirs in submarine mud volcanos. Fluid Dyn. 45 (5), 769–778.

Egorov, A.V., Nigmatulin, R.I., Sagalevich, A.M., Rozhkov, A.N., Tsypkin, G.G., 2011. Investigation of deep water gas hydrates with "MIR" submersibles during 2008–2010 expedition in Lake Baikal. In: Proceedings of the 7th International Conference on Gas Hydrates (ICGH 2011), Edinburgh, Scotland, UK, July 17–21. Paper 243.

Etiope, G., Feyzullayev, A., Baciu, C.L., Milkov, A.V., 2004. Methane emissions from mud volcanoes in eastern Azerbaijan. Geology 32, 465–468.

Etiope, G., Nakada, R., Tanaka, K., Yoshida, N., 2011. Gas seepage from Tokamachi mud volcanoes, onshore Niigata Basin (Japan): origin, post-genetic alterations and CH_4–CO_2 fluxes. Appl. Geochem. 26, 348–359.

Fletcher, H., 1992. Degassing Lake Nyos. Nature 355, 683.

Freeth, S.J., 1990b. Lake Bambuluwe: could it be the source for a third gas disaster in western Cameroon? J. Volcanol. Geotherm. Res. 42, 393–395.

Freeth, S.J., 1990c. The anecdotal evidence, did it help or hinder investigation of the Lake Nyos gas disaster? J. Volcanol. Geotherm. Res. 42, 373–380.

Freeth, S.J., 1992. Incident at Lake Nyos. Sciences 32 (3), 31–36.

Freeth, S.J., 1993. On the problems of translation in the investigation of the Lake Nyos disaster. J. Volcanol. Geotherm. Res. 54 (3–4), 353–356.

Freeth, S.J., Kay, R.L.F., 1987. The Lake Nyos gas disaster. Nature 325, 104–105.

Freeth, S.J., Kling, G.W., Kusakabe, M., Maley, J., Tchoua, F., Tiezte, K., 1990a. Conclusions from the Lake Nyos disaster. Nature 348, 201.

Gardner, J.V., Malikk, M., Walker, S., 2009. Plume 1400 meters high discovered at the seafloor off the northern California margin. EOS Trans. Am. Geophys. Union 90 (32), 275.

Ginsburg, G.D., Milkov, A.V., Soloviev, V.A., Egorov, A.V., Cherkashev, G.A., Vogt, P.R., Crane, K., Lorenson, T.D., Khutorskoy, M.D., 1999. Gas-hydrate accumulation at the Håkon Mosby mud volcano. Geo-Mar. Lett. 19, 57–67.

Hazen, T.C. Dubinsky, E.A., DeSantis, T.Z., et al., 2010. Deep-sea oil plume enriches indigenous oil-degrading bacteria. Science 330, 204–208.

Hjelstuen, B.O., Eldholm, Q., Faleide, J.I., Vogt, P.R., 1999. Regional setting of Håkon Mosby mud volcano, SW Barents Sea margin. Geo-Mar. Lett. 19, 22–28.

Hovland, M., Judd, A.G., 1988. Seabed Pockmarks and Seepages. Impact on Geology, Biology and the Marine Environment. Graham & Trotman Ltd., London, 293 pp.

Hyndman, R.D., Spence, G.D., 1992. A seismic study of methane hydrate marine bottom simulating reflectors. J. Geophys. Res. 97, 6683–6698.

Johnson, J.E., Goldfinger, C., Suess, E., 2003. Geophysical constraints on the surface distribution of authigenic carbonates across the Hydrate Ridge region, Cascadia margin. Mar. Geol. 202, 79–120.

Kanari, S.I., 1989. An inference on the process of gas outburst from Lake Nyos, Cameroon. J. Volcanol. Geotherm. Res. 39, 135–149.

Kennett, J.P., Fackler-Adams, B.N., 2000. Relationship of clathrate instability to sediment deformation in the upper Neogene of California. Geology 28 (3), 215–218.

Klapp, S.A., Enzmann, F., Walz, P., Huthwelker, T., Tuckermann, J., Schwarz, J.-O, Pape, T., Peltzer, E.T., Hester, K.C., Zhang, X., Mokso, R., Wangner, D., Marone, F., Kersten, M., Bohrmann, G., Kuhs, W.F., Stampanoni, M., Brewer, P.G., 2011. Fluid flow in natural gas hydrates revealed by X-ray tomographic microscopy. In: Proceedings of the 7th International Conference on Gas Hydrates (ICGH 2011), Edinburgh, Scotland, UK, July 17–21. Paper 248.

Kling, G.W., Clark, M.A., Comption, H.R., Devine, J.D., Evans, W.C., Humphrey, A.M., Koenigsberg, E.J., Lockwood, J.P., Tuttle, M.L., Wagner, G.N., 1987. The 1986 Lake Nyos gas disaster in Cameroon, West Africa. Science 236, 169–175.

Kling, G.W., Tuttle, M.L., Evans, W.C., 1989. The evolution of thermal structure and water chemistry in Lake Nyos. J. Volcanol. Geotherm. Res. 39, 151–165.

Knittel, K., Boetius, A., 2009. Anaerobic oxidation of methane: progress with an unknown process. Annu. Rev. Microbiol. 63, 311–334.

Kujawinski, E.B., Kido Soule, M.C., Valentine, D.L., Boysen, A.K., Longnecker, K., Redmond, M.C., 2011. Fate of dispersants associated with the Deepwater Horizon oil spill. Environ. Sci. Technol. 45, 1298–1306.

Kusakabe, M., Ohsumi, T., Aramaki, S., 1989. The Lake Nyos gas disaster: chemical and isotopic evidence in waters and dissolved gases from three Cameroonian crater lakes, Nyos, Monoun and Wum. J. Volcanol. Geotherm. Res. 39, 167–185.

Kuzmin, M.I., Geletiy, V.F., Kalmychkov, G., Kuznetsov, F.A., Larionov, E.G., Manakov, A.Yu., Mironov, Yu.I., Smoljakov, B.S., Dyadin, Yu.A., Duchrov, A.D., Bazin, N.M., Mahov, G.M., 2000. The first discovery of the gas hydrates in the sediments of the Lake Baikal. In: Holder, G.D., Bishnoi, P.R. (Eds.), Gas Hydrates: Challenges for the Future (Annals of the New York Academy of Sciences), vol. 912. New York Academy of Sciences, New York, NY, pp. 112–115.

Kvalstad, T.J., Yamamoto, K., Noguchi, S., Uchida, S., Soga, K., 2011. Effect of gas hydrate production on seabed stability in the eastern Nankai Trough area. In: Proceedings of the 7th International Conference on Gas Hydrates (ICGH 2011), Edinburgh, Scotland, UK, July 17–21. Paper 242.

Lane, A., 2005. Overcoming deepwater geohazards in West Africa. In: Proceedings of the Offshore Technology Conference, OTC-17496.

Lanoil, B.D., Sassen, R., La Duc, M.T., Sweet, S.T., Nealson, K.H., 2001. Bacteria and Archaea physically associated with Gulf of Mexico gas hydrates. Appl. Environ. Microbiol. 67 (11), 5143–5153.

Lockwood, 1986. By Jack Lockwood, USGS (public domain), via Wikimedia Commons. <http://commons.wikimedia.org/wiki/File%3ANyos_Lake.jpg>.

Lutken, C.B., Macelloni, L., Sleeper, K., McGee, T., Simonetti, A., Knapp, J.H., Knapp, C.C., Caruso, S., Chanton, J., Lapham, L., Lodi, M., Ingrassia, M., Higley, P., Brunner, C., 2011. New discoveries at Woolsey Mound, MC118, northern Gulf of Mexico. In: Proceedings of the 7th International Conference on Gas Hydrates (ICGH 2011), Edinburgh, Scotland, UK, July 17–21. Paper 626.

MacDonald, I.R., Bender, L., Vardaro, M., Bernard, B., Brooks, J., 2005. Thermal and visual time series at a seafloor gas hydrate deposit on the Gulf of Mexico slope. Earth Planetary Sci. Lett. 233 (1–2), 45–59.

Makogon, Y.F., 1981. Hydrates of Natural Gas. PennWell Publishing Company, Tulsa, OK.

Maksimov, A.O., Sosedko, E.V., 2009. Acoustic manifestations of gas hydrate shelled bubbles. Acoust. Phys. 55 (6), 776–784.

Mazurenko, L.L., Soloviev, V.A., Gardner, J.M., Ivanov, M.K., 2003. Gas hydrates in the Ginsburg and Yuma mud volcano sediments (Moroccan margin): results of chemical isotopic studies of pore water. Mar. Geol. 195, 201–210.

Mazzini, A., Ivanov, M.K., Parnell, J., Stadnikskaia, A., Cronin, B.T., Poludetkina, E., Mazurenko, L., van Weering, T.C.E., 2004. Methane-related authigenic carbonates from the Black Sea: geochemical characterization and relation to seeping fluids. Mar. Geol. 212, 153–181.

McConnell, D.R., Zhang, Z., Boswell, R., 2012. Review of progress in evaluating gas hydrate drilling hazards. Mar. Pet. Geol. 34, 209–223.

Milkov, A.V., 2000. Worldwide distribution of submarine mud volcanoes and associated gas hydrates. Mar. Geol. 167, 29–42.

Milkov, A.V., Sassen, R., Apanasovich, T.V., Dadashev, F.G., 2003. Global gas flux from mud volcanoes: a significant source of fossil methane in the atmosphere and the ocean. Geophys. Res. Lett. 30 (2), 9-1–9-4.

Monastersky, R., 1998. Giant seabed slides may have climate link. Science News, March 28, p. 153.

Nojiri, Y., Kusakabe, M., Hlrabayashi, J.-I., Sato, H., Sano, Y., Shinohara, H., Njine, T., Tanyileke, G., 1990. Gas discharge at Lake Nyos. Nature 346, 322–323.

Obzhirov, A., 2009. Methane fluxes and gas hydrates in the Sea of Okhotsk. In: Fire in the Ice, DOE-NETL Newsletter. Spring Issue, pp. 12–15.

Obzhirov, A., Shakirov, R., Salyuk, A., Suess, E., Biebow, N., Salomatin, A., 2004. Relations between methane venting, geological structure and seismo-tectonics in the Okhotsk Sea. Geo-Mar. Lett. 24, 135–139.

Paull, C.K., Brewer, P.G., Usler, W., Peltzer, E.T., Rehder, G., Clague, D., 2002. An experiment demonstrating marine slumping is a mechanism to transfer methane from seafloor gas hydrate deposits into the upper ocean and atmosphere. Geo-Mar. Lett. 22 (4), 198–203.

Piboule, M., Pourchet, M., Bouchez, R., Amosse, J., Brenac, P., Maley, J., Pinglot, J.F., Vincent, C., 1990. Radiometric studies of Lake Nyos (Cameroon) sediments: evidence of strong mixing and excess ^{210}Pb. J. Volcanol. Geotherm. Res. 42 (4), 363–372.

Pinheiro, L.M., Ivanov, M.K., Sautkin, A., Akhmanov, G., Magalhaes, V.H., Volkonskaya, A., Monteiro, J.H., Somoza, L., Gardner, J., Hamouni, N., Cunha, M.R., 2003. Mud volcanism in the Gulf of Cadiz: results from the TTR-10 cruise. Mar. Geol. 195, 131–151.

Radich, J.G., 2009. Laboratory and theoretical investigations of direct and indirect microbial influences on seafloor gas hydrates. A thesis in partial fulfillment of the requirements for the degree of Master of Science in Chemical Engineering in the Dave C. Swalm School of Chemical Engineering, Mississippi State University, Mississippi State, MS.

Reeburgh, W.S., 1976. Methane consumption in Cariaco Trench waters and sediments. Earth Planetary Sci. Lett. 28, 337–344.

Rehder, G., Leifer, I., Brewer, P.G., Friederich, G., Peltzer, E.T., 2009. Controls on methane bubble dissolution inside and outside the hydrate stability field from open ocean field experiments and numerical modeling. Mar. Chem. 114 (1–2), 19–30.

Rogers, R.E., Yevi, G.Y., 1996. Hydrate theory explains Lake Nyos disaster. In: 2nd International Gas Hydrate Conference Proceedings, Toulouse, France, June 2–6.

Rogers, R.E., Kothapalli, C., Lee, M.S., Woolsey, J.R., 2003. Catalysis of gas hydrates by biosurfactants in seawater-saturated sand/clay. Can. J. Chem. Eng. 81, 1–8.
Rogers, R., Guochang, Z., Dearman, J., Woods, C., 2007. Investigations into surfactant/gas hydrate relationship. J. Pet. Sci. Eng. 56, 82–88.
Ruppel, C., Boswell, R., Jones, E., 2008. Scientific results from Gulf of Mexico Joint Industry Project Leg 1 drilling: introduction and overview. Mar. Pet. Geol. 25, 819–829.
Sano, Y., Kusakabe, M., Hirabayashi, J., Noriri, Y., Shinohara, H., Njine, T., Tanyileke, G., 1990. Helium and carbon fluxes in Lake Nyos, Cameroon: constraint on next gas burst. Earth Planetary Sci. Lett. 99, 303–314.
Sauter, E.J., Muyakshin, S.I., Charlou, J.-L., Schluter, M., Boetius, A., Jerosch, K., Damm, E., Foucher, J.-P., Klages, M., 2006. Methane discharge from a deep-sea submarine mud volcano into the upper water column by gas hydrate-coated methane bubbles. Earth Planetary Sci. Lett. 243, 354–365.
Shakirov, R., Obzhirov, A., 2011. Western Pacific gas hydrate belt. In: Proceedings of the 7th International Conference on Gas Hydrates (ICGH 2011), Edinburgh, Scotland, UK, July 17–21. Paper 781.
Sloan, Jr., E.D., 1990. Clathrate Hydrates of Natural Gas. Marcel Dekker, Inc., New York.
Smith, M., Kou, W., Ahmed, A., Kuzela, R., 2005. The significance of gas hydrate as a geohazard in Gulf of Mexico exploration and production. In: Proceedings of the Offshore Technology Conference, OTC-17655.
Stager, J.C., 1987. Silent death from Cameroon's killer lake. National Geographic, vol. 172, Issue 3, pp. 404–420.
Tietze, K., 1987. Results of the German–Cameroon Research Expedition to Lake Nyos (Cameroon), October/November 1986. Report to the Bundesanstalt fur Geowissenschaften und Rohstofe Hannover, 34 pp.
Tinivella, U., Accaino, F., Vedova, B.D., 2008. Gas hydrates and active mud volcanism on the South Shetland continental margin, Antarctic Peninsula. Geo-Mar. Lett. 28, 97–106.
Valentine, D.L., Kessler, J.D., Redmond, M.C., Mendes, S.D., Heintz, M.B., Farwell, C., Hu, L., Kinnaman, F.S., Yvon-Lewis, S., Du, M., Chan, E.W., Tigreros, F.G., Villanueva, C.J., 2010. Propane respiration jump-starts microbial response to a deep oil spill. Science 330, 208–211.
Vanneste, M., de Batist, M., Golmshtok, A., Kremlev, A., Versteeg, W., 2001. Multi-frequency seismic study of gas hydrate-bearing sediments in Lake Baikal, Siberia. Mar. Geol. 172, 1–21.
Vanneste, M., Poort, J., de Batist, M., Klerkx, J., 2003. Atypical heat-flow near gas hydrate irregularities and cold seeps in the Baikal Rift Zone. Mar. Pet. Geol. 19, 1257–1274.
Werne, J.P., Haese, R.R., Zitter, T., Aloisi, G., Bouloubassi, I., Heijs, S., Fiala-Medioni, A., Pancst, R.D., Damste, J.S.S., de Lange, G., Forney, L.J., Gottschal, J.C., Foucher, J.-P., Mascle, J., Woodside, J., 2004. Life at cold seeps: a synthesis of biogeochemical and ecological data from Kazan mud volcano, Eastern Mediterranean Sea. Chem. Geol. 205, 367–390.
Winters, W.J., 2011. Physical and geotechnical properties of gas-hydrate-bearing sediment from offshore India and the northern Cascadia margin compared to other hydrate reservoirs. In: Proceedings of the 7th International Conference on Gas Hydrates (ICGH 2011), Edinburgh, Scotland, UK, July 17–21. Paper 408.
Winters, W.J., Pecher, I.A., Waite, W.F., Mason, D.H., 2004. Physical properties and rock physics models of sediment containing natural and laboratory-formed methane gas hydrate. Am. Mineral. 89, 1221–1227.
Zhang, G., Rogers, R.E., French, W.T., Lao, W., 2007. Investigation of microbial influences on seafloor gas-hydrate formations. Mar. Chem. 103, 359–369.
Zhong, Y., Rogers, R.E., 2000. Surfactant effects on gas hydrate formation. Chem. Eng. Sci. 55 (19), 4175–4187.

CHAPTER SEVEN

Biogenic Hydrate Provinces

Three biogenic gas hydrate provinces are reviewed in this chapter: (1) Blake Ridge, (2) Cascadia margin, and (3) Nankai Trough. In each case, hydrate-gas content originates predominantly from microbes, as readily seen through the presentation of Bernard plots. Geographic and geologic diversities of the three provinces represent microcosms of universal biogenic hydrate occurrences.

Drilling in Atlantic sediments off the eastern US coast, ODP Leg 164 first reported gas hydrates at Blake Ridge. Although Blake Ridge hydrates are extensive, two disincentives exist for early hydrate-gas production there. First, the deepwater hydrates are primarily dispersed or vein filling with relatively low saturations in sediments. Second, no oil or gas production infrastructure exists in this offshore region.

At Cascadia margin, a subduction zone along the juncture of Juan de Fuca and American Plates off the western US coast, extensive biogenic hydrates have developed from conditions and fluid flows representative of other worldwide subduction zones.

Nankai Trough near Japan's east coast has biogenic hydrates highly saturating permeable sands. Nankai is perhaps the most promising marine province for large-scale gas production, and in fact is poised to first produce commercial volumes of hydrate gas.

7.1 COMPARISON OF BIOGENIC GAS HYDRATE PROVINCES

Comparative reservoir data from the three biogenic hydrate provinces are presented in Table 7.1. Of particular significance in Table 7.1 is the approximately 80% gas hydrate saturation S_{gh} of Nankai because of the parameter's importance for economical gas production. Note that S_{gh} is an unfavorably low less than or equal to 14% in dispersed hydrates of Blake Ridge. Another observation from Table 7.1 is that methane comprises over 98% of occluded gases in these biogenic hydrate reservoirs.

Offshore Gas Hydrates. http://dx.doi.org/10.1016/B978-0-12-802319-8.00007-3

Table 7.1 Comparison of biogenic gas hydrate provinces

	Nankai Trough	Blake Ridge	Cascadia margin
Gas composition, methane (%)	99.37–100.00 (Waseda and Uchida, 2002) MITI well	>98.5 (Dickens et al., 1997)	ODP 146, Site 892: 96.9–98.5 (Suess et al., 1999)
Water depth (m)	945 (Waseda and Uchida, 2002) Landward side	2400–4800	800 South Hydrate Ridge (Tréhu et al., 1999) 600 North Hydrate Ridge (Tréhu et al., 1999)
	600–800 (Nouzé et al., 2002) Eastern Nankai, S.E. flank, Kodaiba Knoll 4841 (Grevemeyer and Villinger, 2001) ODP Leg 131, Site 808	3000 (Ghosh et al., 2010; Dickens et al., 1997) mean	ODP Leg 146: Site 892 – 670 m (MacKay et al., 1994); Site 889 – 1322 m (MacKay et al., 1994)
Well depth (mbsf)	2355 (Waseda and Uchida, 2003) MITI well, December 1999	700–750 (Ghosh et al., 2010; Dickens et al., 1997) ODP 164, Sites 994, 995, 997	Penetrated BSR (MacKay et al., 1994): ODP 146, Sites 892 and 889
Core depth (mbsf)	160–313 (Colwell et al., 2002) MITI well, December 1999	57.85–746.85 (Matsumoto and Borowski, 2000)	ODP Leg 146: 3313
BSR depth (mbsf)	40–200 (Nouzé et al., 2002) Eastern Nankai, S.E. flank, Kodaiba Knoll; parallel BSR 50–100 m below top BSR	200–600 (Dillon and Paull, 1983) at Carolina Rise and Blake Ridge	115 (Tréhu et al., 2003) ODP Leg 204, Site 1249
	300 (Waseda and Uchida, 2002) MITI well, landward side 205 (Hyndman et al., 1992) ODP Leg 131, Site 808		ODP 146, Site 892: 73.9 (MacKay et al., 1994) ODP 146, Site 889: 223.5 (MacKay et al., 1994)

Table 7.1 Comparison of biogenic gas hydrate provinces *(cont.)*

	Nankai Trough	**Blake Ridge**	**Cascadia margin**
Porosity sediments (%)	20		20–30
Hydrate saturation, S_{gh} (%)	80	6–12 (Matsumoto and Borowski, 2000) 10–14 (Ghosh et al., 2010) ODP 164, Sites 994, 995, 997	5–39 50 (Daigle et al., 2011); maximum at Summit of South Hydrate Ridge
Bottom water temperature (°C)	4 (Colwell et al., 2002) 1.6 (Grevemeyer and Villinger, 2001) ODP Leg 131, Site 808	3.5 (Paull et al., 1995) at Blake Ridge Diapir	4.25 (Tryon et al., 2002)
In-place gas (tcf)	50	236	16,492
Mean geothermal gradient (°C/km)	53.6 (Yamano et al., 1982) 40 (Colwell et al., 2002) 35 (Foucher et al., 2002)	41.7 (Yamano et al., 1982)	51.0 (Suess et al., 1999) 55.6 (Tréhu et al., 2004a; Hornbach et al., 2012); 9-well average, ODP Leg 204
Heat flow (mW/m^2)	129 (Hyndman et al., 1992) ODP Leg 131, Site 808		57.4 (Tréhu et al., 2004a; Hornbach et al., 2012) 9-Well average, ODP Leg 204
Thermal conductivity [W/(m K)]	1.08 (Hyndman et al., 1992) mean; ODP Leg 131, Site 808		0.97 (Tréhu et al., 2004a; Hornbach et al., 2012) 9-Well average, ODP Leg 204
$\delta^{13}C_{CH_4}$ (‰)	−71 (Waseda and Uchida, 2002) to −66 MITI well	−65.07 (Paull et al., 1995); Blake Ridge Diapir	−71.5 (Suess et al., 1999) to −62.4

7.2 BLAKE RIDGE AND CAROLINA RISE

Inferred from unpressurized cores, offshore hydrates at Blake Ridge and Carolina Rise are disseminated, morphologically inferred in part because hydrates in the cores rapidly dissociate during retrieval. That is, inherently large surface areas of hydrate particles disperse throughout conductive sand as the core's porous media establish relatively high heat-transfer rates, escalating hydrate dissociation during core retrieval (Paull et al., 1996). If instead, hydrate-filled veins dominate morphology, a similar rapid decomposition occurs, thus requiring more sophisticated means to distinguish morphology.

Small-particle sediments are common at Blake Ridge with some porous media containing clays having less than 10 nm pore spaces. However, secondary porosity may provide larger pore spaces that improve hydrate nucleation and expansion capabilities. Lorenson (2000) reports a supplemental porosity coming from diatoms, foraminifera, and framboidal pyrite found abundantly in the Blake Ridge test area. Also, ferromagnetic particles relate to microbial activity during CH_4 generation, and these nanoparticles may in turn play a role in gas hydrate crystal nucleation. Pyrite and iron monosulfides disperse throughout hydrate intervals at Blake Ridge (Lorenson, 2000).

Pockmarks on the order of 50 m diameter frequent the seafloor at Blake Ridge, indicating past gas/water/sediment/hydrate expulsions so common in hydrate-bearing sediments worldwide (Monastersky, 1996). In accordance with the Cathles Theory, pockmarks dot the seafloor where pressurized gases in fine sediments below bottom of gas hydrate stability (BGHS) breach the hydrate-zone/free-gas interfacial barrier as a first step toward violent expulsion of fluidized mass at the seafloor (Cathles et al., 2010).

At both Blake Ridge and Carolina Rise, bottom-simulating reflectors (BSRs) at 200–600 mbsf indicate possibly extensive gas hydrate occurrences, spreading over nearly 26,000 km^2 of areal extent (Dillon and Paull, 1983).

7.2.1 Blake Ridge Drill Sites 994, 995, 997

During Leg 164 of the Ocean Drilling Program, three Blake Ridge sites (994, 995, and 997) were drilled in water depths ranging from 2781 to 2800 m along a 10 km transect across the Ridge (Collett, 2002; Davie and Buffett, 2002; Monastersky, 1996). Site locations are 330 km offshore

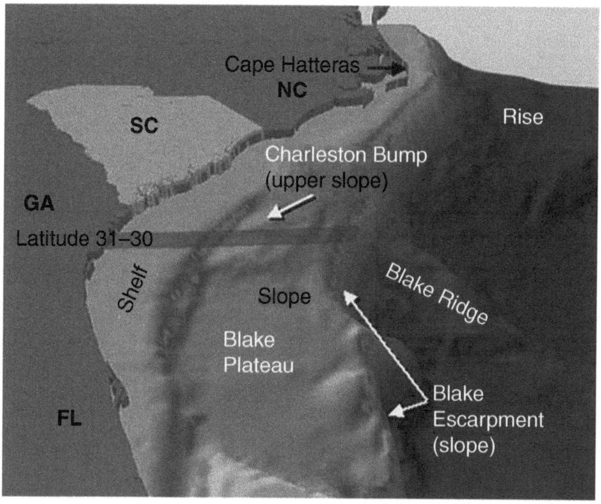

Figure 7.1 ***Location of Blake Ridge gas hydrates.*** *(By NOAA (public domain), via Wikimedia Commons.)*

of Savannah and Charleston, southeastern United States. See the National Oceanic and Atmospheric Administration (NOAA) map of Figure 7.1.

The three wells extend to sediment depths of 700–750 mbsf, sufficiently deep to penetrate BGHS.

One distinct BSR runs below Sites 995, 997, and 994. However, it fades into indistinctness near 3400 mbsl as it approaches Site 994 (Ghosh et al., 2010; Monastersky, 1996; Collett, 2002). Although in close proximity, the three wells along the transect yield a few dissimilar results because of the BSR termination.

At Site 997 hydrates exist throughout the interval 200–450 mbsf, being concentrated at sublevels 230–330 and 370–450 mbsf. Hydrate concentrations check with chloride profiles of pore waters within the hydrate-bearing intervals (Matsumoto, 2000). At BGHS 450 mbsf, free-gas S_g exists in the sediments as 3% of pore spaces. The depth of BGHS and the free-gas concentration check with computer simulations by Davie and Buffett (2002). They estimated a free-gas column below the BSR to be 200 m high.

At the BGHS on the flanks of the Blake Ridge Diapir, methane saturations are estimated to be 169 mM in pore waters having temperatures and pressures of 19°C and 260 bar, respectively (Paull et al., 1995).

Paull et al. (2000a) report rectangular hydrate fragments 3–10 cm long mingling with nanometer-size, dispersed clay platelets, where those clays develop soupy appearances as water reabsorbs after hydrate decomposition during core retrieval.

7.2.2 Blake Ridge Diapir Site 996

7.2.2.1 Diapir Characteristics

Twenty salt diapirs along Blake Ridge and Carolina Rise present an interesting study area of associated gas hydrates (Ghosh et al., 2010). See Figure 7.1.

Van Dover et al. (2003) studied the southernmost Blake Ridge Diapir. The following prominent features were characterized by five shallow boreholes drilled across its dome, four *Alvin* dives, seafloor photography, sonar, and push cores (Paull et al., 1995; Van Dover et al., 2003):

1. Dome diameter is about 5 km.
2. A crater of 50 m diameter and 4 m depth dimples the diapir crest.
3. A gas plume jets 320 m into the overlying water column from a fault along the crest.
4. Mussels, clams, tube worms, and bacterial mats proliferate at methane seeps.
5. Large carbonate blocks lie randomly along the seafloor, associating with gas vents.
6. Gas hydrates accumulate underneath carbonate slabs. But unlike hydrate surface mounds observed on Gulf of Mexico (GOM) seafloors, no outcropping hydrates are reported at Blake Ridge, although the 3.2°C and 21.6 MPa seafloor conditions are well within the hydrate-forming envelope.

Seafloor features at Site 996 are dominated by the salt diapir. Gases released from below the BGHS move upward through faults created by prior movement of the salt mass. Seismic traces show the faults as breaks in the BSR (Paull et al., 1995). As in many hydrate zones of the GOM, salt's high thermal conductivity increases heat flux and salt dissolution increases local pore water salinities. Each phenomenon contributes to concavity in the BGHS profile above the diapir (Taylor et al., 2000).

7.2.2.2 Cratering, Fluid-Venting Mechanics at Blake Ridge Diapir

Data from BSR-penetrating wells at the Blake Ridge Diapir validate calculations by Cathles et al. (2010) relating to gas chimney kinetics, gas pocket height below the BSR, sediment permeability, and gas hydrate saturations. Under gas hydrate saturations S_{gh} of 18% in fine-grained sediments at the BGHS, their calculations require a 28 m free-gas column below BGHS to reach sufficient buoyant force for gas chimney growth. By rule of thumb, when chimney

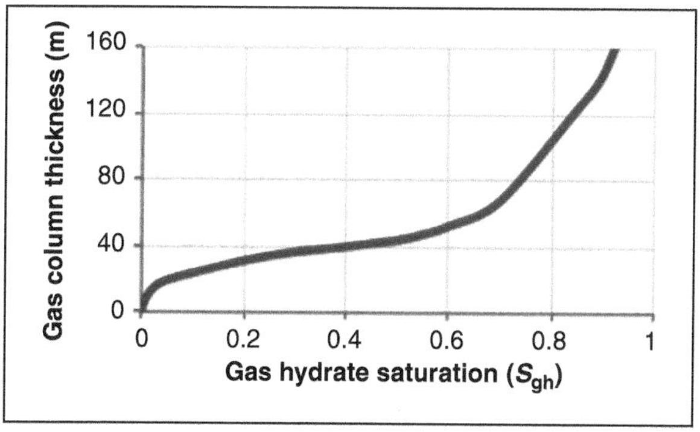

Figure 7.2 *Gas column buildup below BGHS (Cathles et al., 2010).*

extension reaches halfway to the seafloor from BGHS, there begins the final catastrophic event of fluidized water/mud expulsion. A violent expulsion of fluidized mass is thus believed to be the origin of the crater above the Blake Ridge Diapir (Liu and Flemings, 2007; Cathles et al., 2010).

In Figure 7.2 is illustrated the maximum gas column below the BGHS that can be contained by the interfacial barrier of fine sediments before the capillary seal is broken. The theoretical gas column height of Figure 7.2 was derived by Cathles et al. (2010). As gas hydrate saturation increases above the interface, permeability decreases. Consequently, a thicker underlying gas column with its attendant greater buoyant force is required for penetration. Observations of gas column heights reported at three Blake Ridge sites agree with the theoretical curve (Flemings et al., 2003; Liu and Flemings, 2007; Su and Chen, 2007).

7.2.3 Origin of Hydrate Gases at Blake Ridge

Biogenic methane occludes in Blake Ridge hydrates. Carbon isotope analyses give −65.07‰ for $\delta^{13}C$; C_1/C_2 ratios are greater than 1000; these values confirm biogenic origin (Paull et al., 1995). It is noteworthy that biogenic generation of the methane must derive from measured 0.5–1.5% organic carbon contents of sediments (Ghosh et al., 2010).

Gas compositions are similar at Sites 994, 995, and 997, where methane is dominant and carbon dioxide is secondary (Paull et al., 2000a). At Sites 995 and 997, gases released from pressurized cores contained more than 98.5% methane with carbon dioxide the most plentiful residual component (Dickens et al., 1997).

7.2.4 Sulfate Zone and Trends

The sulfate–methane interface (SMI) typically ranges down to limiting depths of 12–35 mbsf on the Carolina Rise at drill sites 994, 995, and 997 (Paull et al., 1995, 2000b; Borowski et al., 1996) Thirty-eight cores collected from Blake Ridge and Carolina Rise by Borowski et al. exhibited linear sulfate depletion with distance below seafloor. The Borowski cores were approximately 12 m long and, although not reaching sulfate zone bottom, extrapolation of the linear concentration gradients allowed the SMI to be identified. These extrapolations place SMI at depths of 10–155 m – the 10 m shallowest depths being from the core of steepest sulfate concentration gradient and the 155 m being from background areas (Borowski et al., 1996).

Where competitive reactions are insignificant, linearity of the sulfate concentration gradient can be seen in the stoichiometry of Equation 7.1. For the anaerobic oxidation of methane (AOM), as in Equation 7.1, the downward sulfate flux at the SMI must equal the upward methane flux at steady state:

$$SO_4^{2-} + CH_4 \rightarrow HS^- + HCO_3^- + H_2O \tag{7.1}$$

Sulfate concentration gradient disparities might indicate nonuniform methane fluxes, where the nonuniformity might be due to two reasons. First, hydrates might develop between BGHS and SMI, acting as capacitors to modulate methane flux. Second, sediment permeability heterogeneities might redistribute methane flux (Borowski et al., 1996).

7.2.5 Gas Hydrate Saturations

Sound wave velocity V_p is affected by whether hydrates exist freely within pores as discrete particles without adhering to confining sand particle surfaces. Correctly assuming that Blake Ridge hydrates do not adhere to sand particle surfaces but primarily disperse in sediment pores, Ghosh et al. (2010) estimated hydrate saturations. They found S_{gh} within the 250 m thickness of hydrate-bearing sediments of the three wells at Sites 994, 995, and 997 to be 10–14% as an average (Ghosh et al., 2010). These findings were corroborated with more details in independent studies determining that hydrates occupy 6.2–8.4% of the sediment pore spaces at Sites 994 and 995 and reach 13.6–25% at Site 997 (Collett, 2002; Matsumoto, 2000). Additionally, Paull et al. (2000a) found 16% hydrate saturations in the interval 230–240 mbsf, decreasing in the deeper intermediate interval, and then maximizing 450 mbsf at 25%.

7.3 CASCADIA MARGIN SUBDUCTION ZONE

Hydrate Ridge, 80 km offshore Oregon, is exemplary of other subduction zones associated with gas hydrates worldwide. See Figure 7.3. Significant methane gas fluxes, having no accompanying oil seeps, perpetuate gas hydrate accumulations along an oblong ridge that peaks about 1.6 km above adjacent seafloor (Bangs et al., 2011; Daigle et al., 2011; Sommers et al., 2007).

Hydrate studies focus on an offshore area extending from Vancouver Island to the northern California coast where at an approximate rate of 40 mm/year the North American Plate overrides the Juan de Fuca Plate (Larrasoaña et al., 2007). Along the line of override, primarily at the 25 km long/15 km wide Hydrate Ridge, a series of gas hydrate events and discoveries have been made. Here, Ocean Drilling Programs supply drilling, coring, and logging data from at least 20 sites.

Methane plumes and flares surge into water columns at the southern as well as the northern perimeter of Hydrate Ridge (Tryon et al., 2002). Extensive carbonate deposits indicate by their patterns long-standing

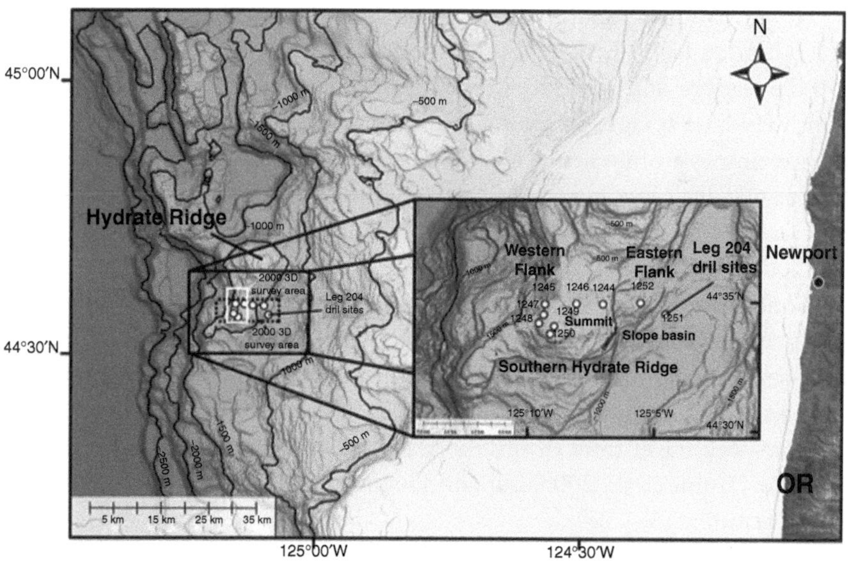

Figure 7.3 ***Location of gas hydrates at North Hydrate Ridge and South Hydrate Ridge (Bangs et al., 2011).*** *(Drawn by Marco D'Emidio, Graphic Information Systems Research Analyst.)*

methane flux, probably cyclic in magnitude, and location over large areas of the ridge.

7.3.1 South Hydrate Ridge

On South Hydrate Ridge, bottom water temperature does not exceed 4.25°C. To put hydrate stability here into perspective, temperatures could rise to 9.8°C but still maintain hydrate equilibrium under the prevailing hydrostatic head of an 800 m water column. That is, a subcooling ΔT of 5–6°C exists to form and stabilize gas hydrates (Dickens and Quinby-Hunt, 1994).

7.3.1.1 Sites 1244–1247, 1251, 1252 on Flank Slopes

Of the nine wells drilled at South Hydrate Ridge during the ODP 204 Program (Daigle et al., 2011), most were drilled on the Ridge flanks. See Figure 7.3. But two sites, Site 1249 and Site 1250, were drilled close to the Summit and the Pinnacle – the two promontories at the Ridge crest. Significantly, the two crest wells are near focused methane flows. SMI differences between crest and flanks reflect methane flux variation of the two locations: SMIs on the flanks register 3–11.5 mbsf, whereas SMIs on the crest reach only a few centimeters amid high methane flows (Larrasoaña et al., 2007).

7.3.1.2 Sites 1249, 1250 on Summit and Pinnacle

On the seafloor and down to 40 mbsf near the Summit and the Pinnacle of South Hydrate Ridge, gas hydrates proliferate and occupy 20–30% of sediment volumes around active gas venting (Tréhu et al., 2003).

Seismic, well logging, and core data from Sites 1249/1250 show interrelationships of hydrates, gas vents, authigenic carbonate deposits, chemosynthetic communities, and gas hydrate outcrops.

Southwest of the Summit, a pinnacle rises 50 m above the seafloor where extensive carbonate accumulations, including 11 wt% $CaCO_3$ in the near-surface sediments, indicate long-term venting over perhaps thousands of years (Bangs et al., 2011; Johnson et al., 2003).

Two-way travel time of seismic waves taken at different sites near the Summit (Tréhu et al., 2003; Liu and Flemings, 2006) indicates some interesting features:

1. Fifty-meter-high authigenic carbonate chimney (Pinnacle) protrudes from the seafloor.
2. Maximum height of seafloor protrusions (Summit) occurs 350 m northeast of the Pinnacle.

3. Strong BSR runs parallel to the seafloor approximately 115–130 mbsf (Tréhu et al., 1999).
4. Wipeout zone and gas chimney rise above a pierced BSR below the Pinnacle.
5. Horizon A, a conduit of high permeability sand, crosses BSR and supplies gas flow.

Liu and Flemings (2006) propose a mechanism consistent with the preceding five features, especially detailing the gas–water–hydrate status as flow proceeds through the hydrate zone until a fraction of the gases vents into the ocean. Their model indicates high gas flows relative to low water flows. The mechanism is explained with the help of Figure 7.4.

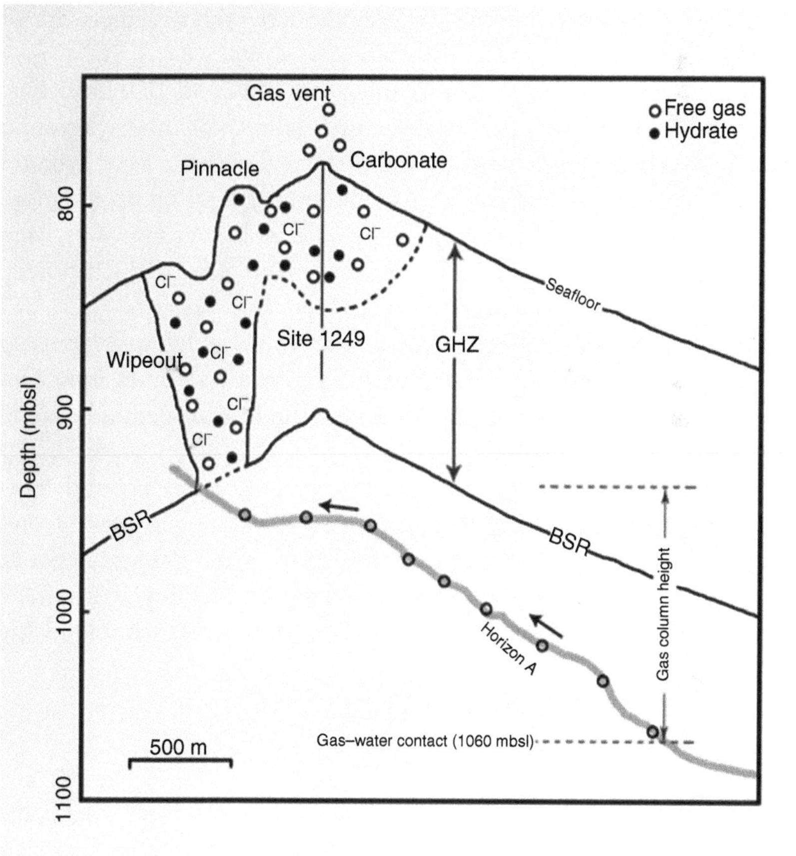

Figure 7.4 ***Model describing gas flow and hydrate formation on South Hydrate Ridge (Liu and Flemings, 2006).***

A porous and permeable fluid channel and reservoir, Horizon A, depicted in Figure 7.4 is composed of 2–4 m thick sand and ash media constrained by relatively impermeable clayey sediments. Horizon A stretches through sediments 150 m below the Summit, trending parallel to the BSR. Directly below the Summit, the channel resides 15–20 m below BSR (Tréhu et al., 2004b; Bangs et al., 2011).

Hydrates begin to form at the intersection of Horizon A with BGHS, which increases flow resistance. As flow becomes blocked, gas column and pressures build until exceeding lithostatic restraints, thus fracturing strata and forming new flow paths to the seafloor. The resulting fractures in fine-grained clay sediments are estimated to be a few centimeters in width (Holland et al., 2008). Simulations by Daigle et al. (2011) find that when gas pressure buildup exceeds vertical effective stresses, gas proceeds from Horizon A to terminal venting at the seafloor through a dynamic network of generated fractures.

Where gas-conducting Horizon A intersects BSR and BGHS in Figure 7.4, salinity of the associated water equals salinity of the background seawater. But salinity increases along the upward flow path: new hydrates exclude salts from their water uptake. As salinity progresses updip, the more saline waters perform as thermodynamic hydrate inhibitors. However, keep in mind that any nearby microbial activities producing bioproducts potentially increase hydrate formation rates (Rogers et al., 2011).

In sediments of the upper hydrate zone, the model of Liu and Flemings (2006) predicts 70% gas hydrate saturations. Verification comes from Site 1249 pressurized cores showing 45% hydrate saturations at depths of 8 and 14 mbsf (Tréhu et al., 2004a).

Methane flux from the seafloor into the water column at South Hydrate Ridge has been measured to be as high as 1000 mol/m^2/day (Torres et al., 2002; Tryon et al., 2002). The permeability of the fracture network must accommodate the high gas flux; fracture permeability is given by Equation 7.2 when no hydrates are present to form resistance within the aperture (Snow, 1968):

$$k = \frac{12a^3}{l} \tag{7.2}$$

where k, permeability; a, fracture aperture (m; 1 mm in subject case); l, interfracture spacing (m; 1 m in subject case).

According to the simulation of Daigle et al. (2011), hydrates forming in fractures reduce permeability dependent on extent of saturation $(1 - S_{gh})^3$.

Liu and Flemings (2006) further postulate that methane and methane-saturated water once exited the Pinnacle, where AOM produced massive carbonate deposits. Permeability progressively declined in the Pinnacle conduits, finally diverting gas flows laterally to vent at the Summit. Fewer carbonate deposits around the Summit vents indicate younger flow paths (Heeschen et al., 2002; Tréhu et al., 1999).

Five interesting observations regarding hydrate formation along BGHS-to-Summit paths come from computer models (Liu and Flemings, 2006; Daigle et al., 2011):

1. Three-phase equilibrium exists not only at BGHS but also along gas flow paths between BSR and Summit vents.
2. Gas–water–hydrate equilibrium exists along flow paths. As hydrates form, water salinity elevates to create more severe equilibrium conditions.
3. Because three-phase equilibrium develops along the entire flow path, hydrates are sensitive to decomposition from bottom water temperature fluctuations as low as 4–8°C.
4. Nonuniform hydrate formation from salinity variations represents an error source for generally estimating offshore hydrate contents.
5. In the Cascadia margin subduction zone, cyclic expulsion of seafloor hydrocarbon gases may last for decades.

7.3.1.3 Episodic Venting at South Hydrate Ridge Crest

Comparisons of 3D seismic surveys over an 8-year period show amplitude increases, signifying interim gas flow rate increases throughout the 100 m updip length of Horizon A below the Summit. The data capture an episodic nature of vent gas flow in hydrate zones (Bangs et al., 2011). Episodic venting was qualitatively verified by visual observations of gas plumes deteriorating over the 8 years to a more subdued venting. Despite the lack of active venting, analyses show that gas still migrated upward in Horizon A. For that time, Daigle et al. (2011) estimate a linear velocity of methane gas in Horizon A of 12.5 m/year with the bulk of gas accumulating below BGHS. Hornbach et al. (2004) suggest that a 50 m gas column height in Horizon A would provide sufficient pressure for BGHS breakthrough to reestablish gas flux toward the Summit, although through new flow paths.

These observations of episodic gas venting at South Hydrate Ridge closely follow theory of gas chimneys forming in fine sediments (Cathles et al., 2010). In fact, the Cathles rule of thumb predicts that fracturing must reach about half the distance from BGHS to seafloor before initiating

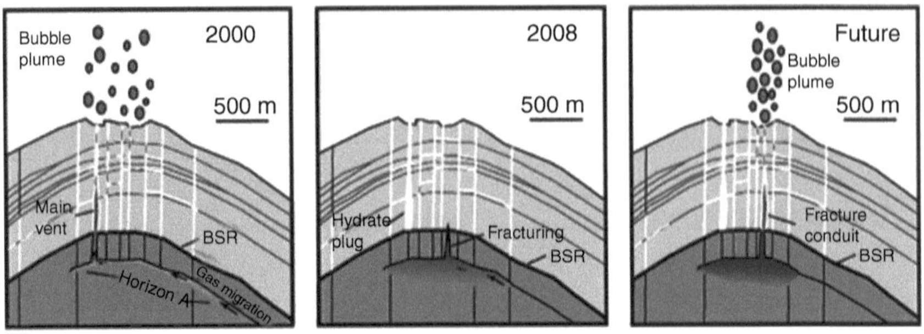

Figure 7.5 ***Episodic venting at South Hydrate Ridge (Bangs et al., 2011).***

pronounced venting, which is indicated to be the case represented by Figure 7.5 (Bangs et al., 2011).

7.3.2 North Hydrate Ridge

The crest of North Hydrate Ridge lies beneath 600 m of water, where bottom water temperature is about 4.25°C. At prevailing hydrostatic pressure, hydrate stability is sustained to a maximum temperature of 7.0°C, establishing a subcooling ΔT of 2.75°C on the crest.

There, anaerobic oxidation of venting methane over long time periods created thick layers of fractured carbonates (Tryon et al., 2002). The elaborateness of the carbonates suggests both focused and long-term methane flows – possibly more intense than at South Hydrate Ridge (Johnson et al., 2003).

Three mud volcanoes lie on the northeastern slope near the Ridge's crest (Johnson et al., 2003). Large, periodic gas expulsions causing gas flares of hydrate-encased bubbles of methane have been documented by submersible, towed camera, and acoustic image profiles.

Distinct BSRs occur throughout the area at depths coincident with calculated BGHS. During ODP Leg 146 at Sites 888–892, cores were extracted from depths below defined BGHS. The first hydrates retrieved from the Cascadia margin came from Site 892 – a high ridge where seismic data delineate a strong BSR. Data and references are presented in Table 7.2.

A distinctive 500 m/s differential in sound wave velocities defines in Table 7.2 the Site 892 BSR interface at 73.9 mbsf (MacKay et al., 1994). Note that the base of the sulfate zone is 20 mbsf. Hydrogen sulfide concentrates at 6000–10,000 ppm in the top hydrate interval within the sulfate zone, but only less than 30 ppm H_2S is in the interval below the sulfate zone (Kastner et al., 1998). The geochemistry of Site 892 is summarized in Table 7.3.

Table 7.2 Site 892 – Leg 146, summary of characteristics

Location	Latitude 44°40.440′N Longitude 125°7.140′E	Cragg et al. (1996)
Water depth (m)	670	MacKay et al. (1994)
BSR depth below sea floor (m)	73.9	MacKay et al. (1994), Housen and Musgrave (1996), Kastner et al. (1998)
Seismic velocities above BSR (m/s)	1750	MacKay et al. (1994)
Seismic velocities below BSR (m/s)	1250	MacKay et al. (1994)
Hydrate intervals (mbsf)	First: 2–19; second: 68 to BSR	Housen and Musgrave (1996)
Geothermal gradient (°C/km)	51	Suess et al. (1999)
Temperature, bottom water (°C)	4.3	Kastner et al. (1998)
Base of sulfate zone	20 (mbsf)	Kastner et al. (1998)
Focused flow, governed by fractured sediments		MacKay et al. (1994)

Table 7.3 Geochemistry of ODP Site 892

H_2O content, hydrate interval	Low Cl^-, 15% dilution	Housen and Musgrave (1996), Kastner et al. (1998)
Occluded gas composition	≥99% methane, 6000–10,000 ppm H_2S top hydrate interval	Kastner et al. (1998)
	≥99% CH_4, <30 ppm H_2S bottom hydrate interval	Kastner et al. (1998)
Sediment composition, hydrate interval	Fine-grained clayey silt High CH_4 and H_2	Clennell et al. (1999); Kastner et al. (1998)
Hydrate interval TOC (wt%)	1.5 1.22	Kastner et al. (1998), Cragg (1996)

The 1.22% mean carbon content in the sediments of Site 892 exceeds that of 0.56% for Site 888, 0.61% for Site 891, and 0.64% for Sites 889/890 (Cragg et al., 1996).

7.3.3 Offshore Vancouver Island, Sites 889/890

Sites 889 and 890 of ODP Leg 146 are located 11 km SSW of Vancouver Island (Cragg et al., 1996). Conditions of the two sites are summarized in

Table 7.4 ODP Sites 889/890

Location	Latitude 48°39.750′N Longitude 126°52.890′E Latitude 48°41.958′N Longitude 126°52.098′E	Cragg et al. (1996)
Water depth (m)	1322	MacKay et al. (1994)
BSR depth (mbsf)	225	MacKay et al. (1994), Housen and Musgrave (1996), Kastner et al. (1998)
Sound wave velocities at BSR		
Above BSR (m/s)	1750	MacKay et al. (1994)
Below BSR (m/s)	1250	
Hydrate-bearing intervals		
First interval (mbsf)	2–19	Housen and Musgrave (1996)
Second interval (mbsf)	68 to BSR	
Geothermal gradient (°C/km)	51	Suess et al. (1999)
Temperature, bottom water (°C)	4.3	Kastner et al. (1998)
Base of sulfate zone (mbsf)	20	Kastner et al. (1998)
Hydrological mode	Slow, dispersed, diffused flow, not fault controlled	MacKay et al. (1994)

Table 7.4, and are listed as a unit because the two sites are in close proximity – 4 km apart – and have similar characteristics.

A strong BSR at 225 mbsf defines BGHS at Sites 889/890 (MacKay et al., 1994; Housen and Musgrave, 1996; Kastner et al., 1998). That is, seismic velocities decrease rapidly beginning 100 m above BSR to the gas-saturated sediments below (Cragg et al., 1996). Perhaps during the last glacier period, an older age BSR existed 70 m below current levels, based on three indicators presented by Housen and Musgrave (1996):

1. Seismic velocities indicate 1–5% free gas in sediments directly below the BSR, possibly a remnant gas from hydrate decomposition (MacKay et al., 1994). MacKay et al. theorize that gas relative permeability, being 3 orders of magnitude less than water relative permeability, restricts gas dissipation.
2. Sediment cores show low chloride content extending below BSR.
3. Magnetic iron sulfides exist through the region of low chloride content.

During nonpressurized core extractions above BGHS at Sites 889/890, core hydrates dissociated, as indicated by depressed temperatures in cores retrieved throughout the 215–224 mbsf interval (Housen and Musgrave, 1996). Clennell et al. (1999) describe the sediments in these retrieved cores as being composed of small-particle clayey silts, assuming soupy textures as they absorb waters from decomposed hydrates.

Compared with five other Pacific Ocean locations having no hydrates, microbial populations at Sites 889/890 increase about 10-fold between 214 and 225 mbsf (Cragg et al., 1996). Bacterial populations in control Site 888, which had no hydrates, were at levels of other Pacific locations beyond gas hydrate occurrences.

7.3.4 Densities of Hydrates, Hydrate Ridge

Using *in situ* hydrates taken from near the surface of Hydrate Ridge, Suess et al. (2002) studied how macropores and micropores affect hydrate bulk density. The near-surface sediment–hydrate conglomerates of Hydrate Ridge, considered to have formed as methane bubbled through the sediment, exhibit a mean bulk density of 0.53 ± 0.18 g/cm^3. However, the range varies from 0.2 to 0.9 g/cm^3.

Both micropores and macropores lower bulk densities of hydrates. Importantly, the low bulk densities increase buoyancy and capability of entraining more sediment when dislodged from the seafloor. Moreover, lower hydrate bulk densities affect compression seismic velocities V_p, impacting gas hydrate estimates.

A network of micropores and macropores exists throughout 79 of the retrieved gas hydrate samples from the Summit, South Hydrate Ridge. By imposing 160 bar pressures on samples, void volumes decreased but insufficiently to attain the 0.91 g/cm^3 theoretical density of methane hydrates. A bulk density of 0.79 ± 0.03 g/cm^3 was reached by Suess et al. by extrapolating compression data to zero cumulative macropore and micropore volumes, yet networks of nanopores less than 1 μm remained to prevent reaching the theoretical value (Suess et al., 2002).

The experiments of Suess et al. showed offshore hydrate densities to be generally less than those theoretically predicted: voids develop during crystal agglomeration, stress fractures permeate the hydrate mass, and crystals adhere imperfectly to neighboring crystals – formation leaves voids/capillaries. Insufficient water or free gas may leave the other component trapped within the sediment matrix (Suess et al., 2002).

An obvious consequence of the lower density is an increased buoyance and an increased capability of entraining sediments. Lower specific gravities of clayey sediments allow increased mineral entrainment with hydrates while retaining overall mass buoyancy. As an example, cores extracted at South Hydrate Ridge exhibited a low 1.6 g/cm^3 sediment density (Riedel et al., 2006).

7.3.5 Magnetic Iron Sulfides Corroborate Double BSRs, Hydrate Ridge

On analyzing Cascadia margin cores, Housen and Musgrave (1996) found unique magnetic properties in intervals where hydrates exist. Abrupt discontinuities of the magnetic characteristics develop above and below the hydrate intervals. Sites 888 and 891, which have no hydrates, exhibit none of the disparate magnetic properties. The data by Housen and Musgrave infer that the properties originate from magnetic iron sulfides associated with bacterial activity at the time of hydrate formation in the interval.

At Sites 889/890, the observed magnetic property is attributed to the presence of small 0.1–1 μm diameter particles of greigite (Fe_3S_4) that have replaced magnetite (Fe_3O_4 composed of $Fe^{2+}Fe_2{}^{3+}O_4$) in the hydrate intervals. Pyrrhotite (Fe_7S_8) may also exist in the gas hydrate intervals. Magnetite occurs above and below the hydrate intervals.

Consider in more detail intervals where hydrates are defined, such as Site 889. Seismic profiles establish 225 mbsf as current BSR depth, but core analyses (Kastner et al., 1995) show low water chlorinity below that depth – suggestive of prior gas hydrate dissociation. Within the interval of low chlorinity, above and below current BSR, most of the magnetite has been replaced by iron sulfides (Musgrave et al., 2006).

Seismic profiles reported by Musgrave et al. (2006) define double BSRs at South Hydrate Ridge along the western flank (substantiated by greigite fines), center ridge, and crest – but not clearly along its eastern flank, where faults apparently dissipated free gas to negate a second seismic BSR. However, magnetic sulfides of fine-grained greigite persist along the eastern flank, as well as western flank, to clearly substantiate a second ancient hydrate lower depth.

7.3.6 Microorganisms Associated with Cascadia Gas Hydrates

At Sites 888 and 891 where gas hydrates do not occur, microbial populations are at the low levels typically found as background in other non–hydrate-bearing zones of the Pacific region. But wherever hydrates reside in Cascadia, such as Sites 889/890/892, microbial populations proliferate to levels as much as 10 times greater (Cragg et al., 1996).

Understandably, an exception develops in the presence of concentrated H_2S, a material toxic to most species except sulfate-reducing bacteria. For example, toxicity of H_2S decreases overall bacterial populations in the upper sulfate reduction zone at Site 892, but below the sulfate reduction zone microbial populations increase greatly (Cragg et al., 1996). Negligible sulfates remain in the lower interval for reduction to H_2S (Suess et al., 1999).

Unlike the sister sites off Vancouver, Site 892 exhibits sediment fractures through which gases permeate. Where the gases vent at seafloor, localized colonies of tube worms and clams concentrate (Cragg et al., 1996).

Near the Summit of South Hydrate Ridge, *Beggiatoa* mats top hydrate mounds created by episodic gas venting. Near the mat periphery 10 m away, clam colonies abound (Tryon et al., 2002). At the crest of the Ridge, organic carbon ranges between 0.9 and 1.6 wt% in the sediments (Larrasoaña et al., 2007).

Based on methane samples from the area, Suess et al. (1999) determined carbon isotopes to have $\delta^{13}C$ values ranging from −71.5 to −62.4‰, as well as high $C_1/(C_2 + C_3)$ ratios greater than 60,000. These analyses confirm methane to be decidedly biogenic, which is graphically verified in the Bernard plot of Figure 7.6.

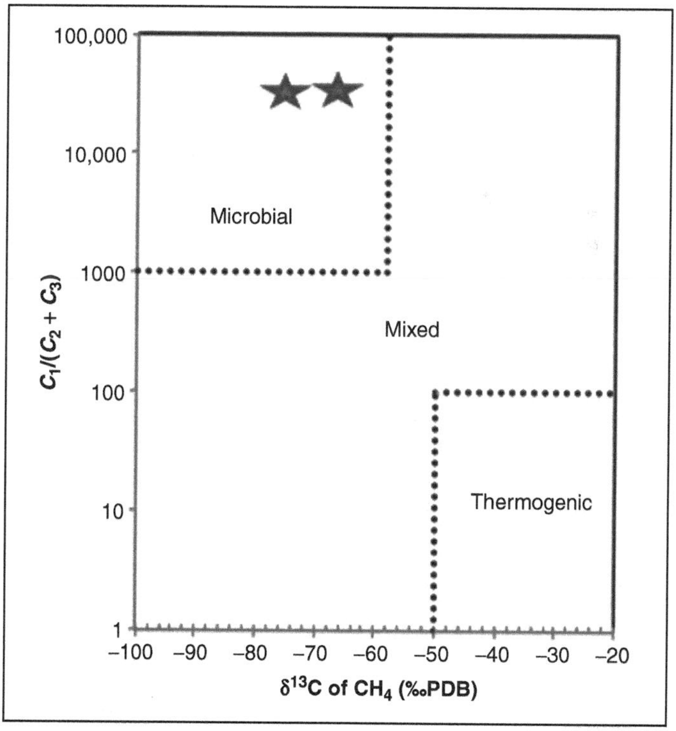

Figure 7.6 ***Biogenic nature of Hydrate Ridge methane (Suess et al., 1999).***

7.4 NANKAI TROUGH

7.4.1 Characteristics

Exceptionally high hydrate saturations in sand reservoirs highlight gas hydrate sites in the eastern Nankai Trough, presenting similarities with Mount Elbert and Mallik reservoirs. Nankai hydrates concentrate in coarse-grained sands sandwiched between layers of finer sediments, where the bounding fine grains contain significantly fewer hydrates and much lower permeabilities (Santamarina and Jang, 2009; Kurihara et al., 2011a,b).

Not only at eastern Nankai Trough but also at multiple sites near Japan, exploratory well drilling and seismic data have identified gas hydrates (Satoh, 2002). Common to the region, BSRs range from 200 to 600 mbsf (Nouzé et al., 2002). Of all Japan's gas hydrate sites identified so far, the most promising for hydrate-gas production is Nankai, where sediments display the most prominent BSRs, water depths are less than 800 m, and sediment porosities are high.

The most important information from Nankai cores is that hydrates saturate 80–90% of the pore space of some sediments (Satoh, 2002). Other selected parameters are summarized in Table 7.5.

A prominent BSR1 runs continuously for 10 km from a sediment depth of 200 mbsf to terminate 40 mbsf near the top of Daiichi-Tenryu Knoll, eastern Nankai Margin, where outgassing to surface begins. A separate BSR (BSR2) runs almost parallel at 50–110 m below BSR1. By assuming a 35°C/km geothermal gradient, calculations with known pressures and gas compositions predict a bottom of the hydrate zone to coincide with the BSR (Foucher et al., 2002).

7.4.2 Gas Source

The approximately 0.5% concentration of organic matter in Nankai sediments has a terrestrial origin (Waseda and Uchida, 2002). Because 0.5% is

Table 7.5 Nankai Trough parameters influencing gas hydrate formation

Water depth (m)	Bottom water temperature (°C)	Thermal gradient (°C/km)	Type of hydrates
<800–1000 m MITI well 945 m	4°C (Colwell et al., 2002)	40°C/km (Colwell et al., 2002) 35°C/km (Foucher et al., 2002)	Biogenic (Colwell et al., 2002; Waseda and Uchida, 2002) Type I

Table 7.6 Timeline of Nankai Trough hydrate development

Date	Activity	Location	References
1999	6 exploration wells drilled. LWD and wireline logging data, cores	Offshore Tokai	Takahashi et al. (2001), Colwell et al. (2002)
2001	2D seismic survey	Eastern Nankai Trough	
2002	3D seismic survey	Eastern Nankai Trough	
2004	32 exploration wells drilled, well log data and cores	Kumano Basin, Daini Atsumi Knoll, and offshore Tokai	Takahashi and Tsuji (2005)
2012 to present	First hydrate production wells	Eastern Nankai Trough	Kurihara et al. (2011a,b)

considered insufficient to account for generation of the large amount of biogenic gas, considerable gas may have migrated from microbial activity external to the gas hydrate zone. Hydrate-gas analyses show a ratio $C_1/(C_2 + C_3) > 4000$ and a $\delta^{13}C = -71$ to $-66‰$, so biogenic gas does prevail in hydrates of the Nankai Trough (Waseda and Uchida, 2002).

7.4.3 Gas Hydrate Wells Drilled, Nankai Trough

Japanese Ministry of International Trade and Industry (MITI) sponsored the first gas hydrate well of Nankai Trough. Drilled to 945 m depth, sediment cores contained hydrates in the interval 185–290 mbsf (Colwell et al., 2002).

In Table 7.6 is a Nankai Trough timeline of important activities.

7.5 OTHER COUNTRIES DEVELOPING GAS HYDRATE RESERVES

7.5.1 Indonesia

Gas hydrates have been verified in the seafloor of deep equatorial waters between Sumatra and Borneo. Hydrates were identified from piston cores taken in the Makassar Strait below 1396–1989 m water depths. The hydrates have a mean methane content of 99.96% and a $\delta^{13}C$ methane value of $-70.6‰$, both parameters indicating decidedly biogenic gases (Sassen and Curiale, 2006).

7.5.2 South Korea: Ulleung Basin

The Ulleung Basin of the East Sea off South Korea is a bowl-shaped basin whose western boundary is the Korean Peninsula and whose northern

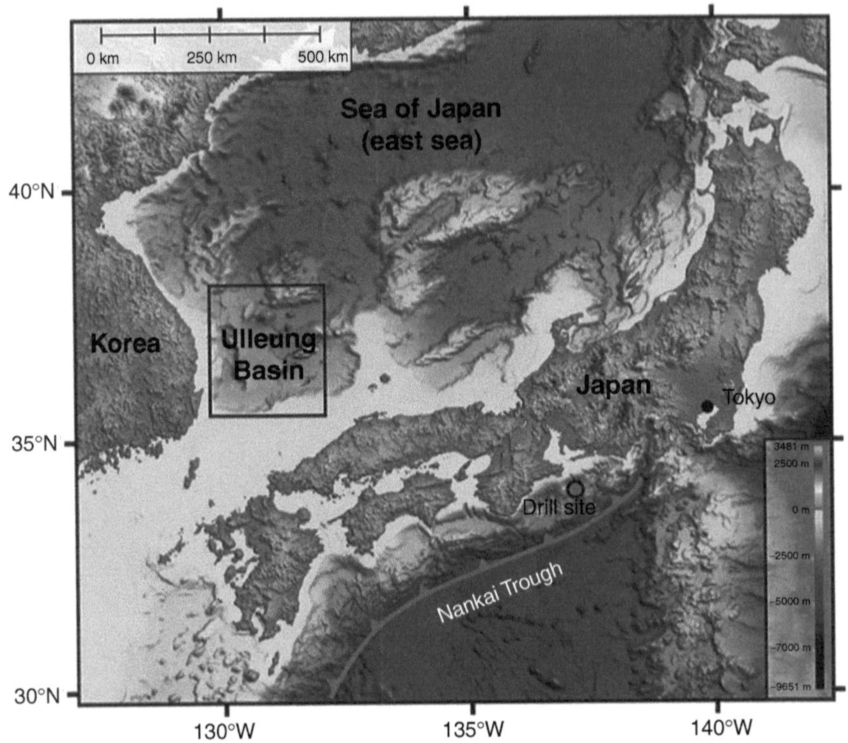

Figure 7.7 ***Map of Ulleung Basin and Nankai Trough.*** *(Graphics credit: Marco D'Emidio.)*

boundary is the Korean Plateau. Water depths are less than 1000 m to the south and greater than 2300 m to the north (Yoo et al., 2011). There is ample geophysical proof of extensive hydrates in the basin. A map is given in Figure 7.7.

Park et al. (2009) report 11–27 vol% hydrate saturations in pressurized cores taken from the Ulleung Basin. A computed tomography (CT)-scan image of the cores shows veins filled with hydrate dominating hydrate morphology. The 3D seismic data show numerous seismic chimneys, BSRs, seafloor pockmarks/mounds, and gas hydrate accumulations – all verified by cores and drilling (Kang et al., 2011). Imagery data indicate pockmarks and possible mounds where seismic chimneys, up to 2 km in width, extend to the seafloor (Yoo et al., 2011; Bahk et al., 2011).

7.5.3 Taiwan

In the southwestern region of offshore Taiwan, numerous seismic surveys show large densities of BSRs (Wang and Liu, 2009). At 325 ± 25 mbsf,

BSRs parallel seafloor contours throughout the survey area. Hydrate saturations are indicated to be 20–25% (Cheng et al., 2011). Free gas is indicated below the BSRs.

During a 4-year program, data from about 270 gravity and piston cores mapped dense accumulations of gas hydrates over an 11,000 km^2 area off southwestern Taiwan under water columns of 700–3500 m.

Some piston cores show biogenic origins of the methane, but a significant number of other analyses show mixtures of biogenic and thermogenic gases. Very few of the analyses indicate wholly thermogenic gases (Chuang et al., 2011).

7.5.4 India: Krishna–Godavari Basin

Gas hydrates occur in the Krishna–Godavari Basin on the eastern continental margin of India, where deep sediments accumulate from the confluence of the Krishna and Godavari river systems (Rao, 2001; Gupta, 2006). It is a passive margin. The hydrates occur in complex networks of fractures, including polygonal faults, within fine-grained muds noticeably depleted of sand.

Hydrates accumulate primarily in self-generated fissures by the classic theory for propagation of fractures in fine sediments dominated by small-particle clays. Near-surface stresses of the Krishna–Godavari muds allow gas bubbles to coalesce and form nodular hydrates. Near the BGHS at ≈160 mbsf, fractures tend to be vertical, i.e., at an angle greater than 60° (Riedel et al., 2010; Abegg et al., 2007).

Gas hydrate saturations in the basin average 25% (Lee and Collett, 2009).

REFERENCES

Abegg, F., Bohrmann, G., Freitag, J., Kuhs, W., 2007. Fabric of gas hydrate in sediments from Hydrate Ridge – results from ODP Leg 204 samples. Geo-Mar. Lett. 27, 269–277.

Bahk, J.-J., Kim, D.-H., Chun, J.-H., Son, B.K., Kim, J.-H., Ryu, B.-J., Torres, M., Schultheiss, P., Collett, T., Riedel, M., Lee, S.R., 2011. Gas hydrate occurrences and their relation to hosting sediment properties: results from UBGH2, East Sea. In: Proceedings of the 7th International Conference on Gas Hydrates (ICGH 2011), Edinburgh, Scotland, UK, July 17–21. Paper 567.

Bangs, N.L.B., Hornbach, M.J., Berndt, C., 2011. The mechanics of intermittent methane venting at South Hydrate Ridge inferred from 4D seismic surveying. Earth Planetary Sci. Lett. 310, 105–112.

Borowski, W.S., Paull, C.K., Ussler, III, W., 1996. Marine pore-water sulfate profiles indicate in situ methane flux from underlying gas hydrate. Geology 24 (7), 655–658.

Cathles, L.M., Su, Z., Chen, D., 2010. The physics of gas chimney and pockmark formation, with implications for assessment of seafloor hazards and gas sequestration. Mar. Pet. Geol. 27, 82–91.

Cheng, W.B., Lin, S.S., Wang, T.K., Lee, C.S., Liu, C.S., 2011. Velocity structure and gas hydrate saturation estimation of active margin off SW Taiwan inferred from seismic tomography. In: Proceedings of the 7th International Conference on Gas Hydrates (ICGH 2011), Edinburgh, Scotland, UK, July 17–21. Paper 16.

Chuang, P.C., Yang, T.F., Chen, N.C., Dale, A., Wallmann, K., Lin, S., Matsumoto, R., Hiruta, A., 2011. Carbon isotopic evidence for coupled sulfate reduction–methane oxidation and the numerical simulation results in offshore SW Taiwan. In: Proceedings of the 7th International Conference on Gas Hydrates (ICGH 2011), Edinburgh, Scotland, UK, July 17–21. Paper 611.

Clennell, M.B., Hovland, M., Booth, J.S., Henry, P., Winters, W.J., 1999. Formation of natural gas hydrates in marine sediments. Conceptual model of gas hydrate growth conditioned by host sediment properties. J. Geophys. Res. 104 (B10), 22985–23003.

Collett, T.S., 2002. Energy resource potential of natural gas hydrates. AAPG Bull. 86 (11), 1971–1992.

Colwell, F.S., Delwiche, M., Blackwelder, D., Cherry, R.S., Mikucki, J., Liu, M.Y., Boone, D.R., Uchida, T., 2002. Evidence of broad thermal tolerance of methanogens in sediments containing gas hydrates. In: Fourth International Conference on Gas Hydrates, Yokohama, pp. 19–24.

Cragg, B.A., Parkes, R.J., Fry, J.C., Weightman, A.J., Maxwell, P.A., 1996. Bacterial populations and processes in sediments containing gas hydrates (ODP Leg 146: Cascadia margin). Earth Planetary Sci. Lett. 139 (3–4), 497–507.

Daigle, H., Bangs, N., Dugan, B., 2011. Transient pressures, hydraulic fracturing, and gas migration at South Hydrate Ridge: geophysical observations and flow modeling. In: Proceedings of the 7th International Conference on Gas Hydrates (ICGH 2011), Edinburgh, Scotland, UK, July 17–21. Paper 321.

Davie, M.K., Buffett, B.A., 2002. A comparison of methane sources using numerical models for hydrate formation. In: Fourth International Conference on Gas Hydrates, Yokohama, pp. 25–30.

Dickens, G.R., Quinby-Hunt, M.S., 1994. Methane hydrate stability in seawater. Geophys. Res. Lett. 21, 2115–2118.

Dickens, D.R., Paull, C.K., Wallace, P., the ODP Leg 164 Scientific Party, 1997. Direct measurement of in situ methane quantities in a large gas-hydrate reservoir. Nature 385 (6615), 426–428.

Dillon, W.P., Paull, C.K., 1983. Marine gas hydrate – II: geophysical evidence. In: Cox, J.L. (Ed.), Natural Gas Hydrates: Properties, Occurrence and Recovery. Butterworth Publishers, Woburn, MA, pp. 73–90.

Flemings, P.B., Liu, X., Winters, W.J., 2003. Critical pressure and multiphase flow in Blake Ridge gas hydrates. Geology 31, 1057–1060.

Foucher, J.P., Nouzé, H., Henry, P., 2002. Observation and tentative interpretation of a double BSR on the Nankai slope. Mar. Geol. 187, 161–175.

Ghosh, R., Sain, K., Ojha, M., 2010. Estimating the amount of gas-hydrate using effective medium theory: a case study in the Blake Ridge. Mar. Geophys. Res. 31, 29–37.

Grevemeyer, I., Villinger, H., 2001. Gas hydrate stability and the assessment of heat flow through continental margins. Geophys. J. Int. 145 (3), 647–660.

Gupta, S.K., 2006. Basin architecture and petroleum system of Krishna Godavari basin, east coast of India. Leading Edge 25 (7), 830–837.

Heeschen, K.U., Tréhu, A.M., Collier, R.W., Suess, E., Rehder, G., 2002. Distribution and height of methane bubble plumes on the Cascadia margin characterized by acoustic imaging. Geophys. Res. Lett. 30, 1–4.

Holland, M., Schultheiss, P., Roberts, J., Druce, M., 2008. Observed gas hydrate morphologies in marine sediments. In: Proceedings of the 6th International Conference on Gas Hydrates (ICGH 2008), Vancouver, Canada.

Hornbach, M.J., Saffer, D.M., Holbrook, W.S., 2004. Critically pressured free-gas reservoirs below gas hydrate provinces. Nature 427, 142–144.

Hornbach, M.J., Bangs, N.L., Berndt, C., 2012. Detecting hydrate and fluid flow from bottom simulating reflector depth anomalies. Geology 40 (3), 227–230.

Housen, B.A., Musgrave, R.J., 1996. Rock-magnetic signature of gas hydrates in accretionary prism sediments. Earth Planetary Sci. Lett. 139, 509–519.

Hyndman, R.D., Foucher, J.P., Yamano, M., Fisher, A., 1992. Deep sea bottom-simulating-reflectors: calibration of the base of the hydrate stability field as used for heat flow estimates. Earth Planetary Sci. Lett. 109, 289–301.

Johnson, J.E., Goldfinger, C., Suess, E., 2003. Geophysical constraints on the surface distribution of authigenic carbonates across the Hydrate Ridge region, Cascadia margin. Mar. Geol. 202, 79–120.

Kang, D.-H., Yoo, D.-G., Ryu, B.-J., Yi, B.-Y., Kang, N.-K., Kim, J.-S., 2011. Relation between strong reflection amplitude anomaly and gas hydrate from 3D seismic data in the Ulleung Basin, East Sea. In: Proceedings of the 7th International Conference on Gas Hydrates (ICGH 2011), Edinburgh, Scotland, UK, July 17–21. Paper 555.

Kastner, M., Kvenvolden, K.A., Whiticar, M.J., Camerlenghi, A., Lorenson, T.D., 1995. Relation between pore fluid chemistry and gas hydrate associated with bottom-simulating reflectors at the Cascadia margin, Sites 889 and 892. In: Carson, B. (Ed.), Proceedings of the Ocean Drilling Program, Scientific results, vol. 146. Ocean Drilling Program, College Station, TX, pp. 175–187.

Kastner, M., Kvenvolden, K.A., Lorenson, T.D., 1998. Chemistry, isotopic composition, and origin of a methane–hydrogen sulfide hydrate at the Cascadia subduction zone. Earth Planetary Sci. Lett. 156, 173–183.

Kurihara, M., Ouchi, H., Sato, A., Yamamoto, K., Noguchi, S., Narita, H., Nagao, J., Masuda, Y., 2011a. Prediction of performances of methane hydrate production tests in the eastern Nankai Trough. In: Proceedings of the 7th International Conference on Gas Hydrates (ICGH 2011), Edinburgh, Scotland, UK, July 17–21. Paper 450.

Kurihara, M., Ouchi, H., Narita, H., Masuda, Y., 2011b. Gas production from methane hydrate reservoirs. In: Proceedings of the 7th International Conference on Gas Hydrates (ICGH 2011), Edinburgh, Scotland, UK, July 17–21. Paper 792.

Larrasoaña, J.C., Roberts, A.P., Musgrave, R.J., Grácia, E., Piñero, E., Vega, M., Martinez-Ruiz, F., 2007. Diagenetic formation of greigite and pyrrhotite in gas hydrate marine sedimentary systems. Earth Planetary Sci. Lett. 261, 350–366.

Lee, M.W., Collett, T.S., 2009. Gas hydrate saturations estimated from fractured reservoir at site NGHP-01, Krishna–Godavari basin, India. J. Geophys. Res. 114, BO7102, doi:10.1029/2008JB006237.

Liu, X., Flemings, P.B., 2006. Passing gas through the hydrate stability zone at southern Hydrate Ridge, offshore Oregon. Earth Planetary Sci. Lett. 241, 211–226.

Liu, X., Flemings, P.B., 2007. Dynamic multiphase flow model of hydrate formation in marine sediments. J. Geophys. Res. 112, B03101, 23 pp.

Lorenson, T.D., 2000. Microscopic character of marine sediment containing disseminated gas hydrate: examples from the Blake Ridge and the Middle America Trench. In: Holder, G.D., Bishnoi, P.R. (Eds.), Gas Hydrates: Challenges for the Future (Annals of the New York Academy of Sciences). New York Academy of Sciences, New York, NY, pp. 189–195.

MacKay, M.E., Jarrard, R.D., Westbrook, G.K., Hyndman, R.D., 1994. Origin of bottom-simulating reflectors: geophysical evidence from the Cascadia accretionary prism. Geology 22, 459–462.

Matsumoto, R., 2000. Methane hydrate estimates from the chloride and oxygen isotopic anomalies, examples from the Blake Ridge and Nankai Trough sediments. In: Holder,

G.D., Bishnoi, P.R. (Eds.), Gas Hydrates: Challenges for the Future (Annals of the New York Academy of Sciences). New York Academy of Sciences, New York, NY, pp. 39–50.
Matsumoto, R., Borowski, W.S., 2000. Gas hydrate estimates from newly determined oxygen isotopic fractionation and $\delta^{18}O$ anomalies of the interstitial waters: Leg 164, Blake Ridge. In: Paull, C.K., Matsumoto, R., Wallace, P.J., Dillon, W.P. (Eds.), 2000 Proceedings of the Ocean Drilling Program Scientific Results, vol. 164. Ocean Drilling Program, Texas A&M University, College Station, TX, pp. 59–66.
Monastersky, R., 1996. The mother lode of natural gas. Science News Online, November 9.
Musgrave, R.J., Bangs, N.L., Larrasoana, J.C., 2006. Rise of the base of the gas hydrate zone since the last glacial recorded by rock magnetism. Geology 34 (2), 117–120.
Nouzé, H., Noble, M., Foucher, J.P., Kuramoto, S., Pascal, G., Lallemant, S., the Shipboard Scientific Party of the SFJ Cruise, 2002. Seismic imaging of a double BSR area on the Nankai Slope. In: Fourth International Conference on Gas Hydrates, Yokohama, pp. 163–168.
Park, K.P., Bahk, J.J., Holland, M., Yun, T.S., Schultheiss, P.J., Santamarina, C., 2009. Improved pressure core analysis provides detailed look at Korean cores. In: Fire in the Ice. DOE-NETL Newsletter. Winter Issue, pp. 18-20.
Paull, C.K., Ussler, III, W., Borowski, W.S., Spiess, F.N., 1995. Methane-rich plumes on the Carolina continental rise: associations with gas hydrates. Geology 23 (1), 89–92.
Paull, C.K., Matsumoto, R., Wallace, P., et al., 1996. Proceedings of the Ocean Drilling Program, Initial Reports 164, College Station, TX.
Paull, C.K., Lorenson, T.D., Dickens, G., Borowski, W.S., Aussler, III, W., Kvenvolden, K., 2000a. Comparisons of in situ and core gas measurements of Leg 164 bore holes. In: Holder, G.D., Bishnoi, P.R. (Eds.), Gas Hydrates: Challenges for the Future (Annals of the New York Academy of Sciences). New York Academy of Sciences, New York, NY, pp. 23–31.
Paull, C.K., Matsumoto, R., Wallace, P.J., Black, N.R., Borowski, W.S., Collett, T.S., Damuth, J.E., Dickens, G.R., Egeberg, P.K., Goodman, K., Hesse, R.F., Hiroki, Y., Holbrook, W.S., Hoskins, H., Ladd, J., Lodolo, E., Lorenson, T.D., Musgrave, R.J., Naehr, T.H., Okada, H., Pierre, C., Ruppel, C.D., Satoh, M., Thiery, R., Watanabe, Y., Wehner, H., Winters, W.J., Wood, W.T., 2000b. Leg 164 overview. In: Proceedings of the Ocean Drilling Program, Scientific Results 164, pp. 3–10.
Rao, G.N., 2001. Sedimentation, stratigraphy, and petroleum potential of the Krishna–Godavari basin, east coast of India. AAPG Bull. 85 (9), 1623–1643.
Riedel, M., Long, P., Liu, C.S., Schultheiss, P., Collett, T., ODP Leg 204 Shipboard Scientific Party, 2006. Physical properties of near-surface sediments at southern Hydrate Ridge: results from ODP Leg 204. In: Trehú, A.M., Bohrmann, G., Torres, M.E., Colwell, F.S. (Eds.), Proceedings of ODP Scientific Results 204. Ocean Drilling Program, College Station, TX.
Riedel, M., Collett, T.S., Kumar, P., Sathe, A.V., Cook, A., 2010. Seismic imaging of a fractured gas hydrate system in the Krishna–Godavari Basin offshore India. Mar. Pet. Geol. 27, 1476–1493.
Rogers, R., Radich, J., Xiong, S., 2011. The multiple roles of microbes in the formation, dissociation and stability of seafloor gas hydrates. In: Proceedings of the 7th International Conference on Gas Hydrates, Edinburgh, Scotland, July 17–21. Paper 358.
Santamarina, J.C., Jang, J., 2009. Gas production from hydrate bearing sediments: geomechanical implications. Fire in the Ice. DOE-NETL Newsletter, vol. 9, no. 4, pp. 18-22.
Sassen, R., Curiale, J.A., 2006. Microbial methane and ethane from gas hydrate nodules of the Makassar Strait, Indonesia. Org. Geochem. 37 (8), 977–980.
Satoh, M., 2002. Distribution and resources of marine natural gas hydrates around Japan. In: Fourth International Conference on Gas Hydrates, Yokohama, pp. 175–178.

Snow, D.T., 1968. Rock fracture spacings, openings, and porosities. Journal of Soil Mechanics & Foundations Div 94, 73–91.

Sommers, S., Gutzmann, E., Pfannkuche, O., 2007. Sediments hosting gas hydrates: oases for metazoan meiofauna. Mar. Ecol. Prog. Ser. 337, 27–37.

Su, Z., Chen, D., 2007. Calculation of methane hydrate solubility in marine environment and its constraints on gas hydrate occurrence. Chin. J. Geophys. 50 (5), 1518–1526.

Suess, E., Torres, M.E., Bohrmann, G., Collier, R.W., Greinert, J., Linke, P., Rehder, G., Tréhu, A., Wallman, K., Winckler, G., Zuleger, E., 1999. Gas hydrate destabilization: enhanced dewatering, benthic material turnover and large methane plumes at the Cascadia convergent margin. Earth Planetary Sci. Lett. 170 (1–2), 1–15.

Suess, E., Bohrmann, G., Rickert, D., Kuhs, W.F., Torres, M.E., Tréhu, A., Linke, P., 2002. Properties and fabric of near-surface methane hydrates at Hydrate Ridge, Cascadia margin. In: Fourth International Conference on Gas Hydrates, Yokohama, Japan.

Takahashi, H., Tsuji, Y., 2005. Multi-well exploration program in 2004 for natural hydrate in the Nankai-Trough offshore Japan. Paper OTC 17162, presented at the Offshore Technology Conference, Houston, TX.

Takahashi, H., Yonezawa, T., Takedomi, Y., 2001. Exploration for natural hydrate in Nankai-Trough wells offshore Japan. Paper OTC 13040, presented at the Offshore Technology Conference, Houston, TX, 2010.

Taylor, M.H., Dillon, W.P., Pecher, I.A., 2000. Trapping and migration of methane associated with the gas hydrate stability zone at the Blake Ridge Diapir: new insights from seismic data. Mar. Geol. 164, 79–89.

Torres, M.E., McManus, J., Hammond, D.E., de Angelis, M.A., Heeschen, K.U., Colbert, S.L., Tryon, M.D., Brown, K.M., Suess, E., 2002. Fluid and chemical fluxes in and out of sediments hosting methane hydrate deposits on Hydrate Ridge, OR, I: hydrological provinces. Earth Planetary Sci. Lett. 201, 525–540.

Tréhu, A.M., Bohrmann, G., Torres, M.E., Moore, G.F., Suess, E., 1999. Temporal and spatial evolution of a gas hydrate-bearing accretionary ridge on the Oregon continental margin. Geology 27, 939–942.

Tréhu, A.M., Long, P.E., Torres, M.E., Bohrmann, G., Rack, F.R., et al., 2004a. Three-dimensional distribution of gas hydrate beneath southern Hydrate Ridge: constraints from ODP Leg 204. Earth Planetary Sci. Lett. 222, 845–862.

Tréhu, A.M., Felemings, P.B., Bangs, N.L., Chevallier, J., Gracia, E., Johnson, J.E., Liu, C.-S., Liu, X., Riedel, M., Torres, M.E., 2004b. Feeding methane vents and gas hydrate deposits at South Hydrate Ridge. Geophys. Res. Lett. 31, L23310.

Tréhu, A.M., Bohrmann, B., Torres, M.E., Colwell, F.S. (Eds.), 2003. Drilling Gas Hydrates on Hydrate Ridge, Cascadia Continental Margin. Proceedings of ODP, Initial Reports, 204. Ocean Drilling Program, College Station, TX.

Tryon, M.D., Broawn, K.M., Torres, M.E., 2002. Fluid and chemical flux in and out of sediments hosting methane hydrate deposits on Hydrate Ridge, OR, II: hydrological processes. Earth Planetary Sci. Lett. 201, 541–557.

Van Dover, C.L., Aharon, P., Bernhard, J.M., Caylor, E., Doerries, M., Flickinger, W., Gilhooly, W., Goffredi, S.K., Knick, K.E., Macko, S.A., Rapoport, S., Raulfs, E.C., Ruppel, C., Salerno, J.L., Seitz, R.D., Sen Bupta, B.K., Shank, T., Turnipseed, M., Vrijenhoek, R., 2003. Blake Ridge methane seeps: characterization of a soft-sediment, chemosynthetically based ecosystem. Deep-Sea Res. I 50 (2), 281–300.

Wang, Y., Liu, C.-S., Taiwan Gas Hydrate Research Team, 2009. Gas hydrate investigation in Taiwan. In: Fire in the Ice, DOE-NETL Newsletter. Winter Issue, pp. 15-17.

Waseda, A., Uchida, T., 2002. Origin of methane in natural gas hydrates from the Mackenzie Delta and Nankai Trough. In: Fourth International Conference on Gas Hydrates, Yokohama, pp. 169–174.

Waseda, A., Uchida, 2003. Gas origin and migration in marine and terrestrial gas hydrate-bearing sediments. In: Proceedings of ICGH7, pp. 42–43.
Yamano, M., Uyeda, S., Aoki, Y., Shihpley, T.H., 1982. Estimates of heat flow derived from gas hydrates. Geology 10, 339–343.
Yoo, D.G., Kang, N.K., Kim, G.Y., Ryu, B.J., Lee, H.Y., Lee, S.R., Yi, B.Y., Lee, G.H., Park, S.C., 2011. Seismic characteristics and identification of gas and gas hydrates in the Ulleung Basin, East Sea off Korea. In: Proceedings of the 7th International Conference on Gas Hydrates (ICGH 2011), Edinburgh, Scotland, UK, July 17–21. Paper 335.

CHAPTER EIGHT

Microbe, Mineral Synergy, and Seafloor Hydrate Nucleation

Hydrate formation is a crystallization process involving gas, liquid, and solid phases. At the pressurized and chilled interface of selected gases with water, crystalline hydrates nucleate within a thermodynamically determined temperature–pressure envelope. Similar to classical nonhydrate crystallizations within liquid solutions of the chemical industry, sites at which gas hydrates nucleate range widely from submicroscopic foreign particles suspended in the liquid phase to ultrasonically generated microbubbles. For example, with limited nucleation sites in purified water and an ultraclean laboratory system, hydrate nucleation may be exceptionally lengthy. But with mineral and bioproduct impurities in interstitial waters of ocean sediments, hydrate nucleation may be exceptionally fast.

Unlike classical crystallization processes, gas hydrate crystallizations require appropriate guest gases saturating bulk water to mesh in a timely manner with locally structured water at nucleation sites, thus introducing a highly stochastic step. However, if nucleating surfaces assist guest gases and structured water to converge locally on a molecular scale, the stochastic nature and induction time of the process are moderated. Such hydrate nucleating assistance was first found in the laboratory to be provided by synthetic anionic surfactants (Kalogerakis et al., 1993; Zhong and Rogers, 2000).

The present chapter introduces synthetic anionic surfactant enhancement of hydrate formations in the laboratory and then extends to similar hydrate enhancement from anionic biosurfactants generated by *in situ* microbes in hydrate zones of ocean sediments. Impacts of anionic biosurfactants on seafloor hydrate formations, as well as mechanisms, are discussed. Further, an indigenous biosurfactant–mineral–hydrate synergy of the seafloor is detailed. Microbes impose direct and indirect influences on seafloor hydrates in extraordinary ways: morphology, formation catalysis, crystal nucleation, hydrate memory of sediments, cell wall inhibition of hydrate formation, adsorption of smectite platelets onto cell walls to protect from hydrate heat of formation, mechanism by which living cells are placed within porous hydrates, protection/growth of microbes within hydrate masses, and magnetic susceptibility of ocean sediments in the hydrate zone.

Offshore Gas Hydrates. http://dx.doi.org/10.1016/B978-0-12-802319-8.00008-5

Experiments are detailed determining microbial effects on hydrate formation in which pure microorganism species from American Type Culture Collection (ATCC) are cultured and compared in tests with cultures of indigenous microbes taken from Gulf of Mexico (GOM) hydrate-zone core samples.

8.1 SYNTHETIC SURFACTANTS AS LABORATORY HYDRATE NUCLEATING AGENTS

Anionic surfactants prevent a barrier film of hydrates from forming at the water–gas interface in a nonstirred laboratory system when the test cell is a metal container. Whereas in a stirred system the active surface area influencing hydrate formation is represented by liquid vortex area, the surfactant-adsorbing surface of a metal test cell represents the active surface area in a quiescent surfactant system. With quiescent systems having anionic surfactant in the water phase, developing hydrate crystals associate and migrate with surfactant nuclei to adsorb on hydrophilic test-cell walls, thus preventing water–gas barrier films and increasing hydrate formation rates. Hydrates collect and build symmetrically on the cell walls.

Vysniauskas and Bishnoi (1985) derived the kinetic Equation 8.1 as a means to calculate rate of hydrate formation as a function of pressure, temperature, degree of subcooling, and water–gas interfacial area. Not only does the equation give insight into parameters affecting kinetics of laboratory hydrate formation, but it also gives insight into seafloor hydrate formation:

$$r = Aa_s \exp\left(\frac{-\Delta E_a}{RT}\right)\exp\left(\frac{-a}{\Delta T^b}\right)P^\gamma \tag{8.1}$$

where r, rate of hydrate formation; ΔT, degree of subcooling (degrees below equilibrium); T, absolute temperature of system; P, pressure of system; γ, overall order of reaction with respect to pressure; A, lumped pre-exponential constant; a_s, surface area of gas–water interface; ΔE_a, activation energy for hydrate formation; R, universal gas constant; a and b, empirical constants.

Although developed for stirred systems, the equation of Vysniauskas and Bishnoi also approximates nonstirred hydrate formation rates in laboratory systems containing anionic surfactants (Zhong and Rogers, 2000). The rapid heat transfer achieved during hydrate formation while mechanically stirring can be equaled if not surpassed by a quiescent system employing anionic surfactants adsorbing on cold, metal test-cell walls. The rapid crystal migration to metal walls continually renews gas–water interfacial areas; conductive metal adsorbing surfaces remove latent heats of hydrate formation. Therefore, high mass-transfer rates and high heat-transfer rates in a quiescent system can generate exceptional hydrate formation rates.

When conditions other than pressure were held approximately constant in a quiescent, anionic surfactant system, the rate of hydrate formation r reduces to Equation 8.2, where C is a constant of the particular system:

$$r = CP^{\gamma} \tag{8.2}$$

In a test series using the gas mixture 90% methane, 6% ethane, and 4% propane in a nonstirred surfactant system, regression analysis gives a γ value of 2.776 ± 0.185. The R^2 value of the regression is 0.99 (Zhong and Rogers, 2000). Vysniauskas and Bishnoi (1985) obtained values of 2.804 ± 0.406 for pure ethane and 2.986 ± 0.467 for pure methane in systems stirred at 500 rpm. In the quiescent surfactant system, therefore, the pressure exponent approaches that of a stirred reactor system, suggesting that the surfactant action performs similar to optimum stirring. That is, mass transfer and heat transfer achieved with the migration of crystal nuclei to adsorb on a thermally conductive metal surface equate to mass and heat transfer rates in a stirred reaction vessel at high revolutions per minute.

Other important properties develop in the surfactant process. Consider a cylindrical, metal test cell where gas supply is replenished to maintain constant pressure during hydrate formation. Water level recedes in the test cell as anionic surfactant–hydrate crystals build on metal surfaces at the gas–water interface. Suppose all free water in the cell before hydrate initiation becomes incorporated into hydrates, including much unreacted free water within hydrate interstitial spaces. Then, as gas is replenished to maintain operating pressure, the interstitial free water continues to react because gas diffuses through the porous hydrate mass. Eventually, all interstitial water reacts, and the test cell fills with hydrates. After approximately 1 h of such laboratory tests, each volume of solid hydrate mass occludes 156 volumes of gas at standard temperature and pressure (STP).

A proof-of-concept facility was constructed from the preceding design features (Zhong and Rogers, 2000; Rogers et al., 2003).

Importantly, then, anionic surfactants catalyze hydrate formation. An effective synthetic anionic surfactant is sodium dodecyl sulfate (SDS), whose structure is given in Figure 8.1.

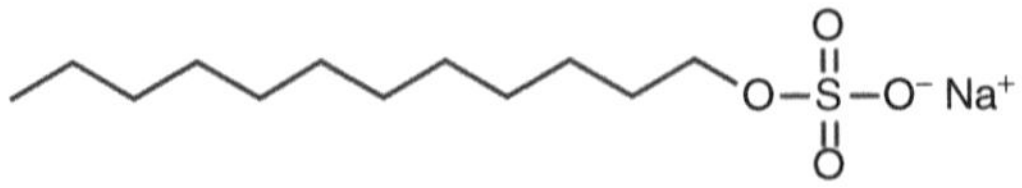

Figure 8.1 ***Sodium dodecyl sulfate structure (Calvero, 2006).***

Anionic surfactants catalyze hydrate formation by absorbing guest gas at the hydrophobic tail while accumulating structured water at the hydrophilic head. Note that water becomes structured on hydrophobic surfaces (Kristiansen et al., 2008). Accordingly, gas and structured water are directed to a common nucleating site rather than doing so randomly.

8.2 BIOSURFACTANTS AS SEAFLOOR HYDRATE NUCLEATING AGENTS

8.2.1 Anionic Biosurfactants Promote Hydrates

Synthetic anionic surfactants boost hydrate formation rates up to several orders of magnitude in the laboratory (Kalogerakis et al., 1993; Zhong and Rogers, 2000). Consider the analog of biosurfactants found naturally in the seafloor.

Microbial activities around near-surface gas hydrates and cold vents of the seafloor are prolific (Lanoil et al., 2001; Cragg et al., 1996) because carbon essential for microbial metabolism abounds there. It should not be surprising that microbes find means to promote storage of carbon essential to their metabolism by catalyzing gas hydrate formation. To do so, the microorganisms interact synergistically with sediment minerals to generate nucleating particles and catalyze hydrate formation by emitting anionic biosurfactants, ensuring long-term carbon access (Rogers et al., 2003). The role anionic biosurfactants play in the seafloor accumulation of hydrates is explained in the following text.

8.2.2 Biosurfactants in Soil Remediation

To access the carbon of water-insoluble hydrocarbons in contaminated soils, waterborne microbes secrete surfactants that solubilize the hydrocarbon. Even if the carbonaceous material is adsorbed on mineral surfaces contacting the water, biosurfactants displace and solubilize the hydrocarbons, making them accessible to the microbes – ergo, soil remediation. Independent of gas hydrate interests, researchers remediate contaminated soils with biosurfactants that are in turn biodegradable (Bai et al., 1997; Banat, 1995; Churchill et al., 1995; Herman et al., 1995). For example, rhamnolipid surface active agent from the bacterium *Pseudomonas aeruginosa* remediates soils contaminated with oil or motor fuels, usually more effectively than synthetic surfactants. The rhamnolipid readily adsorbs on solid surfaces of soil particles to displace hydrocarbons and to solubilize them (Herman et al., 1995; Banat, 1995).

8.2.3 Classifications of Biosurfactants

The building blocks of biosurfactants are based on polysaccharides, amino acids, fatty acids, proteins, and peptides. Surfactants produced by microbes fall within one of five classifications given in Table 8.1 (Kosaric, 1992; Fujii, 1998). The first column of Table 8.1 lists the general microbial species classification, the second column lists cell species chosen to represent each classification, and the third column lists selected common names of representative commercial bioproducts from each cell species. All of the biosurfactants are anionic.

Representative bioproducts were obtained commercially for each of the classifications in Table 8.1 and evaluated in the laboratory for their effect on gas hydrate formation in seawater-saturated porous media (Rogers et al., 2003). For safety or cost reasons, substitutes of related structures were used for the natural products of hydroxylated–cross-linked fatty acids as well as phospholipids.

Table 8.1 Classifications of biosurfactants

Biosurfactant classifications	Microbes	Biosurfactants (commercial) evaluated	References
Hydroxylated and cross-linked fatty acids	*Corynebacterium Lepus*	DL-α-Hydroxystearic acid*	Rosenberg (1986)
Polysaccharide–lipid complexes	1. *Pseudomonas syringae* 2. *Acinetobacter calcoaceticus*	1. Snomax 2. Emulsan	Goodnow et al. (1990), Rosenberg (1993)
Glycolipids	*Pseudomonas aeruginosa*	Rhamnolipid	Fujii (1998), Kosaric (1992)
Lipoprotein–lipopeptides	*Bacillus subtilis*	*Surfactin*	Rosenberg (1986), Kosaric (1992)
Phospholipids	1. *Thiobacillus* 2. *Corynebacterium* species	DMPC* DPPS* POPC*	Fujii (1998), Genzyme (2001)

*Synthetic equivalents.

Identified within cores extracted from near gas hydrate mounds in the northern GOM were *P. aeruginosa* and *Bacillus subtilis* species, source bacteria of rhamnolipid and surfactin biosurfactants listed in Table 8.1 (Lanoil et al., 2001). In addition to the commercial samples of rhamnolipid and surfactin, subsequent tests were made with surfactin extracted from laboratory cultures of *B. subtilis*.

Brief descriptions follow of four biosurfactants representative of classifications listed in Table 8.1. These biosurfactants were evaluated in hydrate tests.

8.2.4 Commercial Anionic Biosurfactants

8.2.4.1 Snomax

Snomax is a biosurfactant used commercially to produce artificial snow on ski slopes. Dissolved in water and sprayed into the air column above the slopes, Snomax promotes snow crystallization at a temperature of about 5 K above water's normal freezing point. The water–Snomax solution solidifies as snow before reaching ground (Amende and Gordon, 1999). The anionic Snomax is produced by the *Pseudomonas syringae* bacterium, falling within the polysaccharide–lipid complex classification (Goodnow et al., 1990).

8.2.4.2 Emulsan

Emulsan does not form micelles. The heteropolysaccharide bioemulsifier structure of emulsan is illustrated in Figure 8.2 (Castro et al., 2008).

The long and slender, high-molecular-weight emulsan can be depicted as tangled thread – an apt description of its randomly coiled appearance. But unfurled emulsan structures depicted in Figure 8.2 illustrate how the

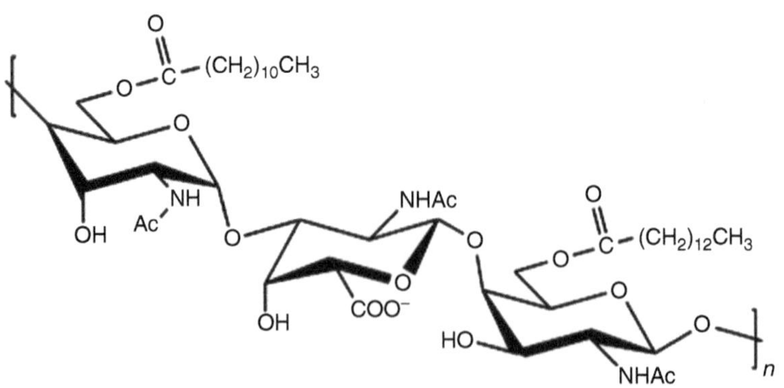

Figure 8.2 ***Emulsan structure.*** *(Drawn by Keith Mead (Castro et al., 2008).)*

many hydrophobic appendages (fatty acid branches) of this long polymeric molecule orient on oil-drop surfaces to create stable oil-in-water emulsions. The long molecule (100/1 length to width ratio) uncoils to spread on the drop surface, covering a high percentage of that surface to prevent coalescence with other oil drops. The comprehensive coverage makes emulsan one of the most effective emulsifying agents known (Gutnick, 1987; Jones-Meehan et al., 1995).

8.2.4.3 Rhamnolipid

Rhamnolipid is a glycolipid produced by *P. aeruginosa* bacterium. It is an anionic biosurfactant, where the ionization of the carboxylic acid group occurs at a pH greater than or equal to 4.0 (Nitschke et al., 2005). Rhamnolipids readily adsorb on porous media surfaces, as found by soil remediation researchers (Herman et al., 1995; Banat, 1995).

The rhamnolipids form micelles at low threshold concentrations; their critical micellar concentration (CMC) in water is at a much lower concentration than that of synthetic surfactant SDS (Thangamani and Shreve, 1994; Herman et al., 1995; Churchill et al., 1995). The micelles of rhamnolipid are as small as 5 nm in diameter (Champion et al., 1995), whereas soil remediation studies estimate that micelles of diameter less than 50 nm will migrate through porous media (Bai et al., 1997; Herman et al., 1995).

Laboratory experiments with a mixture of two rhamnolipids, R1 ($C_{26}H_{48}O_9$) and R2 ($C_{32}H_{58}O_{13}$) having 25% active biosurfactant obtained from Jeneil Corporation (Jeneil, 2001), show that rhamnolipid injected at the column bottom of a hydrate test cell migrates through a chilled, pressurized, packed sand column to catalyze gas hydrate formation at the gas–sand interface near the top of the column (Lee, 2001). This observation is pertinent to ocean sediments where microbes are near an optimum growth temperature that occurs near bottom of gas hydrate stability (BGHS). Biosurfactant produced at BGHS may be entrained with venting gas into the zone above to catalyze hydrate formation far from the source. For example, the biogenic gas stored in hydrates of the Nankai Trough probably migrated from outside the hydrate zone (Collett, 1996).

8.2.4.4 Surfactin

Surfactin is a powerful cyclic lipopeptide anionic surfactant, reducing water surface tension from 72 mN/m to as low as 27 mN/m, far surpassing the surface activity of synthetic surfactants such as sodium lauryl sulfate or SDS (Nakano et al., 1988; Arima et al., 1968; Bernheimer and Avigad, 1970).

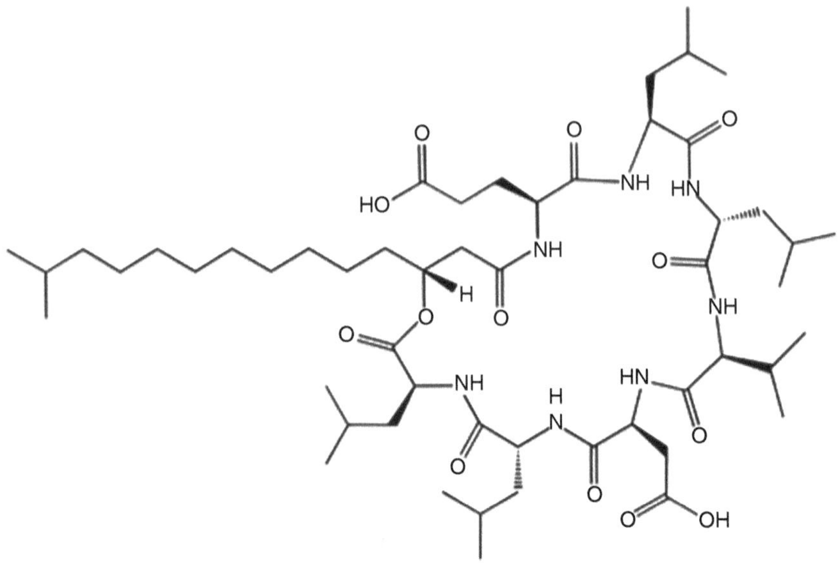

Figure 8.3 ***Surfactin structure (Louisajb, 2011).***

Surfactin is used in the health-care industry as an anti–blood-clotting agent and as a strong antibiotic.

The structure of surfactin is depicted in Figure 8.3. It has a molecular weight near 1036 g/mol. Note that seven amino acids configured in a loop comprise the hydrophilic head of this anionic biosurfactant. A 10- to 15-carbon chain of a fatty acid forms the hydrophobic component. Surfactin forms micelles in surfactin–water solutions at the low concentration of 17.2 ppm – an order of magnitude lower than the synthetic SDS (Heerklotz and Seelig, 2001).

8.2.5 Laboratory Hydrate Catalysis with Biosurfactants

As with synthetic anionic surfactants, the stochastic element of nucleation is reduced when an anionic biosurfactant assembles hydrocarbon gases and structured water around the hydrophobic and hydrophilic moieties, respectively. Water hydrating the anionic surfactant's hydrophilic head becomes structured, enhancing the crystal nucleation site and reducing hydrate induction time. Because hydrocarbon gas is scavenged from solution by the hydrophobic group to create locally high concentrations, hydrate crystallization may commence at these sites even though bulk gas concentration in the water might suggest insufficiency. Anionic biosurfactants in sediments attach to positively charged mineral surfaces. For example, excess positive

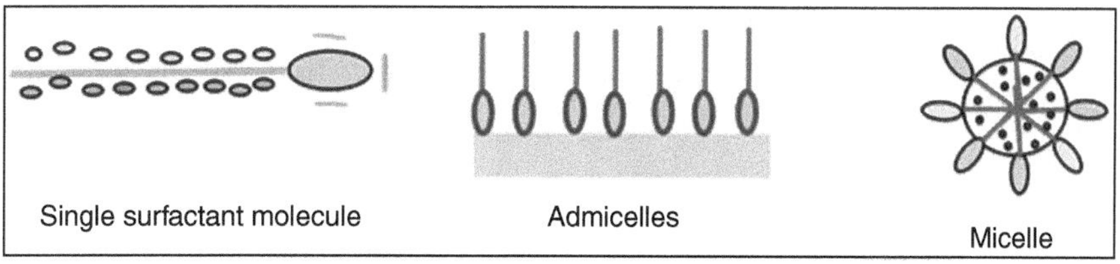

Figure 8.4 ***Three possible surfactant configurations for hydrate nucleation sites.***

charges on the ends of smectite clay platelets adsorb anionic biosurfactant, creating miniscule hydrate nucleation sites.

Three mechanisms are suggested by which anionic biosurfactants promote hydrate crystallization: (1) by an individual surfactant molecule accumulating gas and structured water as a nucleation site, (2) by surfactant molecules adsorbing on solid surfaces to accumulate gas and structured water as nucleation sites, and (3) by surfactant molecules associating in micellar form within the water phase, accumulating gas and structured water as nucleation sites. The three mechanisms are depicted in Figure 8.4.

First, individual anionic biosurfactant molecules congregate gas/structured water at functional groups without forming micelles or admicelles. This appears to be the case for the large-molecular-weight biopolymer emulsan. Periodic hydrophobic and hydrophilic groups along its polymeric saccharide backbone create multiple nucleation centers from which hydrate crystals initiate.

Second, organization of biosurfactant molecules in layers on a positive mineral surface in the form of admicelles might also initiate hydrate crystallization at that interface.

Third, biosurfactants may have very low CMCs in water – e.g., in the case of surfactin, as low as 17.2 ppm in distilled water at 1 atm ambient pressure and 5°C (Heerklotz and Seelig, 2001). Assemblage of individual biosurfactant molecules into micelles could act as hydrate nucleation sites in some cases.

What is the threshold of biosurfactant effectiveness at hydrate-forming conditions in seawater? Pressurized with natural gas (90% methane, 6% ethane, 4% propane), rhamnolipid concentrations in seawater were varied in the laboratory at constant hydrate-forming conditions. Hydrate induction times in seawater show an extremely low minimum threshold concentration for hydrate formation. For example, rhamnolipid less than 20 ppm appears sufficient (Lee, 2001; Kothapalli, 2002). The rhamnolipid used in the experiments contained 25% active ingredient (Jeneil, 2001). The relationship of induction time with rhamnolipid concentration assumes a form typically found with physical properties of surfactant solutions when its CMC is reached. That is, physical property trends such as surface tension, conductivity, and solubility show sharp breaks in value at the CMC threshold.

8.3 SEAFLOOR SYNERGY OF HYDRATES/MINERALS/BIOPRODUCTS

Anionic biosurfactants obtained from commercial sources, as presented in Table 8.1, were evaluated for gas hydrate catalysis. Laboratory tests show that these anionics improve the stochastic process of structured water/hydrocarbon gas association to more rapidly form gas hydrates. The kinetics is especially improved by smectite platelets, a mineral found throughout marine sediments, perhaps forming with biosurfactants important nucleating surfaces for hydrates in ocean sediments.

8.3.1 Bioproducts, Hydrates, Drilling Hazards

A photograph of a test sample immediately after hydrate formation is presented in Figure 8.5. Sand and smectite clay were partitioned in the perforated sample container prior to the test; a smectite base was below the

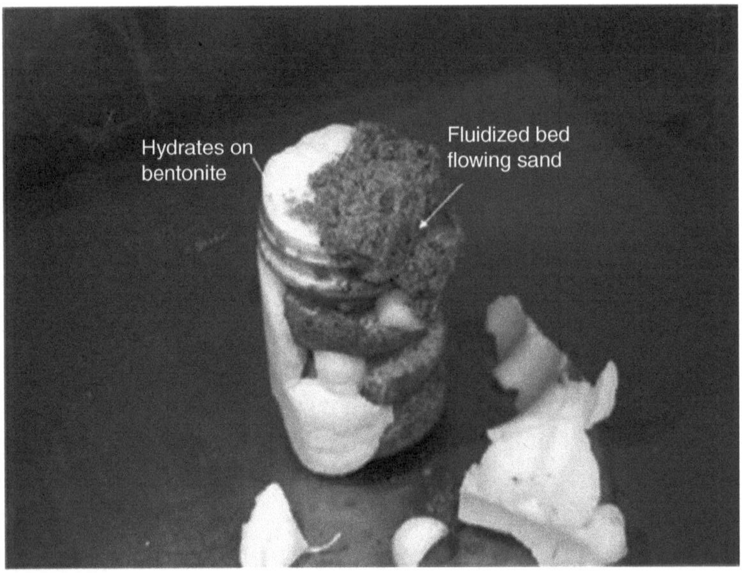

Figure 8.5 ***Rhamnolipid foam ejects sand as fluidized bed on hydrate decomposition (Rogers et al., 2007).***

sand column. Hydrates formed more densely at clay interfaces where clay platelet nucleating particles were more concentrated, distinctly partitioning sand/clay in the sample. There being individual clay platelets throughout the sand, hydrates proceeded to fill pores of the packed sand – without attaching to the sand particles. On hydrate filling of available sand pore spaces and hydrate plugging of gas access ports, nuclei diffused outwardly through hydrate capillaries to form hydrates in the unrestricted space of pressurizing gas exterior to the sample container.

When hydrates were depressurized and decomposition began, sand flowed forcefully out of the sample cup. This occurred because released gas created foam from the water–biosurfactant solution, pressurizing and ejecting unconsolidated sand as a fluidized bed. The anionic biosurfactants have foaming attributes especially stable at lower temperatures where gas hydrates occur.

One shallow drilling hazard in hydrate zones of the GOM derives from hydrate-saturated sands enclosed by impermeable clay strata and having directly below the hydrate a pressurized free gas; the presence of the underlying gas may be masked from seismic detection by the overlying hydrate (McConnell et al., 2012). The preceding laboratory test may be illustrative of this field hazard. Enclosed by impermeable strata, such hydrate/sand masses become vulnerable to heat input from drilling or completing operations that cause hydrate decomposition in the confined space to build extremely high pressures. A surging fluidized bed of sand would threaten drill string damage. Foaming bioproducts within the hydrate mass contribute to pressurization, fluidization, and forceful expulsion of the sand.

8.3.2 Emulsan Impacts Hydrate Morphology

Figure 8.6 presents a view through the transparent top of a hydrate test cell where natural gas pressurizes a chilled porous packing of sand/montmorillonite saturated with distilled water–emulsan solution.

Hydrate morphology differences appear above the two media minerals. Sand is packed on the left side of the container and the smectite montmorillonite (bentonite) on the right. Stringy, broom-like hydrates emanate from the sand; nodules form above the bentonite. In general, hydrate/mineral/bioproduct synergies create hydrate morphology differences (Rogers et al., 2004).

Other naturally occurring as well as synthetically produced polymers influence hydrate morphology. Type I antifreeze protein (AFP) from winter flounder depresses the temperature of catastrophic ice formation in the

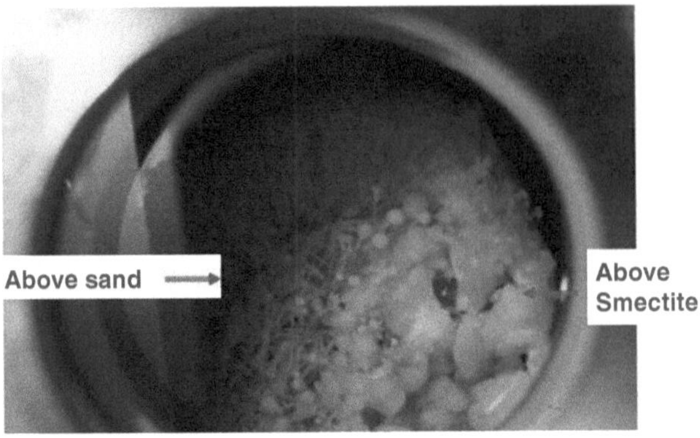

Figure 8.6 ***Hydrate morphology affected by emulsan and mineral combination (Rogers et al., 2004).***

flounder by adsorbing onto the ice crystal. AFP adsorption on a crystal face severely retards crystal growth in that direction. When applied to hydrate experiments, these flounder proteins not only retard gas hydrate formation but also change morphology of the hydrate crystal from three-dimensional octahedron to planar (Zeng et al., 2006).

The synthetic polymer polyvinylcaprolactam (PVCAP), used as a kinetic inhibitor of hydrates in subsea hydrocarbon flow lines, changes hydrate morphology by adsorbing on the developing hydrate crystal (Abay et al., 2011).

8.3.3 Emulsan–Smectite Platelets: Hydrate Nucleation Sites

Further evidence of hydrate preference for smectite surfaces when a seawater–emulsan solution saturates porous media is given in Figure 8.7. Hydrate (white in Figure 8.7) forms more densely wherever smectite concentrates in the packing. Note hydrates protruding through the gas access vents bored through the polypropylene sample container.

When a layered packing of nontronite in Ottawa sand saturated with emulsan–water solution was subjected to hydrate-forming conditions, Dearman et al. (2009) observed gas hydrates preferentially forming around the smectite. Although the gas access perforations became blocked with hydrates and available pore spaces within the packed media were filled, hydrates continued to grow on the outside of the sample container by diffusion of hydrate crystal nuclei through capillaries of hydrates blocking perforations. Nontronite's light green tint made it visually obvious that nontronite was part of the nuclei diffusing through the capillaries. A dynamic light scattering instrument with helium–neon laser, supplemented by

Figure 8.7 ***Smectite clay, emulsan wetted, preferentially promotes gas hydrate (Rogers et al., 2006).***

scanning electron microscope (SEM) photographs, was used to analyze the emulsan–nontronite particles collected from capillary diffusion.

The SEM micrograph of Figure 8.8 shows remnant particles collected from hydrates on the outside, that is, gas side of the perforated sample container. Note the periodic nodular aspects of the particles, which are thought to be smectite clay platelets affixed to periodically spaced functional groups along the unfurled emulsan polymer – each site acts as a hydrate nucleation site. Along the saccharide backbone of the long, linear polymer, hydrophobic fatty acid chains of 12–18 carbons occupy periodic sites as side chains with ester linkages (Zuckerberg et al., 1979). Smectite platelets in SEM views appear to sandwich the sites of the ester linkages, and it is postulated that these alkyl side chains extend from the smectite interlayers, attracting and concentrating methane (Dearman et al., 2009). Meanwhile, the sodium cation in the interlayer of sodium montmorillonite structures the solvated water molecules (Sposito et al., 1983). Since cells of hydrates form within interlayers of smectites (Guggenheim and van Groos, 2003), multiple nucleation sites are postulated to form along the polymer backbone where gas and structured water initiate the formation of gas hydrates.

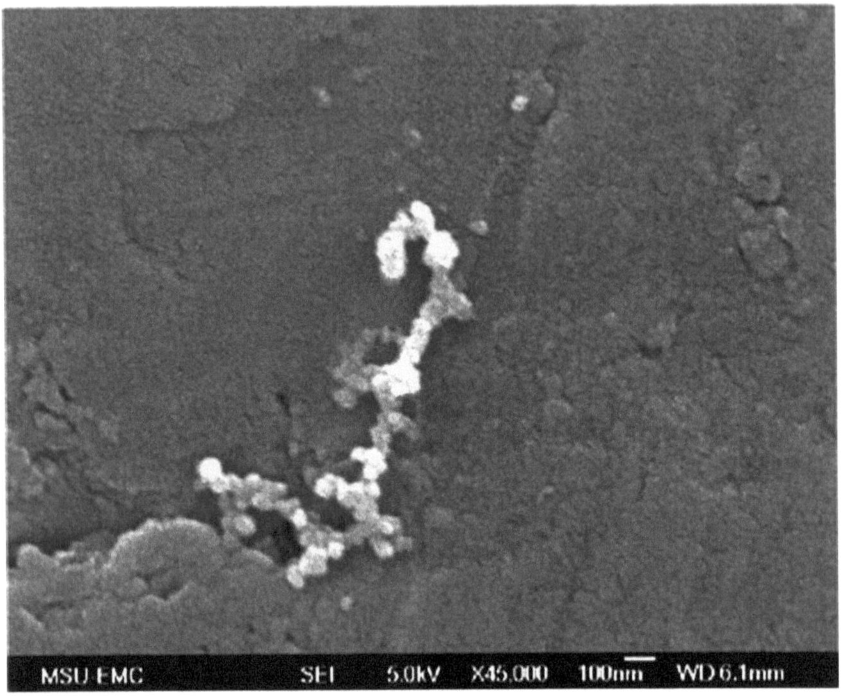

Figure 8.8 ***Association of emulsan and nontronite nuclei having diffused through hydrates to outside sample container; SEM × 45,000 (Dearman et al., 2009).***

By dynamic light scattering (xenon laser), Dearman et al. estimate that emulsan polymer linearity allows polysaccharide–lipid–smectite nuclei to diffuse through hydrate capillaries. For the diffusion to occur, the hydrate capillary diameters must exceed 450 nm, allowing the long slender polymer molecules seen in Figure 8.8 to diffuse lengthwise (Dearman et al., 2009).

The experiments stress the role of smectite in promoting hydrates whenever anionic biosurfactants are present. It is reemphasized that individual smectite platelets appear most important to the hydrate nucleation process rather than bulk clay.

In a possibly related later work, Klapp et al. (2011) discovered by use of high-resolution computed tomography (CT) scan sensitive to submicrometer scales that hollow tubes of greater than 100 μm length by 10–25 μm diameter form throughout hydrate mass early in hydrate dissociation. The Klapp tubes, whose walls are formed of ice, extend through the hydrate mass and allow dissociating gas to diffuse.

8.3.4 Seafloor Mineral Influences on Hydrate Formation

Clay minerals, such as smectites and kaolinites, are prevalent throughout the Earth's crust (Clausen et al., 2009). This is true also in acoustic wipeout zones of the seafloor where microbes are active around venting gas and gas hydrates, and the bioproducts associate with clays.

In conjunction with biosurfactants, smectite clays prove more effective than kaolinite in establishing nuclei for hydrate crystal growth (Woods, 2004; Lee, 2001; Kothapalli, 2002). Nontronite and sodium montmorillonite effectively promote hydrates in the presence of anionic biosurfactants – all occur naturally in the seafloor (Woods, 2004). The smectite clays have large outer surface areas, as well as interlayers that adsorb water or organic matter. The properties help give bentonite, a sodium montmorillonite clay, multiple industrial uses including that of gas adsorbent (Volzone and Ortiga, 2000). Interlayers greatly expand smectite total areas beyond those of kaolinite (Walker et al., 1989), a fact significant for their influence on hydrate nucleation.

Comparative kinetic effects were investigated by Woods (2004) for hydrate nucleation as a function of mineral surface and concentration of rhamnolipid in water solution. Hydrate formation rates were compared when seawater-based solutions containing 0–1000 ppm rhamnolipid saturated packed mineral media containing sodium montmorillonite, sand, or kaolin. When montmorillonite was present, hydrate formation with rhamnolipid concentrations less than 1000 ppm peaked at rates twice those reached with sand or kaolinite present. At 1000 ppm rhamnolipid concentrations in seawater, hydrate formation rates for all three minerals converge, indicating complete coverage of media particle surfaces by each biosurfactant, obviating individual mineral surface effects.

Selectivity occurs in a manner akin to the promotion of gas hydrates by synthetic anionic surfactants adsorbing on the metal walls of a laboratory reaction vessel at the water–gas interface. To investigate the phenomenon further, Kothapalli (2002) determined the relative adsorptions, as measured by solution surface tension changes, of selective anionic biosurfactants (rhamnolipid, emulsan, surfactin, Snomax) saturated with seawater solutions when flowing over packed columns of sand, bentonite, or kaolin. See Table 8.2.

Generally, biosurfactant adsorption maximized over the smectite clay packing of bentonite. Hydrate promotion, as measured by formation rate and induction time, in mineral packings saturated with seawater and the subject biosurfactant, coordinated with the degree of adsorption. This may

Table 8.2 Relative adsorption of biosurfactant from seawater solutions onto mineral surfaces (Kothapalli, 2002)

Biosurfactant	Change in surface tension (mN/m)		
	Sand	Bentonite	Kaolinite
Rhamnolipid	1.1	8.9	3.6
Emulsan	0.5	16.9	1.3
Surfactin	26.9	45.5	39.6
Snomax	9.0	22.5	5.8

Figure 8.9 ***Hydrates preferentially form on surfactin-wetted smectite.*** (a) Before hydrate formation; (b) after hydrate formation (Rogers et al., 2007).

suggest admicelles of biosurfactant collecting on mineral surfaces. Exceptions may indicate orientation anomalies (Kothapalli, 2002).

The mineral selectivity for hydrate formation can be seen in Figure 8.9 where kaolinite, sand, or smectite porous media have been saturated with dilute aqueous surfactin solution; impermeable barriers placed in the test cell separate the three minerals. It is evident that hydrates preferentially form over sodium montmorillonite (bentonite), whereas kaolinite and sand are much less effective.

8.3.5 *B. subtilis* and Seafloor Hydrates

8.3.5.1 *Culturing* B. subtilis

B. subtilis is universally present in soils and sediments; additionally, fragments of its dead cells have extended longevity in soils because of relative chemical inertness (Walker et al., 1989; Ferris et al., 1986, 1988). The *B. subtilis* microbes populate GOM sediments around gas hydrate outcrops – one of many species identified there (Lanoil et al., 2001). As a natural occurrence

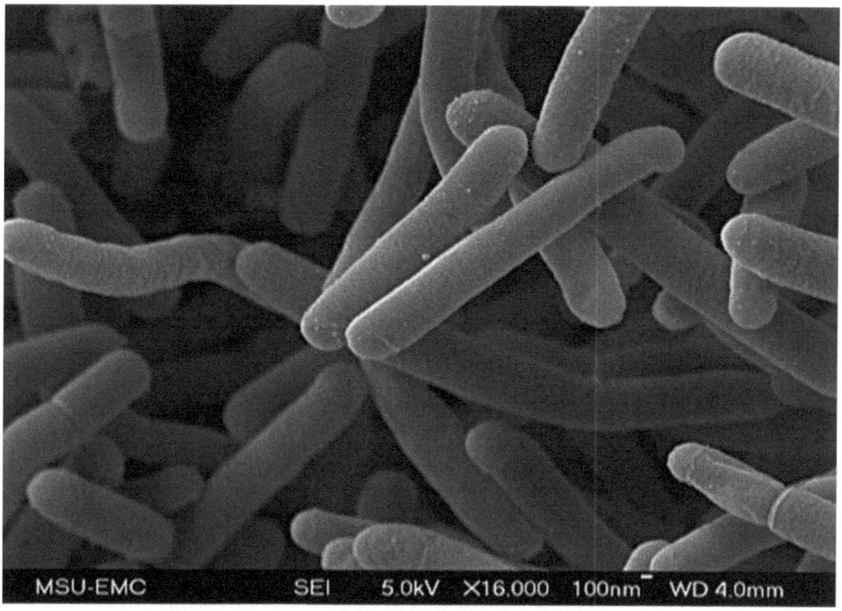

Figure 8.10 ***SEM micrograph of*** **B. subtilis** ***cultured in laboratory (Radich, 2009).***

in the seafloor and because of safety and handling ease, *B. subtilis* is a preferred microbe for representative laboratory cultures to study seafloor gas hydrates. Furthermore, its anionic surface active agent surfactin promotes hydrate formation.

Radich (2009) cultured the ATCC-21332 *B. subtilis* using a mineral medium for optimum surfactin production as advised by Cooper et al. (1981). Samples from the cultures were then viewed by Radich with SEM. One view of these cells at a magnification of ×16,000 is presented in Figure 8.10, where *B. subtilis* exhibit rod shapes with diameters near 300 nm and lengths approaching 1500 nm. To put into perspective, hydrate capillary diameters are thought to be less than or equal to 1500 nm.

To further validate *B. subtilis* as a representative species for seafloor hydrate studies, Zhang et al. (2007) studied their functioning under aerobic or anaerobic conditions with diverse carbon substrates available at low temperatures characteristic of seafloors. Surfactin concentration increase was monitored by periodic measurements of broth surface tension as well as carbon dioxide concentration increase with incubation time. The results show that surfactin forms readily during *B. subtilis* incubation under anaerobic conditions, although the biosurfactant formed more slowly than if produced at higher temperatures under aerobic conditions.

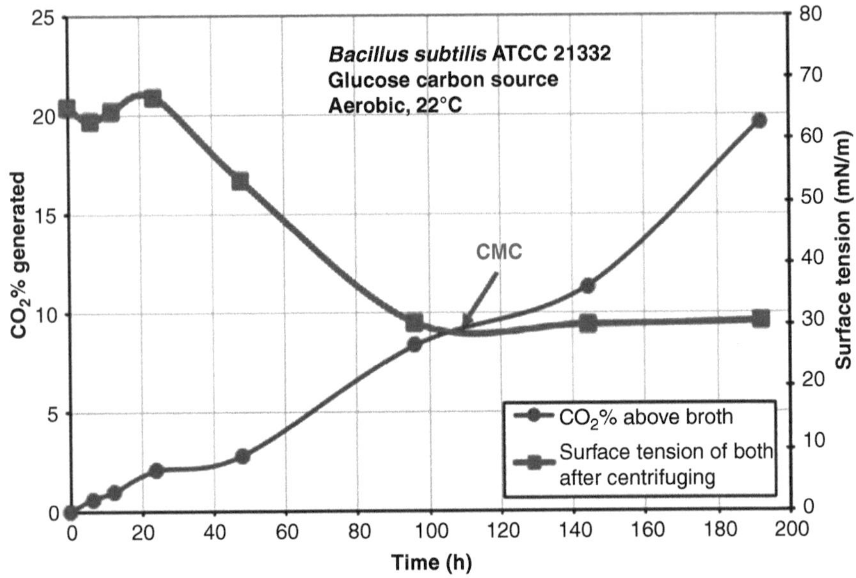

Figure 8.11 **B. subtilis** ***culture rapidly produces surfactin (Rogers et al., 2007).***

In the study by Zhang et al., surface tension of the culture broth decreases as surfactin concentration increases with time until the decrease is arrested at the CMC. At surfactin concentrations above the CMC, surface tension remains constant even though surfactin concentration continues to increase. These observations may be viewed in Figure 8.11, where the CMC is readily spotted.

During the *B. subtilis* culturing, the optimum point at which to harvest manufactured surfactin for future hydrate tests is the CMC, a point occurring 100 h after commencing the culture.

8.3.5.2 Foaming from Decomposing Hydrates Containing Biosurfactants

After culturing *B. subtilis* in a manner approaching seafloor conditions with typically available substrates, laboratory tests determined that when gas percolates through culture solution, as would happen in the hydrate zone of the seafloor, foam forms a gas–liquid emulsion stabilized by biosurfactant (Zhang et al. 2007). The exiting stream is foam composed of water–surfactin–gas. (Note that foam stability increases with cooler solution temperatures.) Passage of the stream through distilled water breaks the foam and efficiently captures the water-soluble surfactin. See Figure 8.12.

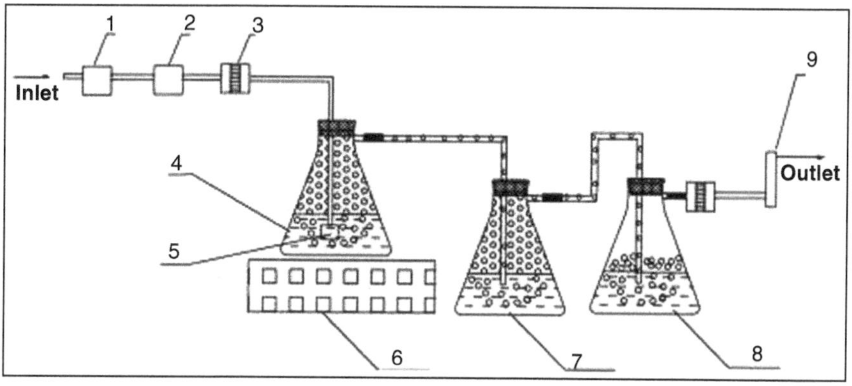

Figure 8.12 ***Separating surfactin from culture broth by bubbling inert gas.*** (1) Silica gel drier, (2) activated carbon, (3) sterile filter (0.1 μm), (4) initial broth, (5) gas distributor, (6) shaker, (7) first distilled water solution collecting biosurfactant, (8) second distilled water solution collecting biosurfactant, and (9) soap-film flowmeter (Zhang et al., 2007).

When natural gas vents through hydrate zones where *in situ* microbes produce biosurfactants, it is hypothesized that thin biofilms surround percolating gas bubbles. For example, if microbes are active at the BGHS, where temperatures are typically near optimum for microbial growth, gases venting through that interface could distribute biosurfactant along venting paths.

As proven in laboratory cultures of *B. subtilis*, as must also happen in seafloor sediments, removal of surfactin from the area of microbial activity prompts the microbes to a greater output of biosurfactant (Cooper et al., 1981). In sediments of wipeout zones, it is likely that entrained biosurfactant adsorbs on mineral surfaces higher in the zone, while greater production by the *in situ* microbes occurs at the biosurfactant source.

8.3.5.3. Surfactin Isoforms, Anaerobic and Aerobic

Under aerobic conditions, *B. subtilis* produces five surfactin compounds having equal molecular weights but different molecular configurations (Wei and Chu, 2002). Under an anaerobic environment, such as exists with seafloor gas hydrates, *B. subtilis* produces surfactin isoforms similar to those produced under aerobic conditions but differing by the proportion of each isoform in the product (Zhang et al., 2007).

The surfactin isoforms influence hydrate morphology. If packed sand in the laboratory is saturated with water (distilled) containing anaerobically derived surfactin from glucose substrate, then nodular hydrate aggregates develop, whereas if the saturating solution contains aerobically derived surfactin, smaller grains of hydrate result (Zhang et al., 2007).

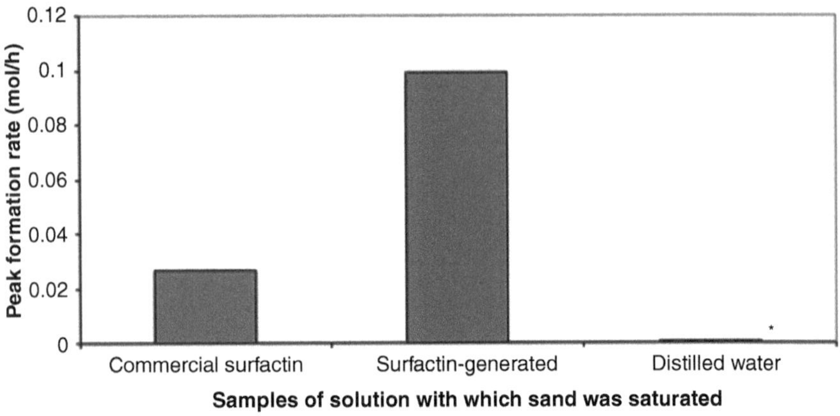

Figure 8.13 ***Surfactin catalysis of hydrates in the laboratory.*** (*) Hydrate had not formed after 72 h at 273.6 K, 2.08 MPa, natural gas.

Surfactin was extracted in the laboratory from cultures at approximate seafloor temperatures and pressures. When water solutions of the extracted surfactin saturated Ottawa sand/bentonite packed samples, hydrate induction times decreased and hydrate formation rates increased. (Distilled water was used in this set of experiments.) Hydrate formation rates from surfactin of the laboratory anaerobic cultures are compared in Figure 8.13 with rates from purchased aerobic surfactin. In the absence of bioproduct, no hydrates formed after 72 h. Under similar hydrate-forming conditions, the anaerobically produced surfactin more readily promoted hydrates.

8.3.5.4. Similar Results: MC-118 Indigenous Microbes and B. subtilis *Cultures*

Laboratory experiments to determine microbial effects on seafloor gas hydrates may be expedited if an ATCC culture such as *B. subtilis* can be used instead of the conglomerate of seafloor *in situ* microbes. *B. subtilis* has been identified as a microbial species in sediments around gas hydrate accumulations in the GOM, accompanying numerous other microbial species in those sediments (Lanoil et al., 2001).

It becomes advantageous to show if *B. subtilis* laboratory cultures affect gas hydrate formation representatively of the large mix of microbes in ocean sediments. Radich (2009) cultured a conglomerate of the many indigenous microbes in sediments near gas hydrate accumulations at MC-118 in the GOM. His anaerobic cultures generated surfactant as a function of incubation time and lowered surface tension similar to ATCC pure *B. subtilis* cultures, although the reduction of surface tension was not as great. Hydrate induction times, formation rates, and extent of reaction comparisons in laboratory incubations and hydrate tests verified a striking similarity between the indigenous microbes and ATCC *B. subtilis* – enough so to warrant *B. subtilis* use in the laboratory to help establish general microbial roles in seafloor hydrates.

8.3.5.5 *Summary of* B. subtilis *Effects*

The *B. subtilis* experiments described in preceding sections have multiple implications for microbial influence on seafloor hydrate formation, which are summarized as follows: (1) *B. subtilis* produces surfactin at seafloor conditions. (2) Hydrocarbon gas coursing through sediments must spread biosurfactants to other regions of the hydrate zone. If optimum conditions for microbial growth and biosurfactant production exist outside the gas hydrate zone, especially just below BGHS, gas may entrain bioproducts throughout the hydrate zone. (3) Temperature increases with depth, improving biosurfactant production rates deeper in the gas hydrate zone. (4) As biosurfactant disperses from bubbling gas and/or adsorbs on surrounding minerals, its concentration decreases at the active microbial site to prompt greater biosurfactant production. (5) Biosurfactants–minerals create nucleation sites that decrease hydrate induction times and increase formation rates. (6) *B. subtilis* cultured in the laboratory gives representative performances of indigenous microbes.

8.4 MICROBIAL SURVIVAL WITHIN HYDRATE MASS

8.4.1 Adsorption of Smectite Platelets on Cell Walls

When microorganisms bind to mineral surfaces, they achieve access to nutrients adsorbed on that surface.

Platelets of smectite clays adsorb onto the cell walls of *B. subtilis*. The data of Walker et al. (1989) illustrate the affinity of smectite to bacteria, where the uptake of *B. subtilis* microbial cell wall material from unit volume of reactant correlates to its adsorption on smectite. The affinity is given by the second-order polynomial Equation 8.3:

$$C_A = -0.248C_W^2 + 0.9993C_W + 0.0121 \qquad (8.3)$$

where C_A, cell adsorption (mg wall/mg smectite); C_W, walls reacted (mg/ml).

Smectite clay platelets, where the smectite is sodium bentonite, adsorb edgewise onto *B. subtilis* cell walls, unless sufficient heavy metals are present to upset the edgewise orientation (Lahav, 1962). The platelet adsorption onto the *B. subtilis* cell walls is greater for smectites than for kaolinite, where the saturation of smectite with *B. subtilis* cell walls reaches 1 mg of wall material adsorbed per 1 mg of clay (Walker et al. 1989). With an edge-on adsorption of the clay platelets on the cell walls, the orientation gives a sandwich effect. This leads to flocculation. However, Lahav (1962) found that electrolytes such as NaCl diminish flocculation because multilayers of clay particles could then adsorb edgewise onto the *B. subtilis* cell walls. The second-order polynomial Equation 8.4 represents the Lahav and Keynan finding for such adsorption of the smectite clay bentonite onto *B. subtilis* in NaCl solution:

$$C_{\text{Abs}} = -0.0207C_{\text{B}}^{2} + 0.936X + 0.1164 \tag{8.4}$$

where C_{Abs}, adsorbed bentonite (μg/million bacteria); C_{B}, equilibrium concentration of bentonite (μg/ml).

Metal ions compete for the same adsorption sites and cause some of the edgewise clay adsorption to flatten, increasing flocculation. Walker et al. (1989) verify the Lahav finding that smectites adsorb onto the cells preferentially in an edgewise fashion.

By an electrophoresis process, Lahav (1962) first showed the edgewise orientation of bentonite (sodium montmorillonite, smectite clay) platelets on *B. subtilis* cells. Although the net charges on sodium bentonite particles are negative, the ends of its platelets are positively charged, and those (+) ends adsorb onto the negatively charged cell walls. When sodium phosphate solution is used in the process, positive charges of the platelet edges are neutralized and bentonite adsorption decreases. The study by Lahav shows that 10 μg of bentonite adsorbs per million of *B. subtilis* cells if in NaCl solution but only 2 μg adsorbs if in sodium phosphate solution.

Lahav found the clay to bacteria affinity to be reversible as given by Equation 8.5:

$$\text{B} + \text{C} \leftrightarrow \text{BC} \tag{8.5}$$

where B, bacterial walls; C, clay particle; BC, bacterium–clay complex.

8.4.2 Adsorption of Metals onto Cell Walls

Both teichoic acid and peptidoglycan have sites that could potentially bind with metals – carboxyl and amino groups in peptidoglycan and phosphodiester units in teichoic acid (Beveridge and Murray, 1980; Doyle et al., 1980; Birdsell et al., 1975). Warth and Strominger (1971) estimate that walls contain about 0.15 μmol of reactive sites per milligram of *B. subtilis* wall mass.

Probably, metal adsorption is due to the presence of peptidoglycan in the cell wall – primarily the carboxyl groups (Walker et al., 1989; Beveridge and Murray, 1980). The affinity of the adsorption sites for metal ions varies, and a competition leads to selectivity in the metals adsorbed. For example, 0.83 μmol of sodium ions adsorbs on 1 mg of *B. subtilis* cell walls (Doyle et al., 1980).

8.4.3 Thermal Protection of Microbial Cells from Hydrate Formation

To put into perspective smectite platelets adsorbing edgewise onto *B. subtilis* cell walls, consider relative sizes of platelets and cells. Notably, bentonite platelets are smaller than the *B. subtilis* cells. Lahav (1962) evaluated 200–500 nm–sized bentonite platelets for their adsorption onto *B. subtilis* cell walls. Radich (2009) determined *B. subtilis* cell dimensions to be 150–200 nm diameter by 1500 nm long.

In his research, Radich proposed that the edgewise adsorptions of smectite platelets onto cell walls help protect the bacteria from exothermic latent heat released during hydrate formation. The action is instrumental in placing bacteria in the protective confines of seafloor hydrate masses, as discussed in the following sections.

The relatively high latent heats released on hydrate formation ΔH_f are presented in Table 8.3.

The significance of this energy release on or near surfaces of microbial cells becomes more apparent by viewing Figure 8.14. Data forming the figure come from a hydrate-generating laboratory test cell where a thermocouple was situated in the gas phase above porous media saturated with biosurfactant–water solution. The temperature spike of 6.5°C indicates

Table 8.3 Hydrate's exothermic heats of formation (Handa, 1986)

Type of hydrate	Heat of formation, ΔH_f (kJ/mol)
Methane, sI	−54.2
Ethane, sI	−71.8
Propane, sII	−129.2
Seafloor, sII	−79

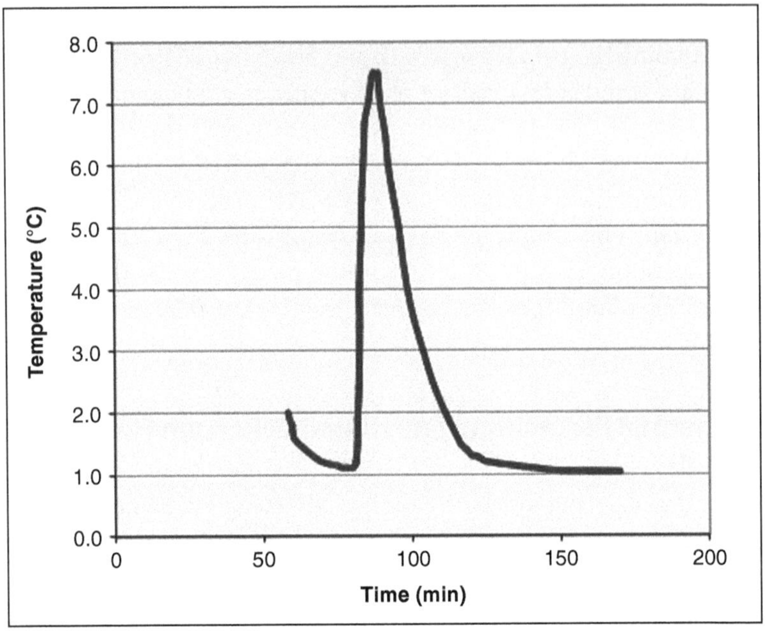

Figure 8.14 ***Temperature spike from exothermic hydrate formation in test cell.***

bulk heat effects from exothermic hydrate formation in the porous media (Radich, 2009). Localized heat effects may be greater.

Therefore, if seafloor hydrates form directly on or near bacterial cells, the microorganisms are subjected to a sudden temperature spike. Can the surge be tolerated?

For the sake of sterilization, medical scientists have long been interested in the susceptibility of bacteria to heat. For this reason, Bender and Marquis (1985) studied the sensitivity of bacterial cells, including *B. subtilis*, to heat by introducing the sensitivity factor *D*. Sensitivity *D* to a given temperature is the time in minutes for 1-log bacterial spore population reduction. They found that 100.3°C imposed on *B. subtilis* gave a *D* of 1-minute; an imposed 93.7°C gave a *D* of 10 minutes.

For survival of the living cells, Radich (2009) suggests a model whereby biosurfactant attaches to the outer edges of adsorbed smectite platelets extending from the cells; the biosurfactant–platelets are hydrate nucleation sites. Thus, forming hydrates surround the microbe(s) with an insulating layer of clay platelets and space protecting cells from liberated heat. The Radich model details means by which microbes are thrust inside

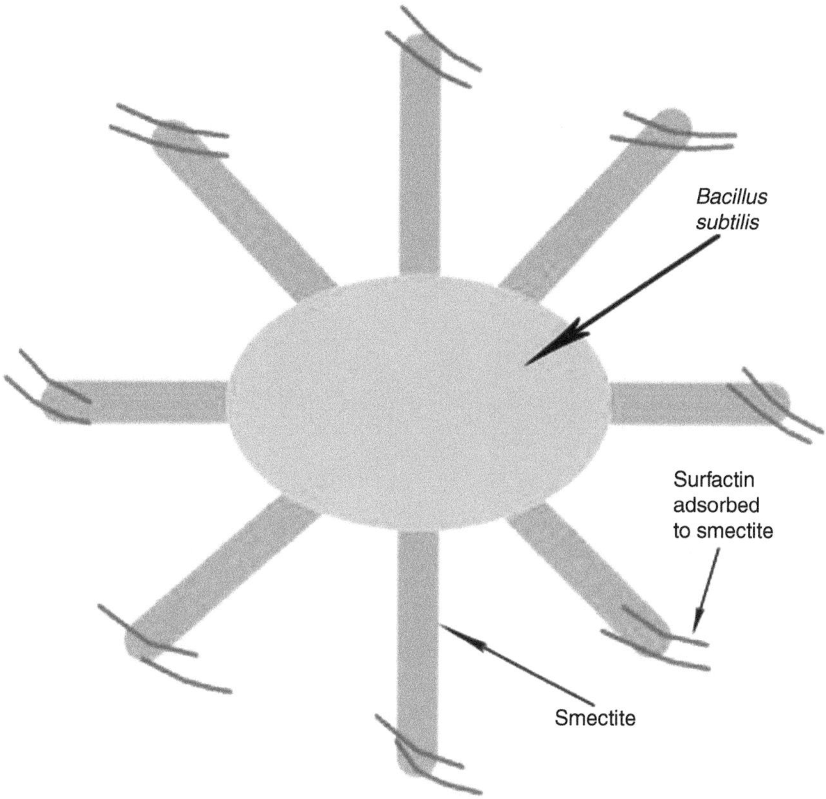

Figure 8.15 ***Model for microbes, smectites, and biosurfactants initiating hydrate nucleation in ocean sediments.*** *(Modified from Radich, 2009.)*

hydrate accumulations in seafloor sediments. The hypothesis is sketched in Figure 8.15.

As partial verification of the surfactant–platelet–microbe configuration in the Radich model, Lahav (1962) found in his nonhydrate experiments that glucose from its water solution attached to the outer ends of smectite platelets adsorbed edgewise onto *B. subtilis* cells.

Within an approximately neutral pH environment, cell walls exhibit net negative surface charges (Walker et al., 1989).

8.4.4 Microbes within Gulf of Mexico Seafloor Samples

In the scientific cruise at MC-118 on September 2006, samples of the hydrate seafloor outcrop were retrieved by the Johnson-Sea-Link submersible's robotic arm as seen in Figure 8.16.

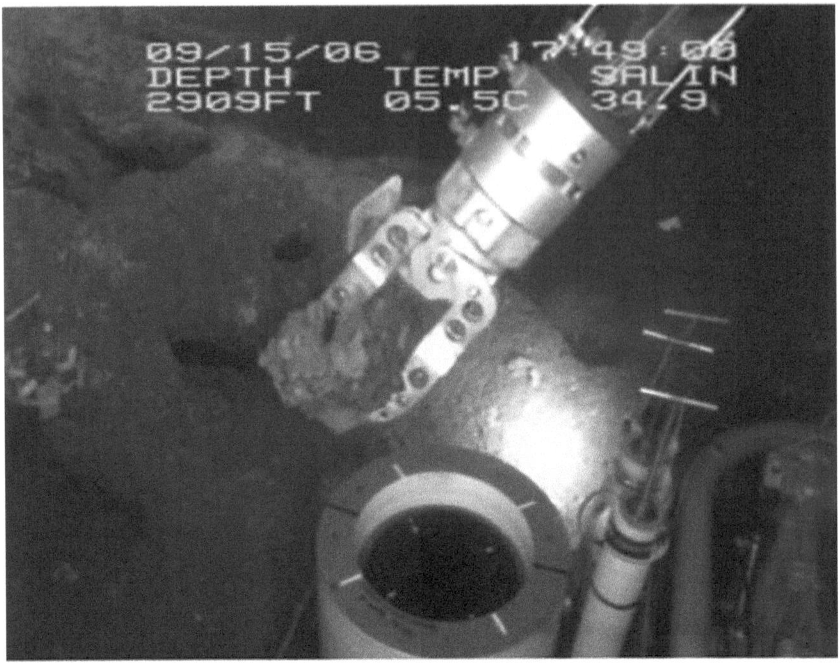

Figure 8.16 ***Manned submersible retrieving hydrate sample, MC-118.*** *(Courtesy: Mississippi Mineral Resources Institute, University of Mississippi. MMRI overseer of Gulf Coast Hydrate Consortium supported by US Department of the Interior, Department of Energy, and National Oceanic and atAtmospheric Administration (NOAA).)*

The hydrate samples, shown being retrieved in Figure 8.16, were decomposed in the laboratory and solid decomposition products viewed by electron microscope as presented in Figure 8.17.

Note the clay platelets oriented edgewise in Figure 8.17. Indigenous microbes are not visible because of the surrounding clay particles. Two other micrographs from the same seafloor hydrate add credence to the conclusion that microbes gained entrance to the seafloor hydrate mass by an association with clay platelets in the initial hydrate nucleation as described by the Radich model.

Figure 8.18 presents another SEM view at a lower magnification of decomposed MC-118 hydrate. Note the biofilm surrounding a section of the residue mass.

No apparent microbes are visible at the ×3700 magnification, but a break in the encasement can be seen in the upper left of the mass, behind which microbial cells are evident. This break is magnified to ×20,000 and reviewed in Figure 8.19.

Figure 8.17 ***MC-118 seafloor gas hydrate decomposition products.*** SEM ×22,000 (Dearman, 2007).

The SEM micrograph of Figure 8.19 reveals a dense population of *in situ* microbes residing behind the break in mineral encasement. The image indicates that encasement protects the microbe colony. The encasement is thought to be composed primarily of smectite clay platelets, possibly cemented with carbonate products of anaerobic oxidation of methane (AOM) inside the hydrate mass.

The seafloor hydrate decomposition residue of Figure 8.20, where the hydrate sample had been chipped from seafloor hydrates, is next compared with a laboratory procedure in which hydrates were formed as follows. *In situ* MC-118 microbes were cultured without agitation and the biosurfactants/cells extracted by centrifuge; hydrates were then formed at 0.5°C/3.14 MPa in a test cell containing samples of extracted biosurfactant/cells, Ottawa sand with 33.3 wt% sodium montmorillonite clay platelets, seawater saturating the porous media, and pressurizing natural gas. An SEM view of residue from the laboratory hydrate is given in Figure 8.20.

Figure 8.18 ***Membrane surrounding mineral and bioproduct agglomerate from MC-118 seafloor hydrates.*** SEM ×3700 (Dearman, 2007).

Note in Figure 8.20 of the laboratory sample that the cultured microbes could not first be found by SEM views until some encasement was removed by spatula, revealing clustered cells quite similar to the counterpart field sample of Figure 8.19. The fixing process for SEM differed for the two micrographs, which accounts for minor differences in shape and size of the encased cells.

8.4.5 Anomaly of Unsaturated Pore Waters Surrounding Massive Hydrates

Seafloor gas hydrate stability depends on maintaining equilibrium values of temperature, pressure, salinity, and saturation of interfacial seawater with the occluded gas species. Measurements of sediment pore waters surrounding near-surface hydrates in the GOM show gas concentrations less than saturation and lower than anticipated to sustain hydrate stability. Yet the hydrates protruding into the seawater remain surprisingly stable (Lapham et al., 2008).

Figure 8.19 ***Inside mineral and bioproduct agglomerate from MC-118 seafloor hydrate residue.*** SEM ×20,000 (Dearman, 2007).

With Raman spectroscopy, Brewer et al. (2011) measured methane concentrations adjacent to seafloor hydrate outcrops. The pore waters held methane concentrations of 28 mM – well below their estimated 50 mM needed for hydrate stability.

Methane concentrations of sediment pore waters may be estimated by the technique of Lorenson et al. (2008), which is given in Equation 8.6:

$$C_{M1} = C_{M2}\frac{21(\rho/m)-1}{24,300\varphi} \tag{8.6}$$

where C_{M1}, methane concentration (millimoles per liter, mM); C_{M2}, methane concentration [parts per million by volume (ppmv)]; ρ, sample density [kg/m^3 (usually taken as 1650 kg/m^3)]; m, mass (kg); φ, porosity.

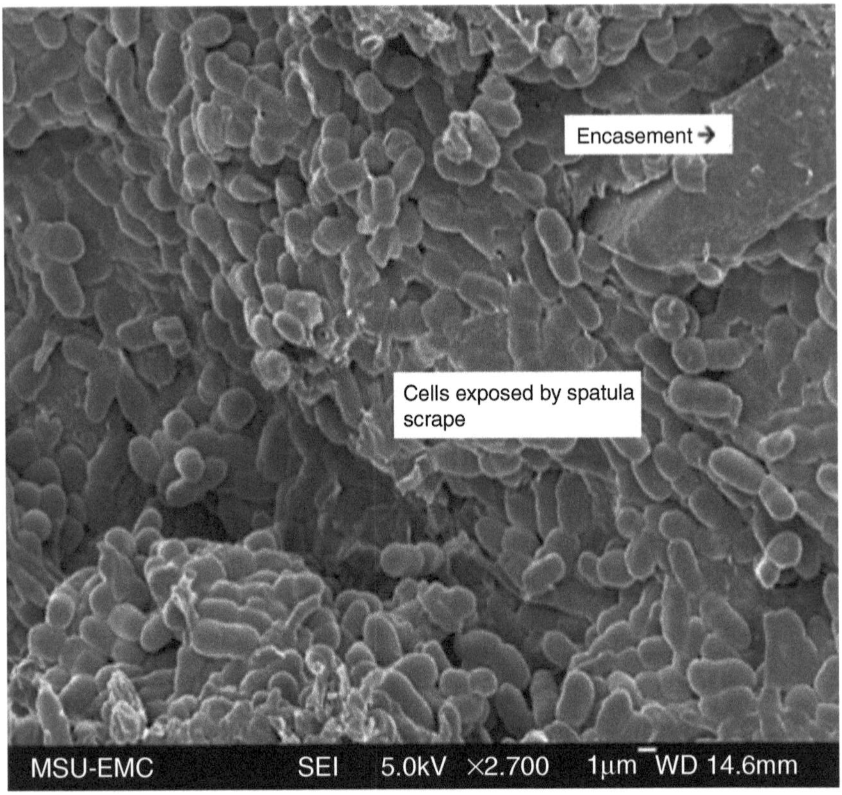

Figure 8.20 ***Laboratory hydrate decomposition residue (×2700) from MC-118*** **in situ** ***microbe culture (Radich, 2009).***

Applying the Lorenson equation to Keathley Canyon site KC151-3 gives dissolved methane concentrations in pore waters of near-surface sediments to be approximately 50 mM, while ranging to 130 mM at depths of 400 mbsf.

8.4.6 Internal Microbial Activity Affects Seafloor Hydrate Stability

Gas hydrate masses are porous associations of hydrate crystals containing discrete interstitial spaces and interconnecting capillaries. Dearman et al. (2009) estimate capillary diameters in laboratory-prepared hydrates to range predominantly between 80 and 450 nm with an upper bound of 1500 nm. These capillaries are large enough to permit passage of seawater-laden nutrients and even hydrate-associated microbes. For example, *B. subtilis* cells from laboratory cultures, as well as other indigenous seafloor microbes, have dimensions permissible for hydrate-capillary diffusion.

Analyses suggest that microbes thrive internally within masses of seafloor hydrate accumulations (Joye et al., 2004; Lanoil et al., 2001; Orcutt et al., 2004; Sassen et al., 1994, 1998, 1999, 2004). Their internal presence may occur by two mechanisms. First, given the accommodating sizes of capillaries, microbes inside seafloor hydrates may have entered by diffusion. Second, and most probable, cells may attain access to the repository of occluded carbon in the hydrate interior according to the microbe/clay platelet/surfactant nucleation hypothesis outlined by Radich et al. (2009).

The Radich hypothesis proposes hydrate stability amid undersaturated seawater to be dually dependent on the kinetics of microbial reactions within the hydrate mass and on reactant/product diffusion rates through the hydrate capillaries rather than solely on gas concentrations of surrounding pore waters. Incremental methane supplied or consumed by these internal reactions helps explain frequently perceived discrepancies of external pore water concentration relative to hydrate stability. The model incorporates internal biochemical reactions, concentration-driven diffusion to the hydrate interior, and an interfacial mass-transfer film at the hydrate–seawater interface.

A wide variety of indigenous microbial species were found in the hydrate samples extracted from MC-118, including both methanogens and methanotrophs (Radich, 2009). To the microbes, the hydrate interior affords protection from external temperature vagaries while providing access to occluded carbon for metabolism and to nutrients via capillary diffusion.

The biochemical reaction that dominates in the hydrate mass depends on where the hydrate mass resides in the seafloor – that is, within the sulfate zone or below the sulfate–methane interface (SMI). Because of diffusion into the hydrate interior according to Fick's first law, the dominant reaction there will relate to the dominant reaction of the external, contacting zone. Large hydrate mounds on the seafloor penetrating both sulfate and carbonate reduction zones, therefore, could be subject to different reactions in compartmentalized mound interiors.

Consider massive gas hydrates at the seafloor surface that protrude into the sulfate zone where the AOM reaction of Equation 8.7 dominates in sediments. Internal to the hydrates, the same reaction will also dominate:

$$CH_4 + SO_4^{2-} \rightarrow HCO_3^- + HS^- + H_2O \qquad (8.7)$$

Internal to the hydrates, dissolved methane in interstitial waters must maintain equilibrium with both occluded methane of contacting hydrate crystals and methane dissolved in external waters. As methanotrophs consume CH_4 of interstitial waters, adjacent hydrate crystals decompose to release methane and external methane diffuses to the interior through hydrate capillaries, overcoming the resistance of a boundary film. The boundary layer resistance helps regulate a proper balance of reaction rates and diffusing rates to maintain equilibrium; a steep methane concentration gradient exists across a very thin interfacial boundary layer. Methane concentrations will be difficult to measure accurately at their hydrate/seawater interface.

As AOM consumes sulfate internally, sulfate is replenished by seawater diffusing through capillaries and assisted by mound-covering *Beggiatoa* mats oxidizing sulfides. If alkaline conditions prevail, carbonates precipitate within interstitial waters – ultimately converting the original hydrate mound to carbonate rubble. Paradoxically, the solid precipitates may help preserve structural size and integrity of the mound.

Below the sulfate zone, methanogens produce methane within the hydrate mass by the dominant biochemical Equation 8.8, assuming bicarbonate is the main substrate reduced by methanogenic Archaea:

$$HCO_3^- + 4H_2 + H^+ \rightarrow CH_4 + 3H_2O \tag{8.8}$$

Again, balances of reaction rates and diffusion rates within hydrate capillaries determine internal compositions and status quo of the hydrate mass. Methane generated internally diffuses to the exterior hydrate–water interface at a rate determined by the diffusion coefficient or is occluded into new hydrates. The hydrate mass stabilizes or expands from produced methane even though surrounding bulk seawater may not be saturated. A hydrate mound may remain stabilized as long as the biogenic reaction rate of Equation 8.8 persists.

Accordingly, multiple scenarios may occur in the hydrate mass interior. Hydrate mass expands, stabilizes, or deteriorates based primarily on reaction rates/diffusion rates, mass position relative to sulfate or carbonate reduction zones, and prevailing pH.

The mathematical details of the Radich model for the microbial influences on seafloor hydrate stability may be reviewed in the publication Radich et al. (2009).

8.5 MAGNETIC PROPERTIES OF HYDRATE ZONES

8.5.1 Microbial Role

Giving seafloor sediments most of their magnetic identity, magnetite is a common mineral of relatively low concentration deposited from volcanoes, rock weathering, and microbes. Blakemore (1975) first reported magnetic particles of microbial origins to be widely distributed in seafloor sediments. Specifically, magnetotactic bacteria produce submicrometer particles of intracellular magnetite, where the iron in the magnetite crystals comprises approximately 3% of cell mass (Housen and Musgrave, 1996; Bazylinski et al., 2007; Canfield and Berner, 1987). To a lesser extent, magnetotactic bacteria produce ferromagnetic greigite particles (Bazylinski and Moskowitz, 1997; Paasche et al., 2004; Lins et al., 2007).

The ferromagnetic biomineral particles produced by the magnetotactics range in size from 35 to 120 nm diameter, which is smaller than that of particles produced from nonbiological sources. The intracellular magnetite becomes attached to membranes in chains of 10–20 crystals within the bacteria (Stolz et al., 1986; Bazylinski et al., 2007; Paasche et al., 2004; Kirschvink and Chang, 1984). Interestingly, magnetotactics migrate under the influence of Earth's magnetic field by means of rotating helical flagella containing membrane-encased, magnetite particles. After cell death and cell lysis, the encapsulating membranes slow dissolution of the ultrafine magnetite particles when the agglomerate attaches to sediment surfaces (Paasche et al., 2004; Kirschvink and Chang, 1984).

Ferromagnetic bioparticles disperse throughout ocean sediments. But where microbial activity is especially high around cold seeps or gas hydrates, magnetite is primarily attributed to biogenic sources. Therefore, hydrate-bearing sediments and sediments previously bearing hydrates have a different magnetic signature. Their magnetic dissimilarity provides insight into the complex ecology of hydrate zones, and the phenomenon also imparts information about hydrate formation mechanisms.

8.5.2 Defining the Ferromagnetic Minerals

Pertinent iron compounds influencing magnetic properties of the hydrate zone and the hydrate accumulations are magnetite, greigite, pyrrhotite, and pyrite.

Equation 8.9 shows pyrite as the terminal point of the reaction series beginning with magnetite (Housen and Musgrave, 1996; Musgrave et al., 2006; Canfield and Berner, 1987):

$$\begin{array}{ccccccc} Fe_3O_4 & \rightarrow & Fe_3S_4 & \rightarrow & Fe_7S_8 & \rightarrow & FeS_2 \\ \text{magnetite} & \rightarrow & \text{greigite} & \rightarrow & \text{pyrrhotite} & \rightarrow & \text{pyrite} \end{array} \quad (8.9)$$

Note in Equation 8.9 that the pyrite end product is the most stable iron sulfide during the progressive conversion in a reducing environment. Also important is the progressively lessening of magnetic intensities while proceeding along the path from magnetite to pyrite. Magnetite is highly magnetic. But pyrite demonstrates low magnetic properties at prevailing conditions in gas hydrate zones.

In order to limit the progressive reaction at greigite or pyrrhotite, short times exposed to H_2S are necessary. Curtailment of the series reaction has occurred within hydrates of the Cascadia margin where ultrafine particles of greigite and pyrrhotite dominate ferromagnetic accumulations (Larrasoana et al., 2007).

Magnetite is a precursor of the other ferromagnetic compounds. It generally determines magnetic properties of ocean sediments because of greater abundance and longevity; remanence, the property measuring magnetization persistence after removal from a magnetic field, is also important (Karlin and Levi, 1985; Anbar, 2004; Stolz et al., 1986).

8.5.3 Sulfide Effects

In marine environments of significant methane flux, where H_2S is produced by AOM, magnetite particle dissolution occurs at a rate given by Equation 8.10 (Canfield and Berner, 1987):

$$\frac{dC_{mag}}{dt} = -k_{rc} C_S^x C_{mag} A_{mag} \quad (8.10)$$

where t, time; k_{rc}, reaction rate constant; C_{mag}, concentration of magnetite (g/g of sediment), C_S, concentration of dissolved sulfide $(mM)^x$; A_{mag}, specific surface area of magnetite (cm^2/g magnetite).

To put magnetite dissolution rates in perspective, Canfield and Berner used k_{rc} of 1.1×10^{-5} and exponential x of 0.5 in Equation 8.10, obtaining some interesting observations:

1. Several hundred years would be required for pyrite to completely replace magnetite if H_2S concentrations are greater than 1 mM.
2. No noticeable conversion of magnetite to pyrite would occur if H_2S concentrations are less than 1 mM.

3. Magnetite would generally have a half-life in ocean sediments of 50–1000 years.

8.5.4 Interpreting Magnetic Characteristics of Gas Hydrate Zones

Remanence, susceptibility, coercivity, and thermal demagnetization are four parameters used to characterize magnetic properties of ferromagnetic minerals. Housen and Musgrave (1996) used the indices D_{JH} defined by Equation 8.11 and D_S defined by Equation 8.12, which combine relevant magnetic parameters, to evaluate significance of magnetic properties in gas hydrate zones:

$$D_{JH} = \frac{J_{rs}/J_s}{H_{cr}/H_c} \tag{8.11}$$

$$D_S = \frac{J_{rs}/k}{H_{cr}} \tag{8.12}$$

where J_s, magnetization saturation; J_{rs}, remanent magnetization saturation; H_{cr}, coercivity of remanence; H_c, coercivity (reverse field intensity for zero magnetization); k, magnetic susceptibility.

If remanence gives a strong measurement, it establishes magnetite as dominant in the sediments – the case above and below hydrate layers in the Cascadia margin. Housen and Musgrave observed high D_{JH} values within strata sandwiching hydrate layers; D_S values maximized within the hydrates. Significantly, ultrafine greigite particles exist as the dominant ferromagnetic material interior to the hydrates.

Double bottom simulating reflectors (BSRs) were identified by Housen and Musgrave (1996) when analyses of sediment samples below current BGHS verified a 70 m lower reach of magnetic properties similar to the current hydrate zone. The ancient BSR presumably dated to the last glacial period.

8.5.5 Mechanism of Magnetic Minerals Accumulating in Hydrates

At Site 892 in Cascadia, gases occluded in hydrates are about 10% H_2S, yet the ferromagnetic composition within the hydrate mass is predominantly greigite (Housen and Musgrave, 1996). Several possibilities arise why the series of reactions of Equation 8.9 do not progress to pyrite:

1. H_2S generated in hydrate interstitial spaces is quickly scavenged into hydrates. Near the seafloor surface, *Beggiatoa* mats covering hydrate masses may assist in scavenging H_2S. These two actions may effectively remove sulfide from direct contact with greigite. The greigite in hydrate interstices becomes isolated from occluded H_2S.
2. There are indications at Cascadia of microbes producing some greigite within the hydrate mass (Larrasoana et al., 2007). Nutrients may diffuse through hydrate capillaries; microbes may be implanted on hydrate nucleation. Continuing microbial activity within the hydrate interstitial spaces becomes feasible.
3. Magnetite particles with membrane coverings act as nucleating sites for hydrates to crystallize, thus putting particles inside hydrate interstitial spaces with limited sulfide access.

8.6 MICROBIAL EFFECTS ON MEMORY OF SEAFLOOR HYDRATE RE-FORMATIONS

8.6.1 Introduction

The remnant water from decomposed conventional hydrates retains a memory that may be erased by time elapse and temperature elevation.

Hydrates exhibit the memory effect in laboratory tests with a wide range of guest gases: carbon dioxide, methane, methane–ethane and other alkane mixtures, xenon, and hydrochlorofluorocarbon (Takeya et al., 2000; Thompson et al., 2006; Lee and Englezos, 2008; Moudrakovski et al., 2001; Ohmura et al., 2003). The phenomenon is also reported after hydrate formation involving tetrahydrofuran solutions (Zeng et al., 2006).

If the memory is not erased, decomposition products re-form hydrates with shorter induction times. However, hydrates extracted from seafloor sediments may perform differently.

Using sediments extracted from MC-118 near current seafloor hydrate occurrences, gas vents, or previous hydrate occurrences, Dearman (2007) evaluated times and temperatures necessary to eliminate memory remnants after laboratory formation/decomposition cycles of hydrates formed in the seafloor samples. In so doing, a unique memory phenomenon was observed for hydrates formed from some seafloor sediments.

8.6.1.1 Hydrate Memory Mechanisms

Three theories are offered to explain the hydrate memory phenomenon of quiescent, ideal systems:

1. Hydrogen-bonded water structuring perseveres in decomposed hydrate remnant liquid. Takeya et al. (2000) suggest that residual structured water is the reason for retained memory during re-formation of CO_2 hydrates on the partially melted surface of ice. Lederhos et al. (1996) propose residual structured water as causative of abbreviated induction times when natural gas hydrates re-form. Moudrakovski et al. (2001) find memory effects in forming xenon hydrates on the surfaces of ice at 223 K and suggest that pentagonal water arrangements from hydrate structure persist after hydrate decomposition; on re-forming hydrate from residual water, induction time diminishes, although formation rate of the re-formed hydrate remains unaffected.
2. Hydrate crystals remain in residual water after bulk hydrate decomposition. Thompson et al. (2006) propose that trace crystalline remnants of decomposed hydrates persist in solution, promoting pentagonal and hexagonal hydrogen-bonded water around the small crystals; these remnant crystals cause faster nucleation during subsequent hydrate formation.
3. Uchida et al. (2000) suggest that residual CO_2 in water from CO_2 hydrate decomposition contributes to faster induction times on hydrate re-formation.

8.6.1.2 Eliminating Laboratory Memory Effects

Subjecting hydrate decomposition liquids to stresses of elevated temperature, vigorous agitation, or extended time eliminates hydrate memory in ideal systems. After stressing, subsequent hydrate induction times return to original higher values.

Some stresses eliminating memory are tabulated in Table 8.4.

Table 8.4 Memory elimination by stressing dissociated hydrate liquids

Type of hydrate	Stress to terminate memory	References
CO_2	298 K for 1 h or 275 K for 2 h	Takeya et al. (2000)
Natural gas	30°C	Lederhos et al. (1996)
Xenon on ice	243 K when decomposed at 223 K	Moudrakovski et al. (2001)
Methane and natural gas	5.5–8.0 K above dissociation temperature	Uchida et al. (2000)
sI and sII	AFP from Type I winter flounder	Zeng et al. (2006)
Tetrahydrofuran (THF)	Higher temperatures	Ohmura et al. (2003)

Hydrates formed from waters of melted ice show shorter induction times, at times orders of magnitude less, than hydrates formed from the melted ice after a delay. Takeya et al. (2000) find that hydrate induction times range from 50 min to over 7200 min on subjecting melted ice to different combinations of time and temperature stresses prior to re-forming hydrates. Impositions of longer time and higher temperature combinations deteriorate orderly hydrogen-bonded structuring left by prior hydrates or ice.

In an attempt to substantiate the proposed causes of hydrate memory in ideal systems, Buchanan et al. (2005) used neutron diffraction to analyze remnant water. They detected no structural effects remaining in the waters after hydrate dissociation near hydrate equilibrium conditions. However, the data were taken after a lengthy time period to reestablish equilibrium P–T – a time lengthy enough that any structured water might have dissipated. Also, the technique's sensitivity limited crystalline detection to greater than or equal to 1%. The data of Buchanan et al. from neutron diffraction did indicate that dissolved gas increases in the remnant water.

8.6.2 Seafloor Sediments – Hydrate Memory Effects

An extraordinary permanent memory effect occurs in a nonideal system when forming gas hydrates from *in situ* connate water and sediments extracted near seafloor hydrocarbon vents and present or past gas hydrate accumulations.

8.6.2.1 Time Effects on Hydrate Memory, Seafloor Sediments

In a series of experiments, Dearman (2007) formed and re-formed hydrates from an MC-118 core sample with GOM *in situ* water where cycle interim times were varied at a constant holding and decomposition temperature of 20°C. Hydrate induction times and formation rates stayed essentially constant at 0.33 ± 0.08 h throughout the series. That is, in the final cycle, decomposition products were subjected to 20°C for about 150 h before final hydrate re-forming. The test results are presented in Figure 8.21. From initial cycle throughout subsequent formation/decomposition cycles constant induction times occur. A permanent memory effect is indicated.

Even during the first cycle of hydrate formation, GOM sediments taken from the proximity of seafloor gas hydrates exhibit short induction times characteristic of structured water. The water remains structured after greater than 150 h at room temperature in the last cycle.

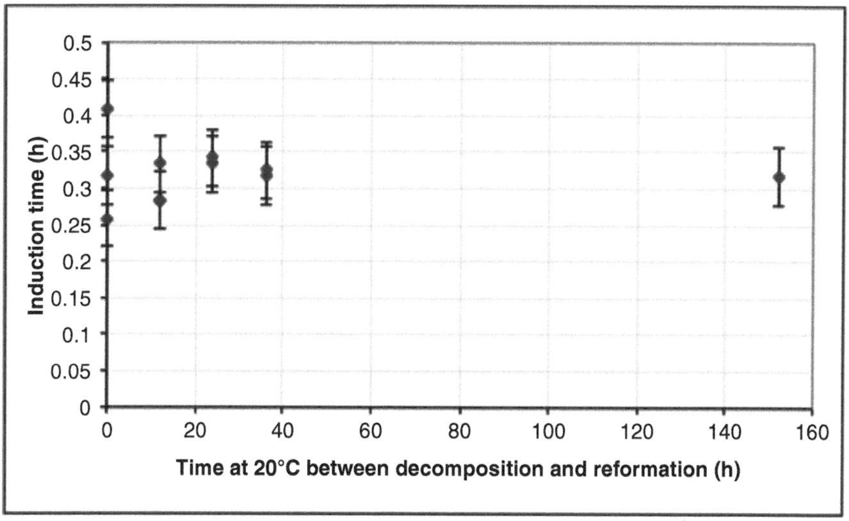

Figure 8.21 ***GOM sediments retain permanent hydrate formation memory (Dearman, 2007).***

In a second set of experiments forming hydrates from sediment samples containing *in situ* water, similar results of constant induction times and formation rates were obtained through a series of hydrate formation–decomposition cycles. These sediments from a different site at that locale in the GOM again gave an induction time of 0.35 ± 0.22 h. Even in the last decomposition phase in which 20°C was maintained for about 46 h, no significant effect on the induction time resulted.

8.6.2.2 Temperature Effects on Hydrate Memory, Seafloor Sediments

Temperatures necessary to erase memories of hydrates formed in seafloor sediments were evaluated with MC-118 sediment samples (Dearman, 2007). Note that the MC-118 site is an acoustic wipeout zone where gas venting, hydrates, and carbonates abound. Eight cycles of natural gas hydrate formation and decomposition were made with sediment samples. Before natural gas hydrate formation in each of the eight cycles, samples were readjusted to 21.1°C and 3.2 MPa and then the test cell was submerged in a 0.58°C cooling bath to re-form hydrates. In successive cycles, hydrates were decomposed at 10°C incrementally higher temperatures.

In the test sequence, no loss of hydrate memory occurred until a decomposition temperature of 51.7°C was imposed on the eighth cycle. After the 51.7°C cycle, no hydrates would form in the subsequent cycle after 20 h under hydrate-forming conditions.

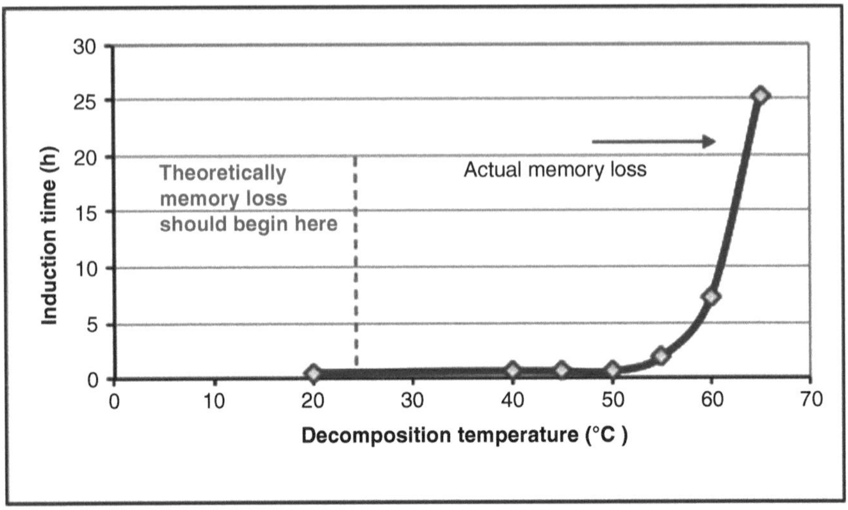

Figure 8.22 ***Hydrate induction times for GOM sediments with indigenous microbes after heating (Xiong, 2009).***

To further study this unusual effect of temperature on seafloor sediment memory, cores from MC-118 were again retrieved near crude oil and natural gas seeps in the vicinity of seafloor gas hydrates. These original sediments contained *in situ* water, indigenous microbes, and oil stains. Two samples were taken from the core and four cycles of identical hydrate formation conditions were imposed on parallel test cells. Then, proceeding through the four cycles in each of the two test cells, hydrates were decomposed at successively higher temperatures for each cycle (Xiong, 2009). The results are presented in Figure 8.22.

All hydrate induction times were low from the beginning, indicating a permanent memory effect in place. Traditionally, water from decomposed hydrates can be expected to retain a memory that can be dissipated with a moderate temperature stress. But this does not occur with core material from a seafloor hydrate zone, which contains *in situ* minerals, microbes, and water. It is seen in Figure 8.22 that hydrates become more difficult to re-form, as measured by longer induction times, only when residue is highly stressed near 52°C. At 50°C the hydrate induction time is 0.57 h, but at 65°C the time has increased to 25.4 h. Surface tensions of *in situ* water at the beginning of the hydrate formation tests were 53.1 mN/m. These resulted from the presence of indigenous biosurfactants. At the end of the tests, the surface tension of the water was 68.1 mN/m.

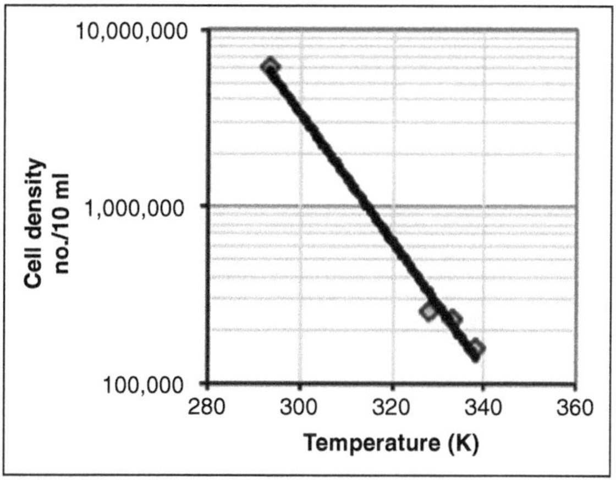

Figure 8.23 ***High temperatures kill indigenous microbes (Xiong, 2009).***

The permanent hydrate memories in hydrate-zone sediments are believed to be caused by indigenous microbes and their bioproducts.

8.6.2.3 Indigenous Microbial Cell Influence on Memory, Seafloor Sediments

Living cell densities of indigenous microbes cultured from MC-118 sediments were determined as a function of rising temperature (Xiong, 2009). Living cells declined exponentially with increasing temperatures; see Figure 8.23. From 20 to 65°C, cells in 10 ml samples declined from 6.10 to 0.167 million.

Figure 8.24 presents the data from another perspective to emphasize the sharp decline of the living cell count on approaching 55°C.

The effects of rising temperatures on cell numbers and cell densities of the indigenous microbes of MC-118 (Xiong, 2009) substantiate data reported for indigenous microbes from Japan Sea sediments cultured at temperatures of 14–65°C (Parkes et al., 1994). Also, thermal cell demise in the Xiong cultures from GOM sediments compares favorably with the kill rate established by Bender and Marquis (1985) for a mixed group of bacteria that included *B. subtilis*.

Comparing the preceding figures, one concludes that indigenous microbes and their bioproducts impart a permanent hydrate memory to current or past hydrate-containing sediments. Dissipation of the permanent memory does not begin until rising temperatures near 50°C. As temperatures approach 65°C, hydrate inhibition properties come to the fore, as seen by the sudden rise in hydrate induction times. That is, teichoic acid and

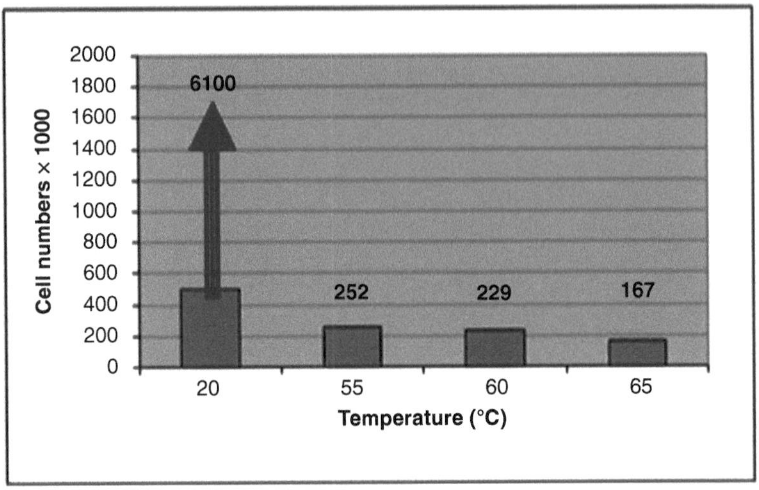

Figure 8.24 ***Cell count of indigenous microbes from GOM sediments of memory tests (Xiong, 2009).***

peptidoglycan structures of the cell walls are fragmented from temperature elevations approaching 65°C, exposing more surface area of these hydrate inhibitors.

From studying sediments extracted near active gas venting or hydrate sites within acoustic wipeout zones, a concept of permanent hydrate memory arises. The permanent memory may derive from microbes and bio-products in those sediments. Historically, gas hydrates once formed and decomposed, hydrate outcrops presently exist in the zone, or shifting gas vents occurred throughout the tested sediments.

8.6.2.4 Antifreeze Protein Influence on Hydrate Memory

It is interesting that permanent memory is reported when hydrates are formed and re-formed in the laboratory with 0.25 mM of a Type I AFP from winter flounder added to the water of a propane hydrate sII system (Zeng et al., 2006). That is, nucleation rates observed in the initial hydrate formation remain the same in ongoing cycles, not being erased by time elapse or moderate temperature elevations.

8.6.2.5 Sediment Memories from Shifting Vents in Wipeout Zones

From the evaluation of cores taken from within the MC-118 acoustic wipeout zone, Lapham et al. (2008) concluded that hydrocarbon gas venting varied over time and space within the zone, triggering corresponding spatial activities of microbes. In other hydrate zones, episodic movements

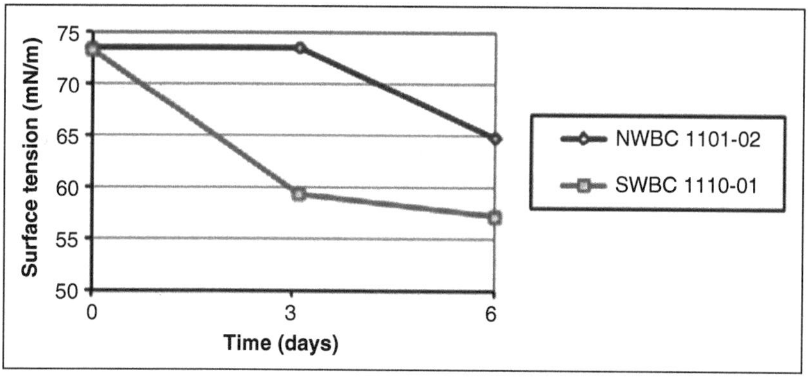

Figure 8.25 ***Microbial longevity in sediments near seafloor gas hydrates (Radich, 2009).***

of gas vents are evident. For example, vent relocations occur in sediments of 1250 m water depth at the submarine mud volcano Håkon Mosby on the Norwegian shelf (Sauter et al., 2006).

Representative samples from two different sites within the MC-118 acoustic wipeout zone were analyzed. Samples were collected near active hydrocarbon vents, and samples were collected near an inactive vent. On culturing indigenous microbes from samples near active gas venting, biosurfactants are produced immediately, but a lag time precedes biosurfactant production at the inactive venting site. See Figure 8.25. Near the inactive vent, spores persist in a dormant state from an earlier active time. Whenever the carbon source is reinstated, whether on the seafloor or in the laboratory culture, microbes activate and biosurfactant production begins after a lag time (Radich, 2009).

This introduces an altogether different memory effect involving the seafloor microbial community in sediments that require activating dormant *in situ* microbes by reintroduction of a carbon source.

REFERENCES

Abay, H.K., Hovring, E., Svartas, T.M., 2011. Does PVCAP promote nucleation of structure II hydrate? In: Proceedings of the 7th International Conference on Gas Hydrates (ICGH 2011), Edinburgh, Scotland, UK, July 17–21. Paper 23.

Amende, B., Gordon, J., 1999. Literature and analyses reports on Snomax. York Snow Inc., Victor, NY.

Anbar, A.D., 2004. Iron stable isotopes: beyond biosignatures. Earth Planetary Sci. Lett. 217, 223–236.

Arima, K., Kakinuma, A., Tamura, G., 1968. Surfactin: a crystalline peptidelipid surfactant produced by *Bacillus subtilis*: isolation, characterization and its inhibition of fibrin clot formation. Biochem. Biophys. Res. Commun. 31, 488–494.

Bai, G., Brusseau, M.L., Miller, R.M., 1997. Influence of a rhamnolipid biosurfactant on the transport of bacteria through a sandy soil. Appl. Environ. Microbiol. 63, 1866–1873.

Banat, I.M., 1995. Biosurfactants production and possible uses in microbial enhanced oil recovery and oil pollution remediation: a review. Bioresour. Technol. 51, 1–12.

Bazylinski, D.A., Moskowitz, B.M., 1997. Microbial biomineralization of magnetic iron minerals: microbiology, magnetism and environmental significance. In: Banfield, J.F., Nealson, K.H. (Eds.), Ribbe, P.H. (Senior Ed.), Reviews in Mineralogy, vol. 35, Geomicrobiology: Interactions Between Microbes and Minerals, Mineralogical Society of America, Washington, D.C. (Chapter 6).

Bazylinski, D.A., Frankel, R.B., Konhauser, K.O., 2007. Modes of biomineralization of magnetite by microbes. Geomicrobiol. J. 24, 465–475.

Bender, G.R., Marquis, R.E., 1985. Spore heat resistance and specific mineralization. Appl. Environ. Microbiol. 50 (6), 1414–1421.

Bernheimer, A.W., Avigad, L.S., 1970. Nature and properties of a cytolytic agent produced by *Bacillus subtilis*. J. Gen. Microbiol. 61 (3), 361–369.

Beveridge, T.J., Murray, R.G.E., 1980. Sites of metal deposition in the cell wall of *Bacillus subtilis*. J. Bacteriol. 141 (2), 876–887.

Birdsell, D.C., Doyle, R.J., Morgenstern, M., 1975. Organization of teichoic acid in the cell wall of *Bacillus subtilis*. J. Bacteriol. 121 (2), 726–734.

Blakemore, R., 1975. Magnetotactic bacteria. Science 190, 377–379.

Brewer, P.G., Ussler, W., Peltzer, E., Walz, P., Kirkwood, W., Hester, K., Xin, Z., 2011. Accurate in situ observation of deep-sea sediment dissolved methane profiles in hydrate bearing provinces. In: Proceedings of the 7th International Conference on Gas Hydrates (ICGH 2011), Edinburgh, Scotland, UK, July 17–21. Paper 48.

Buchanan, P., Soper, A.K., Thompson, H., Westacott, R.E., Creek, J.L., Hobson, G., Koh, C.A., 2005. Search for memory effects in methane hydrate: structure of water before hydrate formation and after hydrate decomposition. J. Chem. Phys. 123 (164507), 1–7.

Calvero, 2006. By Calvero (self-made with ChemDraw) (public domain), via Wikimedia Commons. <http://commons.wikimedia.org/wiki/File%3ASodium_dodecyl_sulfate.svg>.

Canfield, D.E., Berner, R.A., 1987. Dissolution and pyritization of magnetite in anoxic marine sediments. Geochim. Cosmochim. Acta 51, 645–659.

Castro, G., Panilaitis, B., Kaplan, D., 2008. Emulsan, tailorable biopolymer for controlled release. Bioresour. Technol. 99, 4566–4571.

Champion, J.T., Gilkey, J.C., Lamparski, H., Retterrer, J., Miller, R.M., 1995. Electron microscopy of rhamnolipid (biosurfactant) morphology: effects of pH, cadmium, and octadecane. J. Colloid Interface Sci. 170, 569–574.

Churchill, S.A., Griffin, R.A., Jones, L.P., Churchill, P.F., 1995. Biodegradation and bioremediation. J. Environ. Qual. 24, 19–28.

Clausen, P., Andreoni, W., Curioni, A., Hughes, E., Plumer, C.J.G., 2009. Adsorption of low-molecular-weight molecules on a dry clay surface: an *ab initio* study. J. Phys. Chem. C 113, 12293–12300.

Collett, T., 1996. Geologic assessment of the natural gas hydrate resources in the onshore and offshore regions of the United States. In: Proceedings of the 2nd International Conference on Natural Gas Hydrates, Toulouse, France, June 2–6, pp. 499–506.

Cooper, D., MacDonald, C., Duff, S., Kosaric, N., 1981. Enhanced production of surfactin from *Bacillus subtilis* by continuous product removal and metal cation additions. Appl. Environ. Microbiol. 42, 408–412.

Cragg, B.A., Parkes, R.J., Fry, J.C., Weightman, A.J., Rochelle, P.A., Maxwell, J.R., 1996. Bacterial populations and processes in sediments containing gas hydrates (ODP Leg 146: Cascadia margin). Earth Planetary Sci. Lett. 139, 497–507.

Dearman, J.L., 2007. Gas hydrate formation in Gulf of Mexico sediments. Ph.D. Dissertation, Mississippi State University, May, pp. 51–70.

Dearman, J.L., Wilson, W.W., Rogers, R.E., Zhang, G., 2009. Gas-hydrate promotion by smectite–bioproduct interactions. Mar. Chem. 115 (1–2), 21–30.
Doyle, R.J., Matthews, T.H., Streips, U.N., 1980. Chemical basis for selectivity of metal ions by the *Bacillus subtilis* cell wall. J. Bacteriol. 143 (1), 471–480.
Ferris, F.G., Beveridge, T.J., Fyfe, W.S., 1986. Iron-silica crystallite nucleation by bacteria in a geothermal sediment. Nature 320, 609–611.
Ferris, F.G., Fyfe, W.S., Beveridge, T.J., 1988. Bacteria as nucleation sites for authigenic minerals in a metal-contaminated sediment. Chem. Geol. 63, 225–232.
Fujii, T. (Ed.), 1998. Biodetergents. New Products and Applications in Surfactant Technology. Sheffield Academic Press, Sheffield, England, pp. 88–108.
Genzyme Pharmaceuticals, 2001. Synthetic Lipids, Product Data Sheet, Cambridge, MA. <pharmaceuticals@genzyme.com>.
Goodnow, R.A., Harrison, M.D., Morris, J.D., Sweeting, K.B., Laduca, R.J., 1990. Fate of ice nucleation-active *Pseudomonas syringae* strains in Alpine soils and waters and in synthetic snow samples. Appl. Environ. Microbiol. 56, 2223–2227.
Guggenheim, S., van Groos, A.F.K., 2003. New gas-hydrate phase: synthesis and stability of clay-methane intercalate. Geology 31 (7), 653–656.
Gutnick, D.L., 1987. The Emulsan Polymer: Perspectives on a Microbial Capsule as an Industrial Product. Biopolymers. John Wiley & Sons, New York, pp. S223–S240.
Handa, Y., 1986. Compositions, enthalpies of dissociation, and heat capacities in the range 85 to 270 K for clathrate hydrates of methane, ethane, and propane, and enthalpy of dissociation of isobutene hydrate, as determined by a heat-flow calorimeter. J. Chem. Thermodyn. 18 (10), 915–921.
Heerklotz, H., Seelig, J., 2001. Detergent-like action of the antibiotic peptide surfactin on lipid membranes. Biophys. J. 81, 1547–1554.
Herman, D.C., Artiola, J.F., Miller, R.M., 1995. Removal of cadmium, lead, and zinc from soil by a rhamnolipid biosurfactant. Environ. Sci. Technol. 29, 2280–2285.
Housen, B.A., Musgrave, R.J., 1996. Rock-magnetic signature of gas hydrates in accretionary prism sediments. Earth Planetary Sci. Lett. 139, 509–519.
Jeneil Biosurfactant Co., LLC, 2001. Product Data Sheet JBR599, Saukville, WI, August 16.
Jones-Meehan, J., Beard, M., Hind, J., 1995. In situ maintenance of oil/water separators. In: Lardis, A. (Ed.), Annapolis, MD.
Joye, S.B., Boetius, A., Orcutt, B.N., Montoya, J.P., Schulz, H.N., Erickson, M.J., et al., 2004. The anaerobic oxidation of methane and sulfate reduction in sediments from Gulf of Mexico cold seeps. Chem. Geol. 205, 219–238.
Kalogerakis, N., Jamaluddin, A.K.M., Dholabhai, P.D., Bishnoi, P.R., 1993. Effect of surfactants on hydrate formation kinetics, 1993. SPE 25188. In: SPE International Symposium on Oilfield Chemistry, New Orleans, LA, March 2–5.
Karlin, R., Levi, S., 1985. Geochemical and sedimentological control of magnetic properties of hemipelagic sediments. J. Geophys. Res. 90 (B12), 10373–10392.
Kirschvink, J.L., Chang, S.-B.R., 1984. Ultrafine-grained magnetite in deep-sea sediments: possible bacterial magneto-fossils. Geology 12, 559–562.
Klapp, S.A., Enzmann, F., Walz, P., Huthwelker, T., Tuckermann, J., Schwarz, J.-O, Pape, T., Peltzer, E.T., Hester, K.C., Zhang, X., Mokso, R., Wangner, D., Marone, F., Kersten, M., Bohrmann, G., Kuhs, W.F., Stampanoni, M., Brewer, P.G., 2011. Fluid flow in natural gas hydrates revealed by X-ray tomographic microscopy. In: Proceedings of the 7th International Conference on Gas Hydrates (ICGH 2011), Edinburgh, Scotland, UK, July 17–21. Paper 248.
Kosaric, N., 1992. Biosurfactants in industry. Pure Appl. Chem. 64, 1731–1737.
Kothapalli, C.R., 2002. Catalysis of gas hydrates by biosurfactants in seawater-saturated sand/clay. M.S. Thesis, Dave C. Swalm School of Chemical Engineering, Mississippi State University, pp. 43–61.

Kristiansen, E., Pedersen, S.A., Zachariassen, K.E., 2008. Salt-induced enhancement of antifreeze protein activity: a salting-out effect. Cryobiology 57, 122–129.
Lahav, N., 1962. Adsorption of sodium bentonite particles on *Bacillus subtilis*. Plant Soil XVII (2), 191–208.
Lanoil, B.D., Sassen, R., La Duc, M.T., Sweet, S.T., Nealson, K.H., 2001. Bacteria and Archaea physically associated with Gulf of Mexico gas hydrates. Appl. Environ. Microbiol. 67 (11), 5143–5153.
Lapham, L.L., Chanton, J.P., Martens, C.S., Sleeper, K., Woolsey, J.R., 2008. Microbial activity in surficial sediments overlying acoustic wipeout zones at a Gulf of Mexico cold seep. Geochem. Geophys. Geosyst. 9 (abstract).
Larrasoana, J.C., Roberts, A.P., Musgrave, R.J., Gracia, E., Pinero, E., Vega, M., Martinez-Ruiz, F., 2007. Diagenetic formation of greigite and pyrrhotite in gas hydrate marine sedimentary systems. Earth Planetary Sci. Lett. 261, 350–366.
Lederhos, J.P., Long, J.P., Sum, A., Christiansen, R.L., Sloan, Jr., E.D., 1996. Effective kinetic inhibitors for natural gas hydrates. Chem. Eng. Sci. 51 (8), 1221–1229.
Lee, M.S., 2001. The Effects of Biosurfactants on Gas Hydrate Formation in Ocean Sediments. Thesis, Master of Science, Chemical Engineering, Mississippi State University.
Lee, J.D., Englezos, P., 2008. Unusual kinetic inhibitor effects on gas hydrate formation. Chem. Eng. Sci. 61, 1368–1376.
Lins, U., Keim, C.N., Evans, F.F., Farina, M., Buseck, P.R., 2007. Magnetite (Fe_3O_4) and greigite (Fe_3S_4) crystals in multicellular magnetotactic prokaryotes. Geomicrobiol. J. 24, 43–50.
Lorenson, T.D., Claypool, G.E., Dougherty, J.A., 2008. Natural gas geochemistry of sediments drilled on the 2005 Gulf of Mexico JIP cruise. Mar. Pet. Geol. 25, 873–883.
Louisajb, 2011. By Louisajb (talk) 09:11, August 1 (UTC) (own work) (public domain or public domair). <http://commons.wikimedia.org/wiki/File%3ASurfactin.png>.
McConnell, D.R., Zhang, Z., Boswell, R., 2012. Review of progress in evaluating gas hydrate drilling hazards. Mar. Pet. Geol. 34, 209–223.
Moudrakovski, I.L., Sanchez, A.A., Ratcliffe, C.I., Ripmeester, J.A., 2001. Nucleation and growth of hydrates on ice surfaces: new insights from ^{129}Xe NMR experiments with hyperpolarized xenon. J. Phys. Chem. B 105, 12338–12347.
Musgrave, R.J., Bangs, N.L., Larrasoana, J.C., Gracia, E., Hollamby, J.A., Vega, E., 2006. Rise of the base of the gas hydrate zone since the last glacial recorded by rock magnetism. Geology 34 (2), 117–120.
Nakano, M.M., Marahiel, M.A., Zuber, P., 1988. Identification of a genetic locus required for biosynthesis of the lipopeptide antibiotic surfactin in *Bacillus subtilis*. J. Bacteriol. 170, 5662–5668.
Nitschke, M., Costa, S.G.V.A.O., Contiero, J., 2005. Rhamnolipid surfactants: an update on the general aspects of these remarkable biomolecules. Biotechnol. Prog. 21, 1593–1600.
Ohmura, R., Uchida, T., Takeya, S., Nagao, J., Minagawa, H., Ebinuma, T., Narita, H., 2003. Clathrate hydrate formation in (methane + water + methylcyclohexanone) systems: the first phase equilibrium data. J. Chem. Thermodyn. 35, 2045–2054.
Orcutt, B.N., Boetius, A., Lugo, S.K., MacDonald, I.R., Samarkin, V.A., Joye, S.B., 2004. Life at the edges of methane ice: microbial cycling of carbon and sulfur in Gulf of Mexico gas hydrates. Chem. Geol. 205, 239–251.
Paasche, O., Lovlie, R., Dahl, S.O., Bakke, J., Nesje, A., 2004. Bacterial magnetite in lake sediments: late glacial to Holocene climate and sedimentary changes in northern Norway. Earth Planetary Sci. Lett. 223, 319–333.
Parkes, R.J., Cragg, B.A., Bale, S.J., Getliff, J.M., Goodman, K., Rochelle, P.A., Fry, J.C., Weightman, A.J., Harvey, S.M., 1994. Deep bacterial biosphere in Pacific Ocean sediments. Nature 371, 410–413.

Radich, J.G., 2009. Laboratory and theoretical investigations of direct and indirect microbial influences on seafloor gas hydrates. A thesis in partial fulfillment of the requirements for the degree of Master of Science in Chemical Engineering in the Dave C. Swalm School of Chemical Engineering, Mississippi State University, Mississippi State, MS.

Radich, J.G., Rogers, R.E., French, W.T., Zhang, G., 2009. Biochemical reaction and diffusion in seafloor gas hydrate capillaries: implications for gas hydrate stability. Chem. Eng. Sci. 64, 4278–4285.

Rogers, R.E., Kothapalli, C., Lee, M.S., Woolsey, J.R., 2003. Catalysis of gas hydrates by biosurfactants in seawater-saturated sand/clay. Can. J. Chem. Eng. 81, 1–8.

Rogers, R.E., Zhang, G., Kothapalli, C., French, W.T., 2004. Laboratory evidence of microbial–sediment–gas hydrate synergistic interactions in ocean sediments. ISOPE, Paper 2004-JSC-218. Presented at the Fourteenth International Offshore and Polar Engineering Conference, Toulon, France, May 23–28.

Rogers, R.E., Sassen, R., Dearman, J.S., Zhang, G., 2006. Factors prompting seafloor experiments to investigate microbial/hydrate relationships. ISOPE, Paper 2006-TK-01. Presented at the Sixteenth International Offshore and Polar Engineering Conference, San Francisco, CA, May 28 to June 2.

Rogers, R., Zhang, G., Dearman, J., Woods, C., 2007. Investigation into surfactant/gas-hydrate relationship. In: Gas Hydrates and Clathrates, J. Pet. Sci. Eng., 56, 1–3 (special volume).

Rosenberg, E., 1986. Effect of anionic biosurfactant on hexadecane partitioning in multiphase systems. Environ. Sci. Technol. 28, 1993–2000.

Rosenberg, E., 1993. Microbial diversity as a source of useful biopolymers. J. Ind. Microbiol. 11, 131–137.

Sassen, R., MacDonald, I.R., Requejo, A.G., Guinasso, J.N., Kennicutt, J.M., Sweet, S.T., et al., 1994. Organic geochemistry of sediments from chemosynthetic communities, Gulf of Mexico slope. Geo-Mar. Lett. 14, 110–119.

Sassen, R., MacDonald, I.R., Guinasso, N.L., Joye, S., Requejo, A.G., Sweet, S.T., Alcalá-Herrera, J., DeFreitas, D.A., Schink, D.R., 1998. Bacterial methane oxidation in sea-floor gas hydrate: significance to life in extreme environments. Geology 26 (9), 851–854.

Sassen, R., Joye, S., Sweet, S.T., DeFreitas, D.A., Milkov, A.V., MacDonald, I.R., 1999. Thermogenic gas hydrates and hydrocarbon gases in complex chemosynthetic communities, Gulf of Mexico continental slope. Org. Geochem. 30, 485–497.

Sassen, R., Roberts, H., Carney, R., Milkov, A.V., DeFreitas, D.A., Lanoil, B., et al., 2004. Free hydrocarbon gas, gas hydrate, and authigenic minerals in chemosynthetic communities of the northern Gulf of Mexico continental slope: relation to microbial processes. Chem. Geol. 205, 195–217.

Sauter, E.J., Muyakshin, S.I., Charlou, J.-L., Schlüter, M., Boetius, A., Jerosch, K., Damm, E., Foucher, J.-P., Klages, M., 2006. Methane discharge from a deep-sea submarine mud volcano into the upper water column by gas hydrate-coated methane bubbles. Earth Planetary Sci. Lett. 243, 354–365.

Sposito, G., Prost, R., Gaultier, J.P., 1983. Infrared spectroscopic study of adsorbed water on reduced-charge Na/Li-montmorillonite. Clay Clay Miner. 31 (1), 9–16.

Stolz, J.F., Chang, S-B.R., Kirschvink, J.L., 1986. Magnetotactic bacteria and single-domain magnetite in hemipelagic sediments. Nature 321, 849–851.

Takeya, S., Hori, A., Hondowh, T., Uchida, T., 2000. Freezing-memory effect of water on nucleation of CO_2 hydrate crystals. J. Phys. Chem. B 104, 4164–4168.

Thangamani, S., Shreve, G.S., 1994. Effect of anionic biosurfactant on hexadecane partitioning in multiphase systems. Environ. Sci. Technol. 28, 1993–2000.

Thompson, H., Soper, A.K., Buchanan, P., Aldiwan, N., Creek, J.L., Koh, C.A., 2006. Methane hydrate formation and decomposition: structural studies via neutron diffraction and empirical potential structure refinement. J. Chem. Phys. 124, 164508-1–1164508-12.

Uchida, T., Ebinuma, T., Narita, H., 2000. Observations of CO_2-hydrate decomposition and reformation processes. J. Cryst. Growth 217, 189–200.

Volzone, C., Ortiga, J., 2000. O_2, CH_4, and CO_2 gas retentions by acid smectites before and after thermal treatment. J. Mater. Sci. 35, 5291–5294.

Vysniauskas, A., Bishnoi, P.R., 1985. Kinetics of ethane hydrate formation. Chem. Eng. Sci. 40 (2), 299–303.

Walker, S.G., Flemming, C.A., Ferris, F.G., Beveridge, T.J., Bailey, G.W., 1989. Physicochemical interaction of *Escherichia coli* cell envelopes and *Bacillus subtilis* cell walls with two clays and ability of the composite to immobilize heavy metals from solution. Appl. Environ. Microbiol. 55 (11), 2976–2984.

Warth, A.D., Strominger, J.L., 1971. Structure of the peptidoglycan from vegetative cell walls of *Bacillus subtilis*. Biochemistry 10 (24), 4349–4358.

Wei, Y.-H, Chu, I.-M., 2002. Mn^{2+} improves surfactin production by *Bacillus subtilis*. Biotechnol. Lett. 24, 479–482.

Woods, C.E., 2004. Examination of the effects of biosurfactant concentration on natural gas hydrate formation in seafloor porous media. M.S. Thesis, Mississippi State University.

Xiong, S., 2009. Microbially induced and disrupted memory phenomena during gas hydrate occurrences in seafloor sediments. M.S. Thesis, Mississippi State University, August, pp. 46–61.

Zeng, H., Wilson, L.D., Walker, V.K., Ripmeester, J.A., 2006. Effect of antifreeze protein on nucleation, growth and memory of gas hydrates. AIChE J. 52 (9), 3304–3309.

Zhang, G., Rogers, R.E., French, W.T., Lao, W., 2007. Investigation of microbial influences on seafloor gas-hydrate formations. Mar. Chem. 103, 359–369.

Zhong, Y., Rogers, R.E., 2000. Surfactant effects on gas hydrate formation. Chem. Eng. Sci. 55 (19), 4175–4187.

Zuckerberg, A., Diver, A., Peeri, Z., Gutnick, D.L., Rosenberg, E., 1979. Emulsifier of *Arthrobacter RAG-1*: chemical and physical properties. Appl. Environ. Microbiol. 37 (3), 414–420.

CHAPTER NINE

Hydrate Zone Ecology

A complex deepwater ecology, closely associated with cold seeps, thrives around offshore gas hydrates. Microbes proliferate in orders of magnitude greater concentrations around active cold seeps, hydrate masses, and chemosynthetic communities as a consequence of the carbon trove. Here, massive authigenic carbonates and gas hydrates outcrop; fissures in the sediments once venting hydrocarbons fill with hydrates or carbonates to leave evidence of commonly altered flow paths. Populations of tube worms, ice worms, clams, mussels, and microbial mats adjust to the flow of carbon and nutrients within the seeps and to the hydrocarbon gases stored in hydrates. Anaerobic oxidation of methane (AOM) in near-surface sediments saturated with seawater sulfate becomes particularly important to the ecological balance.

Associations within the ecological domain are not benign (Sassen et al., 1993). The dynamic synergy between chemosynthetic communities, gas hydrates, minerals, and hydrocarbon flows throughout the sediments will be discussed in the current chapter.

9.1 SEAFLOOR GAS VENTING

Around seafloor cold seeps, AOM converts a fraction of the methane and stoichiometric amounts of seawater sulfate to hydrogen sulfide and carbonate precipitates. Another fraction of the methane percolating through sediments becomes stored in gas hydrates; in fact, bioproducts from indigenous microbes promote carbon accumulations in hydrates (Lapham et al., 2008). Hydrates, seeps, chemosynthetic communities, and microbes remove or replenish carbon dioxide, hydrogen sulfide, and methane in processes that maintain a sophisticated, seafloor balance.

Chemosynthetic communities locate at water–sediment interfaces of these carbon-rich sites and enter into the methane–sulfate–sulfide–carbon dioxide balance (Boetius and Suess, 2004; Wegener et al., 2008; Sommer et al., 2007). Specific community species oxidize sulfides, where the sulfides come primarily from AOM, and this conversion helps replenish sulfates extracted by AOM from seawater in the near-surface sediments. Therefore, sulfate renewal helps perpetuate methane anaerobic oxidation and the

Offshore Gas Hydrates. http://dx.doi.org/10.1016/B978-0-12-802319-8.00009-7

concomitant carbonate precipitation. One may consider hydrates, microbes, and chemosynthetic communities within the seafloor to be components of a giant gas-emission regulator, modulating methane flux to the atmosphere, storing methane in hydrates, extracting hydrogen sulfide, and storing converted carbon as carbonate precipitates.

Chemosynthetic communities, so named because their entities thrive on chemical energy in the absence of light, were first discovered in 1977 around warm waters venting from the seafloor of Galapagos Rift, Pacific Ocean. The Corliss and Ballard team, descending approximately 2.5 km in the submersible *Alvin*, observed warm water fans extending into the seawater above hydrothermal vents of elevated methane, carbon dioxide, and hydrogen sulfide concentrations (Kennicutt et al., 1985; Corliss and Ballard, 1977). Water temperatures within the fans were found to reach 12–17°C, whereas vents close to the seafloor attained 22°C near tube worm outgrowths (Arp and Childress, 1981).

Eight years after these hydrothermal vents were discovered, Kennicutt et al. (1985) first reported cold seeps in the Gulf of Mexico (GOM) where chemosynthetic communities thrived and gas hydrates outcropped the gas-saturated sediments.

9.2 SOURCES OF VENTING GASES

9.2.1 Thermogenic Gases

Gas hydrates around cold seeps in the GOM usually contain mixtures of biogenic and thermogenic gases. Frequently, fractures develop from shifting of the salt base and of salt diapirs below gas hydrate zones, facilitating gas transport from reservoir traps around the penetrating salt mass or from deeper reservoirs. For example, gases stored in the near-surface hydrate accumulations at Bush Hill in the Green Canyon block come from the Joliet field, a one-time commercial site of conventional oil and gas production (Solomon et al., 2008). During transit to surface vents, leaking hydrocarbon gases from source fields become partially altered microbially with attendant biogenic and thermogenic gas mixing.

9.2.2 Biogenic Sources

Three primary biogeochemical reactions occur in acoustic wipeout zones: methanogenesis, AOM, and sulfate reduction (Aharon and Fu, 2000; Arvidson et al., 2004; Joye et al., 2004; Orcutt et al., 2005).

9.2.2.1 Methanogenesis

Twenty percent of natural gas stored in nature comes from microbial sources, and Whiticar et al. (1986) remind that methane comprises an overwhelming 99% of this biogenic natural gas. Understandably, the creation of methane through methanogenesis is an important phenomenon in the accumulation of seafloor gas hydrates. Even at moderate depths and temperatures, microbial activity contributes to the methane generation because light and volatile fatty acids in the organic matter provide substrates. Acetates lead to methane through the reaction path of Equation 9.1. This is the most direct route to methane, known since 1887 – the fermentation of acetate (Whiticar et al., 1986; Claypool and Kaplan, 1974):

$$CH_3COOH \rightarrow CH_4 + CO_2 \tag{9.1}$$

The second route to methane involves the microbial reduction of carbon dioxide by a different methanogen species (Whiticar et al., 1986; Burdige, 2006; Borowski et al., 1999). See Equation 9.2:

$$4H_2 + CO_2 \rightarrow CH_4 + 2H_2O \tag{9.2}$$

Acetate decomposition to methane is preempted in the sulfate zone of marine sediments by sulfate reducers, which supersede methanogen access to acetate. As a consequence, methanogenesis becomes dominant at depths beyond the sulfate–methane interface (SMI) where sulfate concentrations fall below about 1 mM (Burdige, 2006; Sorensen et al., 1981; Winfrey and Ward, 1983).

Are large enough quantities of biogenic methane available in general for gas hydrate seafloor accumulations? Wellsbury et al. (1997) addressed the question specifically for Blake Ridge, studying the adequacy of substrate below bottom of gas hydrate stability (BGHS) to supply elevated levels of methane through methanogenesis. In the laboratory, Blake Ridge sediments containing organic matter were subjected to temperature increases that would be experienced by progressively deeper seafloor burial. Wellsbury et al. found their laboratory trend of acetate generation checking closely with observed acetate and temperature increases through the Blake Ridge gas hydrate zone and below BGHS. Figure 9.1a, adapted from the Blake Ridge data, shows acetate concentration increases with depth.

Note the approximately 1500 orders of magnitude acetate concentration increase over the interval between seafloor and 700 mbsf. Figure 9.1a and b substantiates sulfate reducer predominance over methanogen conversion of

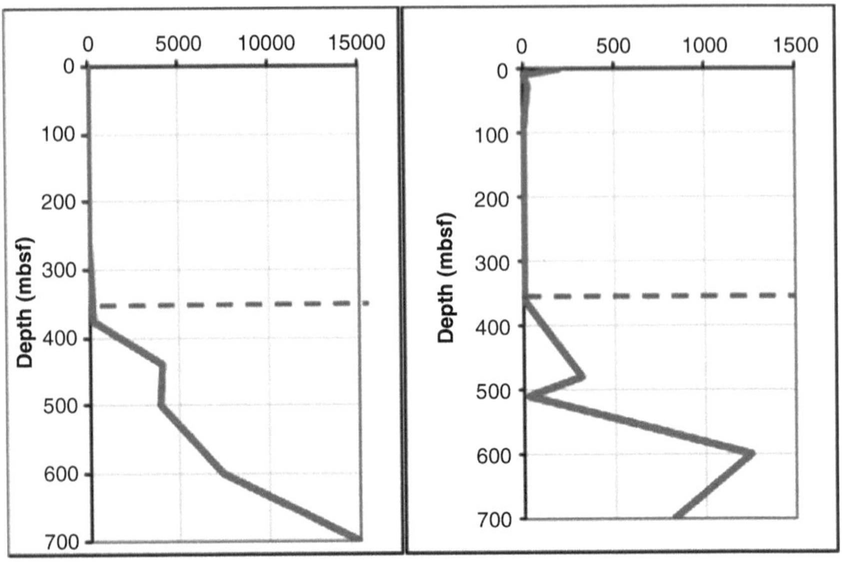

(a) (b)

Figure 9.1 ***Acetate contribution to methanogenesis, deep ocean sediments.*** (a) Acetate concentration (μM); (b) acetate methanogenesis rate (nmol/(ml day)) (Wellsbury et al., 1997).

acetates in the shallow Blake Ridge sediments above SMI. Measurable methanogenesis is nil until depth surpasses the sulfate zone at about 380 mbsf. Figure 9.1b shows acetate being the primary support for methane generation in sediments deeper than about 500 mbsf to the depth of the deepest core at about 700 mbsf and probably beyond (Wellsbury et al., 1997).

Around gas hydrate deposits at 370–500 mbsf, the study by Wellsbury et al. reveals large increases of microbial populations of numerous species, including methanogens and their production of methane directly from acetate. See Figure 9.2a and b.

9.2.2.2 Anaerobic Oxidation of Methane

Typical of all acoustic wipeout zones permeated by hydrocarbon seeps amid gas hydrate accumulations, GOM sulfate reductions around cold seeps were observed by Aharon and Fu (2000) to attain rates reaching 600 times more rapid than reduction rates of sulfate in background waters.

In upper sediments saturated with sulfate at 28 mM, the seawater concentration of sulfate (Burdige, 2006), methane flux promotes the AOM given by Equation 9.3 (Boetius et al., 2000; Orcutt et al., 2005):

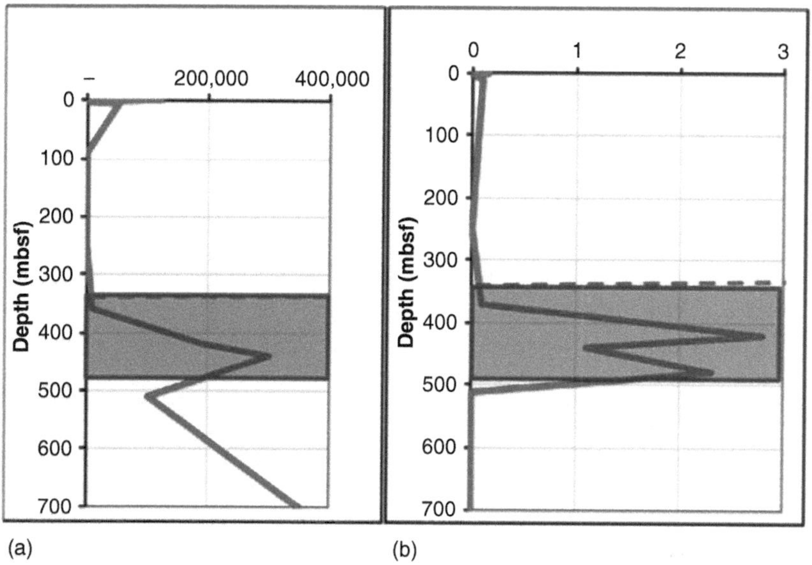

Figure 9.2 ***Acetate contribution to methanogenesis, deep ocean sediments.*** (a) CO_2 rate from acetate (nmol/(ml day); see Equation 9.1); (b) methanogenesis rate from HCO_3 (nmol/(ml day); see Equation 9.2) (Wellsbury et al., 1997).

$$CH_4 + SO_4^{2-} \rightarrow HCO_3^- + HS^- + H_2O \tag{9.3}$$

The overall reaction represented by Equation 9.3 is a composite of the following reactions facilitated by Archaea and sulfate-reducing bacteria (Zhang et al., 2002). In Equation 9.4 methane oxidizes anaerobically to acetate and in Equation 9.5 to carbon dioxide – brought about by Archaea (Ussler et al., 2005):

$$2CH_4 + 2H_2O \rightarrow CH_3COOH + 4H_2 \tag{9.4}$$

$$CH_4 + 2H_2O \rightarrow CO_2 + 4H_2 \tag{9.5}$$

Then, Equations 9.6 and 9.7 show the product of sulfate-reducing bacteria:

$$CH_3COO^- + SO_4^{2-} \rightarrow 2HCO_3^- + HS^- \tag{9.6}$$

$$H^+ + 4H_2 + SO_4^{2-} \rightarrow HS^- + 4H_2O \tag{9.7}$$

Table 9.1 Dimensions of Archaea and sulfate-reducing microbial consortium (Boetius et al., 2000)

Diameter of inner sphere of 100 Archaeal cells	Diameter of single Archaeal cell	Diameter of one sulfate-reducing bacterial cell	Diameter of consortium sphere
2.3 μm	0.5 μm	0.3–0.5 μm	1–11 μm range 3.2 μm average

Archaea and sulfate-reducing bacteria work in consort to accomplish the AOM. Boetius et al. (2000) explain the efficiency of the process, as approximately 100 Archaea cells aggregate to form a dense spherical core while being surrounded by 200 sulfate-reducing bacteria. With the arrangement, intermediates in the reaction have short paths to reach the site of the next reaction. The relative sizes of the anaerobic oxidation consortium are presented in Table 9.1.

The HCO_3^- product in the overall reaction of Equation 9.3 leads to $CaCO_3$ precipitation in alkaline environments, where solid calcium carbonate depositions are assisted by pH increases of pore waters from the sulfate-reducing/methane-oxidizing microbial consortium (Zhang et al., 2002). The HCO_3^-, dissolved inorganic carbon (DIC), reaches a maximum at the bottom of the SMI, and carbonate nodules are usually found below the SMI.

This relationship of DIC, SMI, and carbonate precipitates is illustrated by cores taken from sediments having a 3.00 mbsf SMI beneath a 647 m water column in the Mississippi Canyon. DIC reaches a maximum of 13.5 mM near the maximum negativity of carbon-13 (Ussler et al., 2005). Carbonate nodules occur below the sulfate zone between 3.00 and 4.50 m.

Some random, observed depths of the SMI in gas hydrate provinces are given in Table 9.2.

Borowski et al. (1999) suggest that typical sulfate zones, having negligible methane seeps or gas hydrate deposits, might have SMIs defined at 50 m or more sediment depth. Paull et al. (2005b) corroborate this in a northern GOM scientific cruise on extracting and analyzing 25 cores. Nine cores came from mounds and 16 came from basinal areas of which 435 samples were analyzed. The shallowest SMIs of 35 cmbsf reflect high methane flux, especially around the hydrate mounds or active methane vents. This notably agrees with other worldwide studies finding SMI depths much shallower around gas hydrates or hydrocarbon gas vents. In the basinal areas, or background areas with insignificant methane flux, SMI ranged beyond the 3745 cmbsf sampling limit.

Table 9.2 Comparative SMI of hydrate zones and background values

Location	SMI below seafloor (cm)	Conditions	References
Hydrate Ridge	5	Below bacterial mats	Boetius et al. (2000)
Guaymas Basin	15		Paull et al. (2005a)
Sea of Okhotsk	40	Gas hydrate retrieved	Cho et al. (2005)
	50	Gas hydrate retrieved	
	100	Saturated free methane	
	160	Saturated free methane	
	30–50	Hydrate-bearing cores	Hachikubo et al. (2011)
	200	Gas-rich cores	
Sea of Japan	200	Near extensive gas plumes	Matsumoto et al. (2005)
MDO2-2571 MS Canyon, GOM	300	Cores near gas chimney	Ussler et al. (2005)
India: Krishna–Godavari Basin	1,700	Site NGHP-01-10	Riedel et al. (2010)
Bush Hill; Tunica Mound	600	Cores, mound flanks	Paull et al., 2005b
	900–1,200	Cores, adjacent basins, background	
Carolina Rise and Blake Ridge	1,000	From 5 piston cores	Borowski et al. (1999)
	15,000 extrapolated	Background, no hydrates	

As methane flux increases through sediments, the SMI becomes shallower and the sulfate depletion gradient becomes steeper. For example, Table 9.2 indicates an extremely shallow SMI of 5 cm at Hydrate Ridge (Boetius et al., 2000).

Sulfate gradients in Sea of Okhotsk sediments are more than twice as steep, 0.52–0.62 mM/cm, if gas hydrates are present rather than only pore waters being saturated with methane, 0.14–0.26 mM/cm (Cho et al., 2005).

Sulfate gradients become linear if anaerobic methane oxidation at steady state within the sulfate zone dominates, which occurs in the Blake Ridge hydrate province. Shallow SMIs reflect high methane flux, indirectly implying associated hydrates or free methane gas saturating pore waters (Paull et al., 2005a; Cho et al., 2005). The methane flux may be calculated with Equation 9.8, Fick's first law, when sulfate gradients are linear in steady-state conditions (Borowski et al., 1996):

$$J = D_o \phi^3 \frac{\partial C}{\partial x} \qquad (9.8)$$

where J, sulfate flux; D_o, diffusion coefficient of SO_4^{2-} in water; ϕ, porosity; $\partial C/\partial x$, sulfate concentration gradient.

Borowski et al. (1996) determined sulfate gradients from five Carolina Rise and Blake Ridge cores. The steepest gradient $\partial C/\partial x$ of the five cores was 2.9 mM/m. They used the following values in Equation 9.8 to estimate sulfate flux at the Carolina Rise and Blake Ridge sites:

$D_o = 5.8 \times 10^{-6}$ cm^2/s (Li and Gregory, 1974)

$\phi = 0.70$

$$\frac{\partial C}{\partial x} = 2.9\,\text{mM/m}$$

where M may be stated in units of mol/(1000 cm^3).

Borowski et al. calculated sulfate flux to be $J = 1.8 \times 10^{-3}$ mmol/cm^2/year at Blake Ridge. Methane fluxes over the five cores, each core having been extracted above known bottom-simulating reflectors (BSRs), varied by a factor of 16. Possibly, gas hydrate formations/dissociations contribute to an unsteady-state flow rate or diffusion coefficient variations of the sites account for some differences (Borowski et al., 1996).

9.2.2.3 Hydrogen Sulfide

Sulfides replace sulfates according to Equation 9.3 in a 1:1 molar ratio. The replacement begins at the seafloor surface and extends through the sulfate zone. See Figure 9.3.

Consider a potential hazard while recovering cores containing gas hydrates that may have concentrated H_2S through occlusion: hydrogen sulfide gas is instantly fatal to humans when inhaled at greater than or equal to 300 ppm in air (Smith, 2010).

Although H_2S is produced abundantly in sulfate zones around gas hydrate accumulations and H_2S forms hydrates more easily than most guest gases, the sulfide content in near-surface hydrates is usually surprisingly low. In some cases, chemosynthetic communities may efficiently scavenge sulfides – even oxidize and recycle product sulfates. But there are exceptions.

At ODP Site 892 off the coast of central Oregon, where the base of the sulfate zone is 20 mbsf, the hydrogen sulfide content of decomposed hydrates extracted from within the sulfate zone contained 6000–10,000 ppm by volume of H_2S, compared with less than 30 ppm below the sulfate zone.

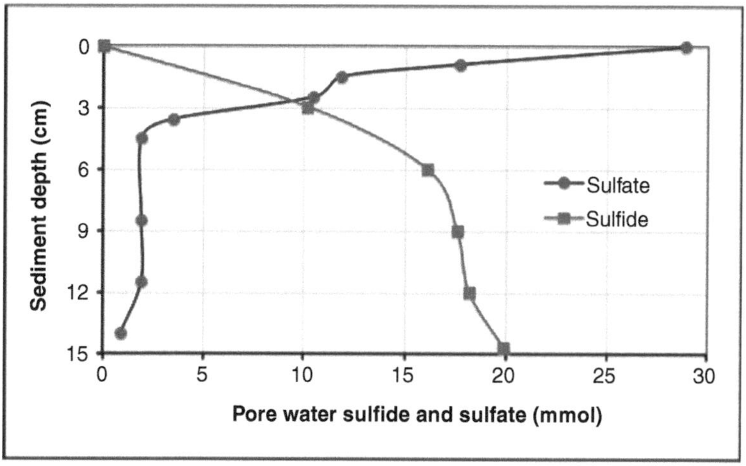

Figure 9.3 ***Reduction of sulfate in the sulfate zone of methane-rich sediments, Hydrate Ridge (Boetius et al., 2000).***

The high H_2S content in recovered hydrates is the first reported (Kastner et al., 1998). Concentrated H_2S in the Site 892 hydrate-containing cores is attributed to advection of methane-rich fluids via faults that replenish methane for AOM sulfide generation, providing sulfide product a direct route to hydrate formation below the scavenging chemosynthetic communities at the seafloor.

High H_2S concentrations with attendant high toxicities generally decrease microbial populations in sediments (Cragg et al., 1996; Brown et al., 1973; van Gemerden, 1993).

Hydrogen sulfide readily forms gas hydrates, forming a clathrate at lower pressures and higher temperatures than methane or carbon dioxide and, in fact, at less stringent conditions than most guest gases (Carroll and Mather, 1991). For example, at 277.6 K, Type I gas hydrates form from hydrogen sulfide at only 0.157 MPa (Selleck et al., 1952).

9.3 MICROBES AFFECT GAS HYDRATES IN DEEP OCEAN SEDIMENTS

Biogenic methane is the predominant hydrate guest gas in offshore hydrate provinces of Cascadia margin, Blake Ridge, and Nankai Trough (Kastner, 2002). Microbial sources extend through the gas hydrate zone and the propitious temperature regimes near BGHS interfaces, but important activity goes surprisingly far below the hydrate zone.

Early in the Ocean Drilling Program, Parkes et al. (1994) analyzed deep cores from five sites taken at widely separated locations in the Pacific Ocean. Temperatures of these deep sediments ranged from 3 to 80°C. Since no gas hydrates were evident to influence microbial cell count, the counts at the five sites are considered a control for widespread comparisons. Cell density data as a function of depth for the Parkes study are given in the Parkes Equation 9.9, a regression line representing the best fit to 299 data points, where the microbial count units are given in cell numbers/cm^3 and depth as mbsf:

$$\log_{10} \text{count} = 8.06 - 0.715 \log_{10} \text{depth} \tag{9.9}$$

Microbial activity trends indicated no discontinuities through measured depths of the control, suggesting probable significant activity continuing beyond the study's 1000 mbsf depth limit.

At Hydrate Ridge, microbial activity near gas hydrates down to BSR gives insight into prevailing biogenic gas contents of the hydrates. Data verify the presence of sulfate reducers as well as methanogens from seafloor to the 225 mbsf BGHS. Within the interval, gas hydrates exist at 215–225 mbsf, where total cell count is 10 times greater than the control area having no gas hydrates (Boetius and Suess, 2004). The greatest increase in bacterial count occurred at about 227.7 mbsf (Cragg et al., 1996).

In further ODP Site 889/890 data evaluations, Barnes et al. (1998) report faster cell growth in laboratory cultures over a large temperature range of 15–45°C of samples from the 222-mbsf depth compared with cultures of microbes from the site's shallow sediments that demonstrated growth over a narrow temperature range of 9–19°C.

Significantly below the 225 mbsf BSR of this ODP Site 889/890, cores at 234 mbsf were extracted from the sediments estimated to be 2–3 million years old. Samples demonstrated substantial microbial activity at 9 m below current hydrate stability depth, including methanogenesis reactions (Marchesi et al., 2001).

9.4 CHEMOSYNTHETIC COMMUNITIES

9.4.1 Chemosynthetic Communities at Venting Sites

Chemosynthetic communities exist on the seafloor around hydrocarbon cold seeps in acoustic wipeout zones. The wipeout zones may exhibit gas chimneys, faults, shifting fractures, and mud volcanoes in their fine-grained sediments as gases percolate through sediments from BGHS to surface. Gas hydrates abound in this backdrop. Aharon and Fu (2000) report that mats of *Beggiatoa* bacteria that oxidize sulfides (thiotrophic) and mussel gills housing methane-oxidizing bacteria (methanotrophic symbionts) typically overlay or reside near outcropping hydrates at oil and gas seeps. Chemosynthetic communities are found at water depths greater than about 400 m, the approximate critical minimal depth for gas hydrate stability (Carney, 1994), and have been documented to water depths of 1240–2740 m (Roberts et al., 2010a).

9.4.2 *Beggiatoa* Mats

Bacterial mats spread in the midst of seafloor gas seeps and over hydrate accumulations in the GOM, Cascadia margin, and Blake Ridge. The sulfide-oxidizing bacteria *Beggiatoa* spp. cover tops of mounds at Hydrate Ridge in the Cascadia margin, where orange and white mat–clad hydrate outcrops are reported by Boetius and Suess (2004). In similar circumstances, *Thioploca* spp. form yellowish mats and are found along the continental shelf of Chile and Peru (Fossing et al., 1995).

The *Beggiatoa* are large and filamentous. Zhang et al. (2002) report those at GC 234 in the GOM to have individual cell diameters of 4–5 μm. At Hydrate Ridge, one of many *Beggiatoa* mats measures 0.3 mm thick and extends another centimeter into the sediments below – covering 25 m^2 of surface area. A thin surface sample of these Hydrate Ridge mats yields 500 g *Beggiatoa* biomass per square meter of surface area, although the mass reduces to 25 g of carbon because of voluminous vacuoles within the bacteria (Boetius and Suess, 2004).

The mats at Blake Ridge spread 10–80 cm in width and appear to attain 1–2 mm thickness (Van Dover et al., 2003). The Blake Ridge *Thiovulum* species detailed by Fossing et al. (1995) have filament bundles composed of 7 cm individual filament lengths; 15-cm long sheaths housing filament bundles orient horizontally atop sediments and may form an estimated 2-cm thick mats.

Mats play an important role in the ecosystem surrounding offshore gas hydrates. Primarily, the sulfide-oxidizing *Beggiatoa* composing the mats help replenish sulfate in sediment pore waters to sustain AOM. Within a single long sheath of the *Thiovulum* species, found at Blake Ridge, a bundle of 100 filaments slide freely along the sheath's interior length – vertically extending nitrate-laden filaments 6–10 cm down into the sediments and also a like distance into the water column to scavenge nitrates. In this manner *Beggiatoa* convey nitrates 40 times faster than diffusion, storing the nitrates in large vacuoles to later access for sulfide oxidation and for metabolism energy (Boetius and Suess, 2004; Fossing et al., 1995). Nitrates may be stored to concentrations 20,000 times greater than that in seawater from which they are extracted. Equally efficient is the oxidation of sulfides in the oxygen-depleted sediments below (Fossing et al., 1995).

Suess et al. (1999) made an interesting observation on extracting two cores from South Hydrate Ridge. In each core gas hydrates were present 5–10 cm below sediment surface, but a *Beggiatoa* mat was present on the surface of only one core. Nitrate concentrations within the interstitial pore waters of the two cores were quite different: 2850 μM of nitrate when mat covered the core but less than 1 μM of nitrate when no mat was present. High nitrate content reflected vacuole effects (Boetius and Suess, 2004).

Fossing et al. (1995) suggest the stoichiometry in Equations 9.10 and 9.11 as *Beggiatoa* utilize nitrates to oxidize sulfides:

$$2NO_3^- + 5HS^- + 7H^+ \rightarrow N_2 + 5S + 6H_2O \quad (9.10)$$

After the preceding reaction, sulfates are replenished in the sediments via oxidation of elemental sulfur shown in Equation 9.11:

$$6NO_3^- + 5S + 2H_2O \rightarrow 3N_2 + 5SO_4^{2-} + 4H^+ \quad (9.11)$$

Trace concentrations of oxygen in bottom waters directly above mats penetrate only a few millimeters into the mats before being completely extracted by the *Beggiatoa*, leaving sediments directly below absent of oxygen and containing 18,000 μmol/l sulfides (Sommer et al., 2007; Boetius and Suess, 2004).

Not surprisingly, therefore, *Beggiatoa* mats flourish above extensive AOM, where fractures and faulting systems provide high methane fluxes or hydrates provide methane accumulations. In the GOM, heavier hydrocarbons supplied by seeps may supplement the methane (Boetius et al., 2000; Van Dover et al., 2003; Sommer et al., 2007; Zhang et al., 2005).

9.4.3 Tube Worms

9.4.3.1 Hydrothermal Vent Sites

Tube worm–populated hydrothermal vents were first discovered in 1977 (Kennicutt, 1985; Corliss and Ballard, 1977). Into waters above hydrothermal vents, warm water convection currents distribute sulfides, where vestimentiferan tube worms extend their cane-like structures meters above the seafloor into the sulfide environment. At the upper extremity of the canes, anterior plumes extend vertically into the warm water column, ingesting sulfides through the gills and conveyed through their vascular systems. Hydrogen sulfide concentrations in the water above hydrothermal vents may reach 111,000 μM (Naganuma et al., 2005).

9.4.3.2 Cold-Seep Vent Sites

Near gas hydrate seafloor protrusions at Bush Hill, tube worm–populated cold seeps were discovered in 1984 on the continental slope of the northern GOM (Kennicutt et al., 1985). Tube worm colonies frequent seafloor cold seeps of acoustic wipeout zones. Long-term cold seeps are common in the GOM, as well as worldwide (MacDonald et al., 2003).

As a striking contrast to hydrothermal vents, hydrogen sulfide concentrations in seawater above cold seeps are insufficient to support worm colonies by absorbing H_2S solely through anterior plumes. Julian et al. (1999) determined H_2S concentrations to be less than 1 μM/l near the gills of tube worms growing above GOM cold hydrocarbon seeps. Instead, sulfides are produced by AOM in the pore waters of seafloor sediments near vents, necessitating vestimentiferan tube worms to scavenge H_2S through an extensive root system that spreads into underlying sediments. Mean sulfide concentrations in GOM sediment pore waters below tube worm bushes range up to 1800 μM/l over the interval from surface to 70 cm below the surface (Julian et al., 1999).

Vestimentiferan tube worms associate with cold hydrocarbon seeps to play an important role in the complex ecology of near-surface seafloors in hydrate zones. Of the numerous vestimentiferan species, the most common is *Lamellibrachia* spp. These are reported in the northern GOM at 400 m to greater than 1000 m water depths (Harmer et al., 2008; Freytag et al., 2001; Kennicutt et al., 1985; Fisher et al., 2007). See Figure 9.4. The vestimentiferan tube worms at hydrocarbon cold seeps are estimated to live 170–250 years.

At cold seeps, methane vents through sediments saturated with sulfates. AOM sulfide product is extracted by tube worm roots to feed symbiont microbes sequestered within the tube-like body. By chemosynthesis, the microbes use the sulfide as an energy source to fixate carbon from CO_2 (Dattagupta et al., 2006; Julian et al., 1999).

Symbionts provide nutrition to the vestimentiferan worm host, which in turn provides sulfide, oxygen, and carbon dioxide to the symbionts. Tubular

Figure 9.4 Lamellibrachia *tube worms, Gulf of Mexico (NOAA Photo Library, 2010).*

sections of the worms extend 2 m or more into the water column (Freytag et al., 2001).

For their growth, tube worms depend on symbiotic microbes to supply nutrition. A mature worm has no mouth, gut, or anus (Dattagupta et al., 2006; Jones, 1981; Kimura et al., 2003), although the larvae do (Harmer et al., 2008). The endosymbiotic bacteria originate in the surrounding environment, so that each individual worm becomes implanted with supportive microbes (Harmer et al., 2008). The symbionts are contained within a trophosome, which is a large sac. A sophisticated vascular system (Julian et al., 1999) circulates H_2S and up to 11 ml O_2 per 100 ml of blood to the trophosome. Because hemoglobin in its blood has an extremely high affinity for sulfide, hydrogen sulfide can be extracted from sediments having very low sulfide concentrations (Freytag et al., 2001). Microbes oxidize sulfide within the trophosome, and the sac may contain up to 3.7×10^9 bacteria per 1 g of the sac housing them (Cavanaugh et al., 1981; Arp and Childress, 1981). The bacterial species present may be thiotrophic and/or methanotrophic, but with *Lamellibrachia* spp. its symbionts are usually only thiotrophic (Naganuma et al., 2005).

(a)
(b)
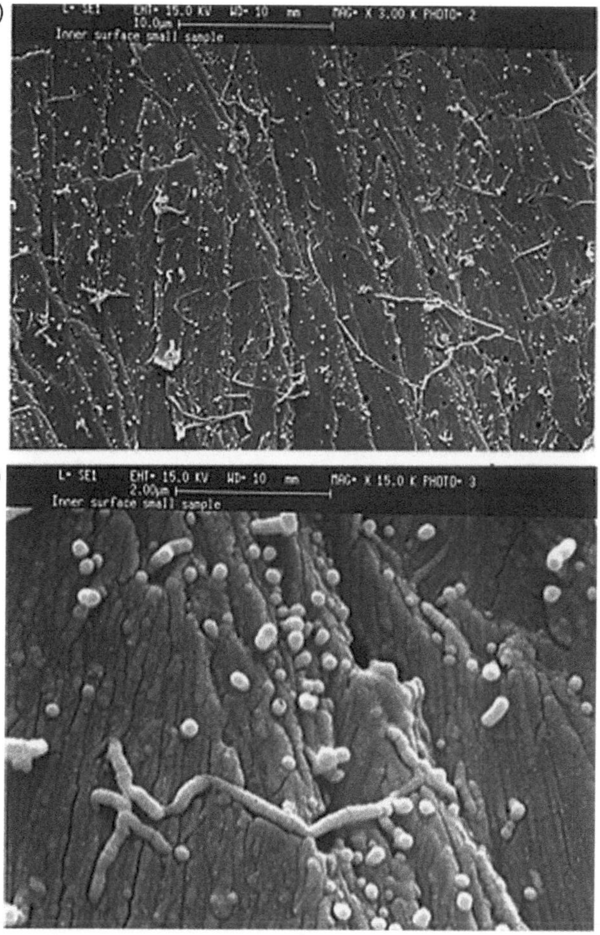

Figure 9.5 ***Interior walls of vestimentiferan tube worm from Bush Hill.*** (a) SEM ×3000 magnification; (b) SEM ×15,000 magnification. *(SEM by R.E. Rogers, MSU Laboratory; tube worm samples provided by R. Sassen and H. Roberts.)*

9.4.3.3 Tube Worm Wall Structure

The relatively inflexible, opaque tubes of the worms provide structure and support, while also offering protection from predators. The tube walls are laminates of composite materials, composed of β-chitin crystallites as the discontinuous phase in a matrix of proteins containing the amino acids glycine, cysteine, asparagines, arginine, and tyrosine (Julian et al., 1999). In Figure 9.5a and b is presented electron micrographs with progressive magnifications of the interior wall surface of a vestimentiferan tube worm retrieved from Bush Hill. The scanning electron microscope (SEM)

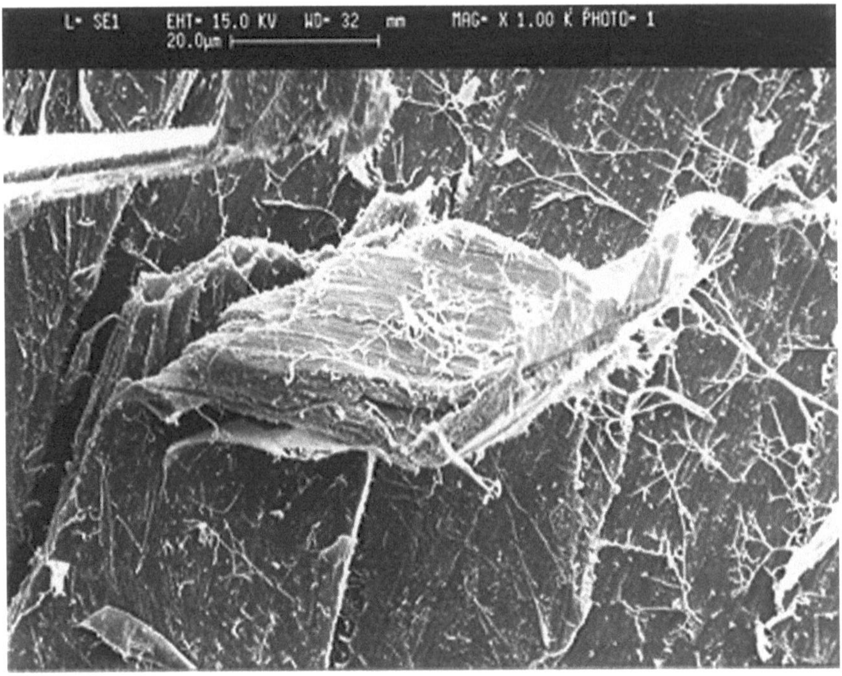

Figure 9.6 ***Break in vestimentiferan tube worm walls reveals structure.*** SEM ×15,000. *(SEM by R.E. Rogers, MSU Laboratory; tube worm samples provided by R. Sassen and H. Roberts.)*

micrographs were taken at the Mississippi State University (MSU) Hydrate Laboratory, samples courtesy of H. Roberts and R. Sassen. Note numerous microbes, including *Beggiatoa*, scattered along the surfaces. Interestingly, seawater-saturated porous media containing sections of these tube worm walls and subjected to hydrate-forming conditions in the laboratory formed gas hydrates throughout the packed sample. But no hydrates formed on the worm segments, reflecting inhibitive effects of amino acids composing the walls and inhibitive effects of microbial cell wall materials.

Note the random occurrence of microbes on the interior walls of the tube worm presented in the SEM micrograph of Figure 9.5. The cell distribution confirms a prediction by Julian et al. (1999) of random bacterial occurrences on a tube worm's wall matrix.

Figure 9.6 reveals features of tube worm wall materials, especially the manual break showing laminations and composite nature of its structure. The micrograph reveals a laminate structure resembling plywood, an analogy first suggested by Gaill et al. (1992).

9.4.3.4 Root System

Rising into the water column from a massive system of intertwined roots are tube worms forming thickets on the order of 1000 single organisms. To sustain the high density, a steady supply of sulfides must be available over long time periods. The sulfide supply comes by means of the roots absorbing AOM sulfide by-product (Cordes et al., 2005; Dattagupta et al., 2006; Julian et al., 1999; Freytag et al., 2001). In fact, the root system becomes extensive to adequately supply symbionts with sulfide and to maintain vestimentiferan metabolism. The roots may extend 1 m or more below sediment surface (Dattagupta et al., 2006). Measuring individual tentacle lengths becomes problematic because of root entanglements and extraction (Freytag et al., 2001). With increasing distance from the tube trunk, the flexible root tentacles maintain an approximately constant diameter of 1.4 mm with 0.07 mm wall thickness. These dimensions compare with trunk wall thicknesses of 0.40 mm (Julian et al., 1999).

The roots grow deeper into sediments if methane flux decreases. Root bends have been reported to be suggestive of root change in direction in order to follow the SMI, but upward root growths may be physically limited by carbonate crusts deposited from earlier SMI changes (Naganuma et al., 2005). Freytag et al. (2001) report that tube worm roots so effectively scavenge toxic H_2S from near-surface sediments that other faunas are protected. The scavenging effect must help account for a surprisingly low concentration of H_2S in surface hydrates – countering an expectation based on the high affinity of gas hydrates for hydrogen sulfide and the substantial production of hydrogen sulfide by AOM.

Sulfide in the sediments may become limited by either unavailability of methane or unavailability of sulfate. In most cases the diffusion rate of ocean waters into sediments is slower than the sulfide generation rate. *Beggiatoa* help replenish pore water sulfate by their oxidation of hydrogen sulfide. See Equations 9.10 and 9.11. To help perpetuate sulfate feed supply, vestimentiferans return sulfate converted from sulfide, distributing the sulfate into the sediments by their extensive root system (Dattagupta et al., 2006; Julian et al., 1999; Freytag et al., 2001). Additional sulfate enters through the anterior plume whereupon the tube worm moves seawater and its sulfate, like a peristaltic pump, through trunk and roots into sediments (Julian et al., 1999).

The comprehensive scavenging and redistribution capability of the vestimentiferans is understandable from the photograph of their root system given in Figure 9.7.

Figure 9.7 ***Root system of vestimentiferan tube worms.*** Public domain. *(By Gulf of Mexico 2002, NOAA/OER National Oceanic and Atmospheric Administration–Ocean Exploration and Research.)*

9.4.3.5 Ultra-Deepwater Occurrences

Roberts et al. (2010b) report the deepest occurring tube worms found to date at Green Canyon Block 600 of the GOM, where the vestimentiferans thrive at 1250 m water depth fed by cold seeps alongside carbonate deposits.

9.4.4 Mussels, Clams, Ice Worms

Vesicomyid clams, genus *Calyptogena* spp., and solemyid bivalve mollusk *Acharax* spp. populate the southern peak of Hydrate Ridge (Sommer et al., 2007; Torres et al., 2002). Boetius and Suess (2004) detail relative seafloor coverage at one site as follows: (1) *Beggiatoa* mats 1–25 m^2, (2) vesicomyid clam fields less than 25 m^2, and (3) bivalve mollusk beds indeterminate because they reside 5–30 cm below the sediment surface.

Situated over an area of seafloor with high sulfide flux reaching 63 mmol/m^2/day, *Beggiatoa* mats are at the center of Hydrate Ridge communities. Situated at seep edges, vesicomyid clams reduce sulfide flux rates to 18 mmol/m^2/day. Further outside the seep periphery, buried *Acharax* bivalve mollusks hover over the lowest sulfide flux (Sommer et al., 2007; Torres et al., 2002).

Each community member exhibits a specific efficiency in filtering the methane. Methane retention/conversion efficiencies were measured at Hydrate Ridge as follows: (1) *Beggiatoa* mats in high-flux areas allow only 50% of the transient methane to reach the water column above; (2) vesicomyid clams allow less than 15% of the methane to escape; (3) *Acharax* bivalve beds allow essentially no methane to pass into the water column above them (Boetius and Suess, 2004).

Mussels found in the GOM by Brooks et al. (1987) support only methanotroph symbionts, which supply the carbon source for mussel tissues (Brooks et al., 1987).

Polychaete ice worms *Hesiocaeca methanicola* burrow into gas hydrates to feed on bacteria sequestered within the hydrate mass. Unlike tube worms, polychaetes have guts and do not depend on symbionts to sustain life, but get nutrition by consuming sulfide-oxidizing bacteria. Fisher et al. (2000) found oxygen consumption to be 2.9–67 μg O_2/min per gram net weight of the polychaete worms.

Sulfide-oxidizing *Beggiatoa* were observed to populate outcropping hydrate surfaces with polychaete ice worms burrowing into the hydrate surface (Orcutt et al., 2004). Ice worm densities on hydrate surfaces have been measured to reach 2500 worms per square meter of hydrate surface area at the time of discovery, whereas a month later the densities had increased to 3000/m^2. In most cases the worms are 2–4 cm long and each creates a hole in the hydrate surface of of approximately 4 cm depth × 3 cm width × 2 cm height. As worm densities increase, larger hydrate surface areas favor more rapid hydrate decompositions (Fisher et al., 2000).

9.4.5 Chemosynthetic Communities in Hydrate Zones, Gulf of Mexico

Chemosynthetic communities were first discovered on the GOM continental slope in less than 1000 m water depths and here early studies concentrated in the shallower gas hydrate zones (Paull et al., 1984). Later, Roberts et al. (2010a) found that the communities extend into deep waters of potential hydrate exploration sites near the leading edge of the Sigsbee Escarpment; sites were subsequently explored by the group with deep-diving Alvin and Jason II remotely operated vehicles (ROVs). The deepwater communities were typically found along lines of faults where crude oil and natural gas surfaced as cold seeps in waters 1240–2740 m deep. For example, Roberts et al. (2010a) observed carbonate and chemosynthetic

Figure 9.8 ***Linear surface fairway along a 1240 m deep GOM fault where carbonates, vestimentiferan tube worms, mussels, and clams reside along cold seep (Roberts et al., 2010a).***

communities similar to those on the upper slope along a fault line in 2740 m water depth at Alaminos Canyon-818 site where crude oil and natural gas migrate.

As another example, view the photograph of deep chemosynthetic communities in Figure 9.8 (Roberts et al., 2010a). There, vestimentiferan tube worms grow amidst carbonate rubble in 1240 m deep waters of Green Canyon, exemplary of thousands of cold seeps occurring in the northern GOM along the continental slope.

9.5 CARBONATE DEPOSITS

Prolific authigenic carbonate deposits occur universally in offshore hydrate zones (Boetius and Suess, 2004). The carbonates derive as a product of AOM, where bicarbonate ion by-products react with calcium, magnesium, and strontium cations in alkaline seawater to form carbonate precipitates. Prevalence of calcium ions leads to predominance of an aragonite carbonate phase (Luff and Wallman, 2003).

Near the seafloor surface, AOM transforms methane as given in Equation 9.12:

$$CH_4 + SO_4^{2-} \rightarrow HCO_3^- + HS^- + H_2O \qquad (9.12)$$

Note that bicarbonate and sulfide ions are produced in equimolar amounts during AOM (Boetius et al., 2000; Orcutt et al., 2005). The bicarbonate ions increase alkalinity of adjacent waters, helping facilitate carbonate precipitation as given in Equation 9.13 (Luff and Wallman, 2003):

$$Ca^{2+} + 2HCO_3^- \Leftrightarrow CaCO_3\,(s) + CO_2 + H_2O \tag{9.13}$$

Carbonates occur at Blake Ridge in some instances as overhanging solid slabs at the water–sediment interface, where hydrates form underneath the slabs as a consequence of venting methane bubbles striking slab undersides. Carbonates at Blake Ridge also exhibit forms such as orifice and chimney-like configurations performing as conduits for hydrocarbon gases (Van Dover et al., 2003). Such configurations are massive in the Cascadia margin, where the carbonate chimney form is noted to reach as high as 40 m above the seafloor (Boetius and Suess, 2004). Massive authigenic carbonates also occur in the North Sea around cold methane seeps at Gullfaks and Tommeliten fields (Wegener et al., 2008).

REFERENCES

Aharon, P., Fu, B., 2000. Microbial sulfate reduction rates and sulfur and oxygen isotope fractionations at oil and gas seeps in deepwater Gulf of Mexico. Geochim. Cosmochim. Acta 64 (2), 233–246.

Arp, A.J., Childress, J.J., 1981. Blood function in the hydrothermal vent vestimentiferan tube worm. Science 213, 342–344.

Arvidson, R.S., Mose, J.W., Joye, S.B., 2004. The sulfur biogeochemistry of chemosynthetic cold seep communities, Gulf of Mexico, USA. Mar. Chem. 87, 97–119.

Barnes, S.P., Bradbrook, S.D., Cragg, B.A., Marchesi, J.R., Weightman, A.J., Fry, J.C., Parkes, R.J., 1998. Isolation of sulfate-reducing bacteria from deep sediment layers of the Pacific Ocean. Geomicrobiol. J. 15 (2), 67–83.

Boetius, A., Suess, E., 2004. Hydrate Ridge: a natural laboratory for the study of microbial life fueled by methane from near-surface gas hydrates. Chem. Geol. 205, 291–310.

Boetius, A., Ravenschiag, K., Schubert, C.J., Rickert, D., Widdei, F., Gleseke, A., Amman, R., Jergensen, B.B., Witte, U., Pfannkuche, O., 2000. A marine microbial consortium apparently mediating anaerobic oxidation of methane. Nature 407, 623–626.

Borowski, W.S., Paull, C.K., Ussler, III, W., 1996. Marine pore-water sulfate profiles indicate in situ methane flux from underlying gas hydrate. Geology 24 (7), 655–658.

Borowski, W.S., Paull, C.K., Ussler, III, W., 1999. Global and local variations of interstitial sulfate gradients in deep-water, continental margin sediments; sensitivity to underlying methane and gas hydrates. Mar. Geol. 159, 131–154.

Brooks, J.M., Kennicutt, II, M.C., Fisher, C.R., Mack, S.A., Cole, K., Childres, J.J., Bidigare, R.R., Vetter, R.D., 1987. Deep-sea hydrocarbon seep communities: evidence for energy and nutritional carbon sources. Science 238 (4830), 1138–1142.

Brown, D.E., Groves, G.R., Miller, J.D.A., 1973. pH and Eh control of cultures of sulfate-reducing bacteria. J. Appl. Chem. Biotechnol. 23, 141–149.

Burdige, D.J., 2006. Geochemistry of Marine Sediments. Princeton University Press, Princeton, Oxford.

Carney, R.S., 1994. Consideration of the oasis analogy for chemosynthetic communities at Gulf of Mexico hydrocarbon vents. Geo-Mar. Lett. 14, 149–159.

Carroll, J.J., Mather, A.E., 1991. Phase equilibrium in the system water–hydrogen sulphide, hydrate-forming conditions. Can. J. Chem. Eng. 69, 1206–1212.

Cavanaugh, C.M., Gardiner, S.L., Jones, M.L., Jannasch, H.W., Waterbury, J.B., 1981. Prokaryotic cells in the hydrothermal vent tube worm *Riftia pachyptila* Jones: possible chemoautotrophic symbionts. Science 213, 340–342.

Cho, J.H., Jeong, K.S., Kim, K.H., Obzhirov, A., Jin, Y.K., Shoji, H., 2005. Migration patterns of sulfate reduction–methane oxidation interface in hydrate-bearing and free-gas saturated sediments in the Sea of Okhotsk. In: Proceedings of the Fifth International Conference on Gas Hydrates, Trondheim, Norway.

Claypool, G.E., Kaplan, I.R., 1974. The origin and distribution of methane in sediments. In: Kaplan, I.R. (Ed.), Natural Gases in Marine Sediments. Plenum, New York, pp. 99–139.

Cordes, E.E., Arthur, M.A., Shea, K., Arvidson, R.S., Fisher, C.R., 2005. Modeling the mutualistic interactions between tubeworms and microbial consortia. PLoS Biol. 3 (3), 497–506.

Corliss, J.B., Ballard, R.D., 1977. Oasis of life in the cold abyss. National Geographic Magazine, vol. 152, pp. 440-453.

Cragg, B.A., Parkes, R.J., Fry, J.C., Weightman, A.J., Rochelle, P.A., Maxwell, J.R., 1996. Bacterial populations and processes in sediments containing gas hydrates (ODP Leg 146: Cascadia margin). Earth Planetary Sci. Lett. 139, 497–507.

Dattagupta, S., Miles, L.L., Barnabei, M.S., Fisher, C.R., 2006. The hydrocarbon seep tubeworm *Lamellibrachia luymesi* primarily eliminates sulfate and hydrogen ions across its roots to conserve energy and ensure sulfide supply. J. Exp. Biol. 209, 3795–3805.

Fisher, C.R., MacDonald, I.R., Sassen, R., Young, C.M., Macko, S.A., Hourdez, S., Carney, R.S., Joye, S., McMullin, E., 2000. Methane ice worms: *Hesiocaeca methanicola* colonizing fossil fuel reserves. Naturwissenschaften 87, 184–187.

Fisher, C.R., Roberts, H.H., Cordes, E.E., Bernard, B.B., 2007. Cold seeps and associated communities of the Gulf of Mexico. Ocean Explor. 20 (4), 118–129.

Fossing, H., Gallardo, V.A., Jorgensen, B.B., Huttel, M., Nielsen, L.P., Schulz, H., Canfield, D.E., Ulloa, O., 1995. Concentration and transport of nitrate by the mat forming sulphur bacterium *Thioploca*. Nature 374 (6524), 713–715.

Freytag, J.K., Girguis, P.R., Bergquist, D.C., Andras, J.P., Childress, J.J., Fisher, C.R., 2001. A paradox resolved: sulfide acquisition by roots of seep tubeworms sustains net chemoautotrophy. Proc. Natl. Acad. Sci. U. S. A. 98, 13408–13413.

Gaill, F., Persson, J., Sugiyama, J., Vuong, R., Chenzy, H., 1992. The chitin system in the tubes of deep-sea hydrothermal vent worms. J. Struct. Biol. 109, 116–128.

Hachikubo, A., Tatsumi, K., Sakagami, H., Minami, H., Yamashita, S., Takahashi, N., Shoji, H., Jin, Y.K., Vereshchagina, O., Obzhirov, A., 2011. Molecular and isotopic compositions of hydrate-bound hydrocarbons in subsurface sediments from offshore Sakhalin Island, Sea of Okhotsk. In: Proceedings of the 7th International Conference on Gas Hydrates (ICGH 2011), Edinburgh, Scotland, UK, July 17–21. Paper 14.

Harmer, T.L., Rotjan, R.D., Nussbaumer, A.D., Bright, M., Ng, A.W., DeChaine, E.G., Cavanaugh, C.M., 2008. Free-living tube worm endosymbionts found at deep-sea vents. Appl. Environ. Microbiol. 74 (12), 3895–3898.

Jones, M.L., 1981. *Riftia pachyptila* Jones: observations on the vestimentiferan worm from the Galápagos Rift. Science 213 (4505), 333–336.

Joye, S.B., Boetius, A., Orcutt, B.N., Montoya, J.P., Schulz, H.N., Erickson, M.J., Lugo, S.K., 2004. The anaerobic oxidation of methane and sulfate reduction in sediments from Gulf of Mexico cold seeps. Chem. Geol. 205, 219–238.

Julian, D., Gaill, F., Wood, E., Arp, A.J., Fisher, C.R., 1999. Roots as a site of hydrogen sulfide uptake in the hydrocarbon seep vestimentiferan *Lamellibrachia* sp. J. Exp. Biol. 202, 2245–2257.

Kastner, M., 2002. Gas hydrates in convergent margins: formation, occurrence, geochemistry and global significance. Paull, W.P., Dillon, W.P. (Eds.), Natural Gas Hydrates: Occurrence, Distribution, and Dynamics, 124, American Geophysical Union, Washington, DC, pp. 67–86.

Kastner, M., Kvenvolden, K.A., Lorenson, T.D., 1998. Chemistry, isotopic composition, and origin of a methane–hydrogen sulfide hydrate at the Cascadia subduction zone. Earth Planetary Sci. Lett. 156, 173–183.

Kennicutt, II, M.C., Brooks, J.M., Bidigare, R.R., Fay, R.R., Wade, T.L., McDonald, T.J., 1985. Vent-type taxa in a hydrocarbon seep region on the Louisiana slope. Nature 317, 351–353.

Kimura, H., Higashide, Y., Naganuma, T., 2003. Endosymbiotic microflora of the vestimentiferan tubeworm (*Lamellibrachia* sp.) from a bathyal cold seep. Mar. Biotechnol. 5, 593–603.

Lapham, L.L., Chanton, J.P., Martens, C.S., Sleeper, K., Woolsey, J.R., 2008. Microbial activity in surficial sediments overlying acoustic wipeout zones at a Gulf of Mexico cold seep. Geochem. Geophys. Geosyst. 9 (6), 1–17.

Li, Y.-H., Gregory, S., 1974. Diffusion of ions in sea water and in deep-sea sediments. Geochimica et Cosmochimica Acta 38, 703–714.

Luff, R., Wallman, K., 2003. Fluid flow, methane fluxes, carbonate precipitation and biogeochemical turnover in gas hydrate-bearing sediments at Hydrate Ridge, Cascadia margin: numerical modeling and mass balances. Geochim. Cosmochim. Acta 67 (18), 3403–3421.

MacDonald, I.R., Sager, W.W., Peccini, M.B., 2003. Gas hydrate and chemosynthetic biota in mounded bathymetry at mid-slope hydrocarbon seeps: northern Gulf of Mexico. Mar. Geol. 198, 133–158.

Marchesi, J.R., Weightman, A.J., Cragg, B.A., Parkes, R.J., Fry, J.C., 2001. Methanogen and bacterial diversity and distribution in deep gas hydrate sediments from the Cascadia margin as revealed by 16S rRNA molecular analysis. FEMS Microbiol. Ecol. 34, 221–228.

Matsumoto, R., Okuda, Y., Aoyama, C., Hiruta, A., Ishida, Y., Sunamura, M., Numamani, H., Tomaru, H., Snyder, G., Komatsubara, J., Takeuchi, R., Hiromatsu, M., 2005. Methane plumes over a marine gas hydrate system in the eastern margin of Japan Sea: a possible mechanism for the transportation of subsurface methane to shallow waters. In: Fifth International Conference on Gas Hydrates, Trondheim, Norway, pp. 749–754.

Naganuma, T., Elsaied, H.E., Hoshii, D., Kimura, H., 2005. Bacterial endosymbioses of gutless tube-dwelling worms in nonhydrothermal vent habitats. Mar. Biotechnol. 7, 416–428.

NOAA Photo Library, 2010. By image courtesy of Expedition to the Deep Slope 2007, NOAA-OE. NOAA Photo Library HTML. <http://commons.wikimedia.org/wiki?File%3AExpl1767_-_Flickr_-_NOAA_Photo_Library.jpg>.

Orcutt, B.N., Boetius, A., Lugo, S.K., MacDonald, I.R., Samarkin, V.A., Joye, S.B., 2004. Life at the edge of methane ice: microbial cycling of carbon and sulfur in Gulf of Mexico gas hydrates. Chem. Geol. 205, 239–251.

Orcutt, B.N., Boetius, A., Elvert, M., Samarkin, V.A., Joye, S.B., 2005. Molecular biogeochemistry of sulfate reduction, methanogenesis, and the anaerobic oxidation of methane at Gulf of Mexico cold seeps. Geochim. Cosmochim. Acta 69 (17), 4267–4281.

Parkes, R.J., Cragg, B.A., Bale, S.J., Getliff, J.M., Goodman, K., Rochelle, P.A., Fry, J.C., Weightman, A.J., 1994. Deep bacterial biosphere in Pacific Ocean sediments. Nature 371, 410–413.

Paull, C., Hecker, B., Commeau, R., Freeman-Lynde, R.P., Neumann, C., Corso, W., Golubic, S., Hook, J., Sikes, E., Curray, J., 1984. Biological communities at the Florida Escarpment resemble hydrothermal vent taxa. Science 226, 965–967.

Paull, C.K., Ussler III, W., Peltzer, E.T., Brewer, P., Keaten, R., Mitts, P., 2005a. Carbon signatures entombed in methane-soaked gas-hydrate-bearing sediments; preliminary results. In: Fifth International Conference on Gas Hydrates, Trondheim, Norway, pp. 699–708.

Paull, C.K., Ussler, III, W., Lorenson, T., Winters, W., Dougherty, J., 2005b. Geochemical constraints on the distribution of gas hydrates in the Gulf of Mexico. Geo-Mar. Lett. 25, 273–280.

Riedel, M., Collett, T.S., Kumar, P., Sathe, A.V., Cook, A., 2010. Seismic imaging of a fractured gas hydrate system in the Krishna–Godavari Basin offshore India. Mar. Pet. Geol. 27, 1476–1493.

Roberts, H.H., Shedd, W., Hunt, Jr., J., 2010a. Dive site geology: DSV Alvin (2006) and ROV Jason II (2007) dives to the middle-lower continental slope, northern Gulf of Mexico. Deep-Sea Res. II 57, 1837–1858.

Roberts, H.H., Feng, D., Joye, S.B., 2010b. Cold-seep carbonates of the middle and lower continental slope, northern Gulf of Mexico. Deep-Sea Res. II 57, 2040–2054.

Sassen, R., Roberts, H.H., Aharon, P., Larkin, J., Chinn, E.W., Carney, R., 1993. Chemosynthetic bacterial mats at cold hydrocarbon seeps, Gulf of Mexico continental slope. Org. Geochem. 20, 77–89.

Selleck, F.T., Carmichael, L.T., Sage, B.H., 1952. Phase behavior in the hydrogen sulfide–water system. Ind. Eng. Chem. 44, 2219–2226.

Smith, R.P., 2010. A short history of hydrogen sulfide. Am. Sci. 98 (1), 6–9.

Solomon, E.A., Kastner, M., Jannasch, H., Robertson, G., Weinstein, Y., 2008. Dynamic fluid flow and chemical fluxes associated with a seafloor gas hydrate deposit on the northern Gulf of Mexico slope. Earth Planetary Sci. Lett. 270, 95–105.

Sommer, S., Gutzmann, E., Pfannkuche, O., 2007. Sediments hosting gas hydrates: oases for metazoan meiofauna. Mar. Ecol. Prog. Ser. 337, 27–37.

Sorensen, J., Christensen, D., Jorgensen, B.B., 1981. Volatile fatty acids and hydrogen as substrates for sulfate-reducing bacteria in anaerobic marine sediment. Appl. Environ. Microbiol. 42 (1), 5–11.

Suess, E., Torres, M.E., Bohrmann, G., Collier, R.W., Greinert, J., Linke, P., Rehder, G., Trehu, A., Wallmann, K., Winckler, G., Zuleger, E., 1999. Gas hydrate destabilization: enhanced dewatering, benthic material turnover and large methane plumes at the Cascadia convergent margin. Earth Planetary Sci. Lett. 170, 1–15.

Torres, M.E., McManus, J., Hammond, D.E., de Angelis, M.A., Heeschen, K.U., Colbert, S.L., Tryon, M.D., Brown, K.M., Suess, E., 2002. Fluid and chemical fluxes in and out of sediments hosting methane hydrate deposits on Hydrate Ridge, OR, I: hydrological provinces. Earth Planetary Sci. Lett. 201, 525–540.

Ussler III, W., Paull, C.K., Chen, Y., Matsumoto, R., Lorenson, T.D., Winters, W.J., 2005. The consequences of methane oxidation at the sulfate–methane interface in a methane-rich core from the northern Gulf of Mexico. In: Proceedings of the Fifth International Conference on Gas Hydrates, Trondheim, Norway, June 13–16, pp. 1056–1062. Paper 3053.

Van Dover, C.L., Aharon, P., Bernhard, J.M., Caylor, E., Doerries, M., Flickinger, W., Gilhooly, W., Goffredi, S.K., Knick, K.E., Macko, S.A., Rapporport, S., Raulfs, E.C., Ruppel, C., Salerno, J.L., Seitza, R.D., Sen Gupta, B.K., Shank, T., Turnipseed, M., Vrijenhoek, R., 2003. Blake Ridge methane seeps: characterization of a soft-sediment, chemosynthetically based ecosystem. Deep-Sea Res. I 50, 281–300.

van Gemerden, H., 1993. Microbial mats; a joint venture. Mar. Geol. 113, 3–25.

Wegener, G., Shovitri, M., Knittel, K., Niemann, H., Hovland, M., Boetius, A., 2008. Biogeochemical processes and microbial diversity of the Gullfaks and Tommeliten methane seeps (Northern North Sea). Biogeosciences 5, 1127–1144.

Wellsbury, P., Goodman, K., Barath, T., Cragg, B.A., Barnes, S.P., Parkes, R.J., 1997. Deep marine biosphere fuelled by increasing organic matter availability during burial and heating. Nature 388, 573–576.

Whiticar, M.J., Faber, E., Schoell, M., 1986. Biogenic methane formation in marine and freshwater environments: CO_2 reduction vs acetate fermentation – isotope evidence. Geochim. Cosmochim. Acta 50, 693–709.

Winfrey, M.R., Ward, D.M., 1983. Substrates for sulfate reduction and methane production in intertidal sediments. Appl. Environ. Microbiol. 45 (1), 193–199.

Zhang, C.L., Li, Y., Wall, J.D., Larsen, L., Sassen, R., Huang, Y., Wang, Y., Peacock, A., White, D.C., Horita, J., Cole, D.R., 2002. Lipid and carbon isotopic evidence of methane-oxidizing and sulfate reducing bacteria in association with gas hydrates from the Gulf of Mexico. Geology 30 (3), 239–242.

Zhang, C.L., Huang, Z., Cantu, J., Pancost, R.D., Brigmon, R.L., Lyons, T.W., Sassen, R., 2005. Lipid biomarkers and carbon isotope signatures of a microbial (*Beggiatoa*) mat associated with gas hydrates in the Gulf of Mexico. Appl. Environ. Microbiol. 71 (4), 2106–2112.

CHAPTER TEN

Martian Hydrate Feasibility: Extending Extreme Seafloor Environments

Do living microbes exist on Mars? Gas hydrates form in extreme environments of Earth's ocean sediments and thrive with assistance of hydrate/mineral/microbe synergies. Recent studies of seafloor hydrate formation, decomposition, and longevity offer insights when applied to hydrate feasibilities under Martian conditions. Thriving in the protective confines of hydrate pores and capillaries in Earth's ocean sediments, microbes metabolize on carbon of occluded gases and on nutrients of capillary diffusion fluids. Under the harsh conditions of Mars, such a protective environment may be necessary for microbes to survive over long periods.

Numerous recent discoveries amplify knowledge of Martian conditions to support this approach. This chapter details pertinent data relayed by spacecraft probes delineating materials and conditions in the Martian atmosphere, surface, and subsurface: availability of water, hydrate-forming gases, pressure/temperature combinations, and minerals promoting hydrate formation. All requirements seem to be in place for hydrate formation.

Finally, some erosional and crater features of Mars relative to somewhat similar features above hydrate zones on Earth seafloors are discussed in this chapter.

10.1 INTRODUCTION

10.1.1 Mars Spacecraft Launch History

After six failures, the first successful Mars spacecraft probe, Mariner 4, was launched in 1964 by the United States. During flyby of the planet, Mariner 4 transmitted 21 images of Martian surfaces.

Data transmitted by Mariner 4 and by subsequent probes allow more accurate evaluations of possible Martian gas hydrate occurrences. This becomes important because hydrates on Mars would mean profound consequences regarding fuel and water for planet exploration and colonization. Additionally, hydrates impact the ongoing question of whether microbes

Offshore Gas Hydrates. http://dx.doi.org/10.1016/B978-0-12-802319-8.00010-3

exist on Mars, where porous and permeable hydrates in the subsurface could provide microbes the protection and a carbon source while accommodating nutrient diffusion for long-term cell survival, as gas hydrates do in Earth's deep ocean sediments.

In Table 10.1 is a summary compiled from Jet Propulsion Laboratory/National Aeronautics and Space Administration (NASA) information of Mars launches.

Table 10.1 Summary of spacecraft launches to Mars

Launch date	Name	Country	Result	Comments
1960	Korabl 4	USSR (flyby)	Failure	Did not reach Earth orbit
1960	Korabl 5	USSR (flyby)	Failure	Did not reach Earth orbit
1962	Korabl 11	USSR (flyby)	Failure	Failed in Earth orbit
1962	Mars 1	USSR (flyby)	Failure	Radio failed
1962	Korabl 13	USSR (flyby)	Failure	Failed in Earth orbit
1964	Mariner 3	US (flyby)	Failure	Shroud did not jettison
1964	Mariner 4	US (flyby)	Success	Returned 21 images
1964	Zond 2	USSR (flyby)	Failure	Radio failed
1969	Mars 1969A	USSR	Failure	Launch vehicle failed
1969	Mars 1969B	USSR	Failure	Launch vehicle failed
1969	Mariner 6	US (flyby)	Success	Returned 75 images
1969	Mariner 7	US (flyby)	Success	Returned 126 images
1971	Mariner 8	US	Failure	Launch failed
1971	Kosmos 419	USSR	Failure	Earth orbit only
1971	Mars 2 Orbiter/Lander	USSR	Failure	Orbiter arrived. No useful data Lander destroyed
1971	Mars 3 Orbiter/Lander	USSR	Success	Orbiter got 8-month data Lander got 20-s data
1971	Mariner 9	US	Success	Returned 7329 images
1973	Mars 4	USSR	Failure	Flew past Mars
1973	Mars 5	USSR	Success	Returned 60 images. Lasted 9 days
1973	Mars 6 Orbiter/Lander	USSR	Success/failure	Some data produced, but Lander failed on descent
1973	Mars 7 Lander	USSR	Failure	Missed planet; now in solar orbit

Table 10.1 Summary of spacecraft launches to Mars *(cont.)*

Launch date	Name	Country	Result	Comments
1975	Viking 1 Orbiter/Lander	US	Success	First successful landing on Mars
1975	Viking 2 Orbiter/Lander	US	Success	Returned 16,000 images, data from atmospheric and soil experiments
1988	Phobos 1, Orbiter	USSR	Failure	Lost en route to Mars
1988	Phobos 2 Orbiter/Lander	USSR	Failure	Lost
1992	Mars Observer	US	Failure	Lost prior to Mars arrival
1996	Mars Global Surveyor	US	Success	More images than all Mars missions
1996	Mars 96	USSR	Failure	Launch vehicle failure
1996	Mars Pathfinder	US	Success	Experiments 5× longer than expected
1998	Nozomi	Japan	Failure	No orbit insertion; fuel problems
1998	Mars Climate Orbiter	US	Failure	Lost on arrival
1999	Mars Polar Lander	US	Failure	Lost on arrival
1999	Deep Space 2 Probes	US	Failure	Lost on arrival
2001	Mars Odyssey	US	Success	High-resolution images of Mars
2003	Mars Express Orbiter/Beagle 2	European Space Agency	Success/ failure	Orbiter detailed images of Mars Lander lost on arrival
2003	Mars Exploration Rover-Spirit	US	Success	Operating lifetime ×15 designed longevity
2003	Mars Exploration Rover-Opportunity	US	Success	Operating lifetime ×15 designed longevity
2005	Mars Reconnaissance Orbiter	US	Success	Returned >26 TB of data
2007	Phoenix Mars Lander	US	Success	Returned >25 GB of data
2011	Curiosity Robotic Rover	US	Success	Landed August 2012, Gale Crater. To test for methane, minerals, microbes, supportive microbial environment

Credit: NASA/JPL-Caltech.

10.1.2 Martian Geologic Periods

Three periods characterize Martian geologic history. These periods are defined by comparing impact-crater density on Mar's surfaces with impact-crater density on Earth's moon surfaces, the latter being dated by carbon-13 isotopes from samples astronauts returned to Earth (Max et al., 2011). Ordered from oldest to youngest, the Martian periods (Bibring et al., 2006) are designated as follows:

Greater than 4500–3700 mya: Noachian
3700–3000 mya: Hesperian
3000 mya to present: Amazonian

10.2 ATMOSPHERIC COMPOSITION

10.2.1 Overall Composition

The five primary gases that compose the Martian atmosphere are given in Figure 10.1. Composition was measured by mass spectrometer on Mars Rover inside Gale Crater. Most noteworthy is the overwhelming presence of carbon dioxide and significant amounts of nitrogen and argon (Moroz, 1998).

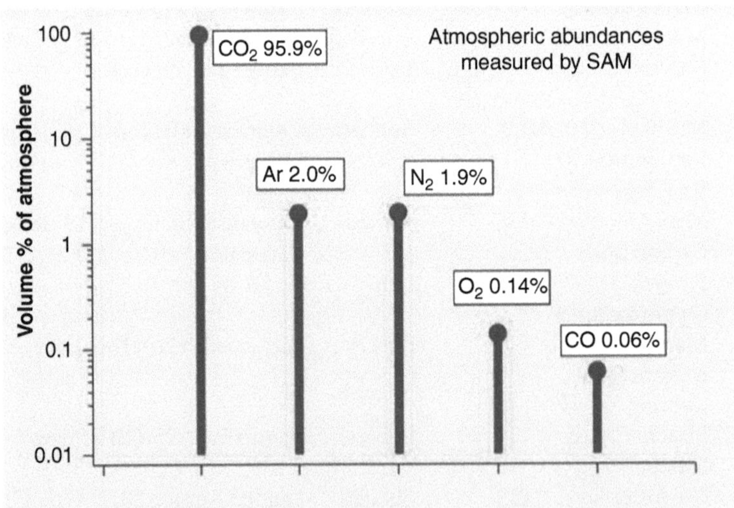

Figure 10.1 ***Martian atmosphere composition; public domain.*** *(Image credit: NASA/JPL-Caltech, Sample Analysis at Mars (SAM)/GSFC; http://www.jpl.nasa.gov/spaceimages/details.php?id=PIA13536; http://photojournal.jpl.nasa.gov/catalog/PIA16460.jpg.)*

Table 10.2 Important trace constituents of Mars' atmosphere*

Element	Concentration	References
Methane	10.5 ppb <50 ppb; 15 ppb average	Krasnopolsky et al. (2004), Formisano et al. (2004), Mumma et al. (2004), Herri and Chassefiére (2012)
Sulfur dioxide	1–2 ppb	Krasnopolsky (2005)
Hydrogen	15 ppm 40–50 ppm	Krasnopolsky and Feldman (2001)
Water vapor	0.03 mol%	Mumma et al. (2009)

*Other minor components include neon, krypton, formaldehyde, xenon, ozone, and nitric oxide.

Trace elements measured in the Martian atmosphere become significant in a discussion of possible gas hydrates on the planet. See Table 10.2.

As would be expected, the water vapor measured as 0.03 mol% in the Martian atmosphere is notably small compared with that in Earth's atmosphere (Baker, 2001; Mumma et al., 2009). But there is a seasonal net transfer of sublimed water vapor from the northern polar cap to the colder and permanent southern polar cap because of a cold trap effect – a basic procedure familiar in laboratories (Richardson and Wilson, 2002).

The Mars' atmosphere has a mean molecular weight of 43.3 g/mol, reflecting the dominance of CO_2, which compares with the molecular weight of 29 g/mol of Earth's atmosphere, reflecting the dominance of O_2 and N_2. The year-around average of atmospheric pressure at the surface of Mars is accepted to be 6.1 mbar (Krasnopolsky, 2005).

With the exception of neon and hydrogen, each trace gas constituent of the Martian atmosphere listed in Table 10.2 forms gas hydrates in the laboratory. Primary gas constituents of the Martian atmosphere shown in Figure 10.1 are hydrate formers.

10.2.2 Sulfur Dioxide, Hydrogen Sulfide

Inorganic reactions and biological reactions on Earth often link sulfur to carbon and in so doing govern the molecular distributions of the two elements (King and McLennan, 2010). For example, the anaerobic oxidation of methane in the presence of seawater sulfates is an important microbial reaction in the gas hydrate zones of Earth's seafloors. It is known, also, that sulfates are abundant in the Martian regolith (Arvidson et al., 2005).

Sulfur dioxide and hydrogen sulfide greatly ameliorate the *P–T* requirements for hydrate stability. The presence of either sulfur dioxide or hydrogen sulfide enhances hydrate formation and stability. For example, the stringency of pressure–temperature combinations to form sulfur dioxide hydrates is dramatically lowered. That is, at 271 K a pressure of only about 0.04 MPa is required to form SO_2 hydrates (Tamman and Krige, 1925; Mohammadi and Richon, 2011; Chinworth and Katz, 1947).

To illustrate the efficacy of H_2S and SO_2 in forming hydrates, Elwood Madden et al. (2011) calculate that 10% H_2S mixed with methane gas would form hydrates in the Martian subsurface stable to within 6 m of its equatorial surface. The sulfide hydrate affinity is so great that hydrates scavenge the free gas.

Sulfur dioxide concentrations of 1–2 ppb have been detected in the Martian atmosphere, as compared with methane concentrations of 10.5 ppb. Krasnopolsky (2005) found that three different models establish an average 2-year SO_2 lifetime in the Martian atmosphere, whereas global mixing time is 0.5 year. Krasnopolsky estimates a 17,000 ton/year annual SO_2 net production on the planet.

What is the source of SO_2 production on Mars? Table 10.3 gives a summary of estimated sources. These estimates of SO_2 production inexplicably fall far short of the amount observed in the atmosphere.

Volcanic eruptions on Earth typically contain SO_2 as a main component of the expelled gases. Although gas compositions vary, SO_2 usually substantially exceeds CH_4. On Earth the ratio of CH_4/SO_2 in volcano eruptions ranges from 0.0001 to 0.001 (Nakagawa et al. 2009; Krasnopolsky, 2005).

Table 10.3 Sulfur dioxide mass balance for Mars' atmosphere (Krasnopolsky, 2005)

Source	Production (ton/year)	Comments
Volcano eruptions	0	No current volcanic activity or hot spots No plate tectonics on Mars
Seepage, ultraviolet (UV) on sulfates	1	Might come from geothermal or hydrothermal water seepage Solar UV decomposition of sulfate minerals on surface
Comet impacts	70	Based on annual comet impacts and comet sulfur content
Meteorites, interplanetary dust	330	Based on delivered carbon and sulfur/carbon ratio
Total SO_2 estimated	≈400	
Total SO_2 observed=	≈17,000	Based on data from Texas Echelon Cross Echelle Spectrograph at NASA Infrared Telescope Facility

Gaillard and Seaillet (2009) estimate that Mars' volcanic gases should contain 10–100 times more SO_2 than the volcanic gas emissions on Earth, but instead the CH_4/SO_2 in the Mars' atmosphere ranges in ratio from 5/1 to 10/1 – orders of magnitude larger in methane than on Earth.

Since methane is commonly associated with microbial activity or geologic activity, the mass imbalance of sulfur dioxide on Mars lends some credence to a biogenic methane source in the subsurface (Nakagawa et al., 2009).

The presence of SO_2 on Mars presages the following: (1) sulfur dioxide (similar effect as H_2S) would form the most stable gas hydrates; (2) SO_2 could persist in ultrastable gas hydrates from ancient times; (3) any increase of SO_2 or H_2S content of hydrate-forming gases would bring the hydrate zone closer to surface and would extend the hydrate zone downward.

Several researchers think the most recent Martian volcanic activity occurred between 2 and 100 million years ago (Neukum et al., 2004; Christensen et al., 2003), necessitating an improbable slow diffusion of sulfur dioxide through the regolith to supply observed atmospheric concentrations (Weiss et al., 2000).

The observed inexplicable SO_2 concentration might be explained if gas hydrates containing SO_2 formed in the subsurface. First, suppose gases emitted in Mars' period of volcanic activity were occluded and preserved in subsurface gas hydrates – especially considering the high stability of SO_2-containing hydrates as intimated by Figure 10.2 (Mohammadi and Richon, 2011; Mohammadi et al., 2005). A second consideration is that hydrates formed from a gas phase of methane–sulfur dioxide would exhibit a much higher sulfur dioxide concentration in the occluded gases.

The gas separation property of hydrates can perhaps be applied to the separation of Martian volcanic gases, if those gases when emitted were highly concentrated in SO_2. The last hydrate layer formed from the volcanic gases, depleted in SO_2, would decompose first and emit gases into the atmosphere of higher than expected methane-to-sulfur ratio.

10.2.3 Trace Gases

Gases present only in trace amounts in the Martian atmosphere form hydrates as pure gases or in mixtures. This is illustrated by the three selections of ozone, formaldehyde, and carbon monoxide. Muromachi et al. (2010) prepared ozone gas hydrates in their laboratory. These researchers reported extended stability of ozone in gas hydrate form lasting greater than 20 days. Second, formaldehyde gas hydrates have been prepared in the laboratory

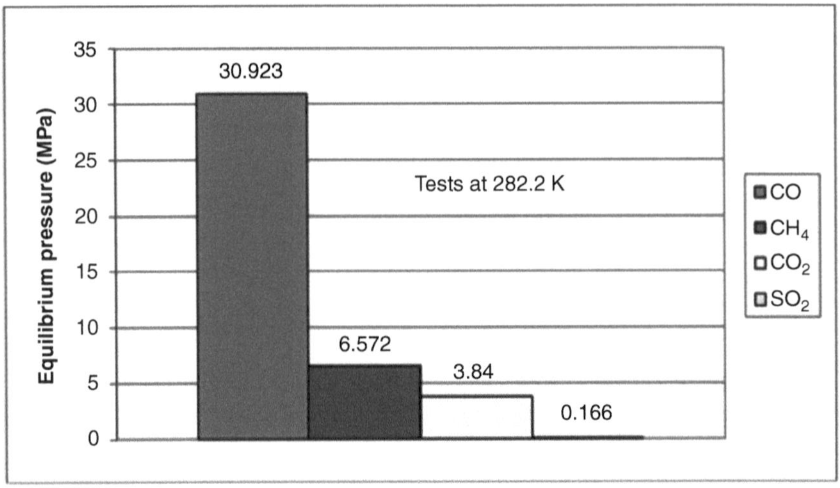

Figure 10.2 ***Comparison of hydrate-forming ease at 282.2 K (Mohammadi and Richon, 2011; Mohammadi et al., 2005).***

(Ripmeester et al., 1996) using a technique of vapor deposition of the formaldehyde with water on a cold surface. Third, carbon monoxide forms Type I gas hydrates according to the hydrate–liquid–vapor equilibrium.

Although difficult to form, carbon monoxide hydrates have been prepared in the laboratory by Davidson et al. (1987). Mohammadi et al. (2005) further prepared CO hydrates and hydrates of gaseous mixtures that included CO, developing an analytical model for CO hydrate P-T equilibrium. Earlier, Miller (1961) predicted CO hydrates to exist in comets, planets, and interstellar dust.

Carbon dioxide represents 95.3% of the current Martian atmosphere, so that it would be expected to dominate any gas hydrate contemporary formation. For example, in a simulation where hydrate formation was evaluated assuming pressurizing gases to be the three primary components of current Martian atmosphere, greater than 99% of the gases occluded in the hydrate was carbon dioxide with only traces of nitrogen and argon (Herri and Chassefiére, 2012).

10.3 SURFACE TEMPERATURES AND PRESSURES ON MARS

10.3.1 Temperatures

Surface temperatures, pressures, and atmospheric gas compositions bolster the feasibility of gas hydrate formation on Mars. Carbon dioxide, the dominant atmospheric gas, is a hydrate former. Selective areas of the planet have

Table 10.4 Martian temperatures

Location	Season	Temperature (K)	Comment	References
Surface, global mean	All seasons	200	Global mean	Duxbury et al. (2004), Max and Clifford (2000), Fairen et al. (2009), Krasnopolsky et al. (2004)
Latitudes around poles	All seasons	170	Mean	Prieto-Ballesteros et al. (2006)
Tharsis volcano region	Summer	276	Surface daytime maximum	Krasnopolsky (2005)
Midlatitude	Summer	230/180	Surface, day/night	Weiss et al. (2000)
Southern polar cap, high altitude	Winter	148–143	Permanent CO_2 ice	Herri et al. (2011), Herri and Chassefiére (2012)
Southern midlatitudes, 32–48°S	Summer afternoon	250–300 K	Prevalent around Re-curring slope lineae (RSL)	McEwen et al. (2011)

pressure/temperature combinations near the surface, or directly below, conducive to hydrate formation.

The global mean temperature at the surface of Mars is 200 K (Duxbury et al., 2004; Max and Clifford, 2000; Fairen et al., 2009; Krasnopolsky et al., 2004). However, local Martian surface temperatures vary widely. These temperatures range from 148 to 276 K, depending on locations, seasons, and altitudes.

Representative Martian surface temperatures are presented in Table 10.4, where the large variations are observed because the listing covers seasonal changes, various altitudes, equatorial regions, and polar regions. Temperatures may also be affected by suspended dust (Leovy, 2001).

10.3.2 Planet Obliquity Effects

Orbital obliquity affects long-term temperatures of the Martian regolith. In fact, the greatest impact on surface temperatures and pressures come from the obliquity of the planet's orbit (Haberle et al., 2003).

Obliquity relates to a planet's plane of orbit. As an orbiting planet spins on its axis, obliquity is the angle between a perpendicular to its orbital plane and its spin axis – the tilt of its axis. Currently, the obliquity of Mars is 25.2°, giving contemporary temperatures of Table 10.4. Higher angles cause warmer polar temperatures (Thomas, 2011). On a geologic time scale, Mars is thought to have relatively frequent changes of obliquity, which become an important influence on subsurface thermal gradients and dissociation of gas hydrates. One estimate sets the present Martian obliquity period at 120,000 years (Perron and Huybers, 2009).

Using data from instruments on the Mars Reconnaissance Orbiter, Phillips et al. (2011) discovered a carbon dioxide solid phase in the near-surface of the South Pole having a volume of about 10,000 km^3 – enough if released to increase the current CO_2 mass in the atmosphere by about 80%. The release could be facilitated by a change in the planet's obliquity that increased polar temperatures.

When obliquity increases, sun rays striking the surface increase and surface temperatures elevate. This in turn increases regolith thermal gradients and endangers stability of any hydrates in the subsurface. Equilibrium subsurface temperatures, and subsequently equilibrium thermal gradients, of a given obliquity period may be calculated by Equation 10.1 (Root and Elwood Madden, 2012):

$$T = T_O + \frac{h}{k} d \qquad (10.1)$$

where T, temperature at depth (K); T_O, temperature at surface, latitude dependent (K); h, geothermal heat flow (W/m^2); k, thermal conductivity (W/m); d, depth (m).

In Equation 10.1 temperature T_o applies to temperature on the surface at a specific latitude. The composite thermal conductivity k must compensate for conductivity variations of different mineral matrices and matrix fills.

10.3.3 Atmospheric Pressures

Earth's atmospheric pressure at sea level is more than 100 times greater than on Mars (Baker, 2001). By use of the international telescope located on Mauna Kea, Hawaii, a 6.1 mbar Martian atmospheric pressure at the surface was verified as the median, annual atmospheric pressure (Krasnopolsky, 2005; Krasnopolsky et al., 2004), a pressure close to being in equilibrium with solid CO_2 at temperatures prevalent within Martian polar regions (Thomas, 2011).

However, Herri and Chassefiére (2012) estimate that median pressures fluctuate by 0.6–0.9 kPa because of seasonal variations and surface elevations. Annually, atmospheric pressure may vary by 20% because of CO_2 condensation at the poles (Leovy, 2001), which is verified by mathematical models. Further, if obliquity reduced to a low threshold value, CO_2 would condense *en masse* at the poles in a "cold finger" effect (Phillips et al., 2011; Newman et al., 2005; Thomas, 2011).

10.4 MARTIAN WATER

10.4.1 Water Detection

A package of instruments carried by Mars Odyssey in 2001 included the Gamma-Ray Subsystem, High-Energy Neutron Detector, and Neutron Spectrometer to determine whether water exists within the top 1 m of the Mars' surface. (Cosmic rays strike the surface of Mars and generate neutrons from the nuclei of soil components. The generated neutron flux is moderated on striking hydrogen atoms in water molecules, whereupon the water emits gamma rays. Therefore, measurements of hydrogen as well as gamma ray concentrations give accurate determination of water presence.) Accordingly, Boynton et al. (2002) estimated water extent based on a desiccated soil composition at the Pathfinder landing site. They noted that concentrations of hydrogen increase nearer the poles, indicating water ice in the subsurface below a desiccated thin surface.

Another verification of water is reported by Byrne et al. (2009). Five impact craters were spotted via Context Camera. Subsequently, the High-Resolution Imaging Science Experiment distinguished within the five craters' excavated regolith the presence of water ice from which a model predicted nominal ice sublimation rates during the 140–200 days of monitoring.

Also, indirect verification of Martian water comes from data of the European 2003 Mars Express Orbiter and the US 2003 Mars Exploration Rover Opportunity. These probes, together as well as independently, identified phyllosilicates and hydrated minerals of sulfates (Bibring et al., 2006; Poulet et al., 2005). The presence of water would have been necessary for these minerals to form.

Other instruments contribute to the Martian water base of knowledge. Data transmitted by the Mars Reconnaissance Orbiter in 2008 give additional independent verification of the presence of water (Byrne et al., 2009). The OMEGA instrument package on the Mars Express Orbiter can distinguish different water phases: vapor, absorbed water, hydrated water in minerals, or ice (Bibring et al., 2006).

Curiosity Rover data from the Gale Crater show rounded pebbles of a size and sorting consistent with water transport during a brief but highly abrasive process where particle transport extended a few kilometers during fluvial deposition (Williams et al., 2013).

10.4.2 Observations of Subsurface Water

The five craters discovered by the Mars Reconnaissance Orbiter in 2008 give interesting insight into subsurface ice, as the impact excavation descends below regolith to depths of solid, rigid materials. For example, two of the craters exhibit flat bottoms, and those bottoms are etched with 7–8 m polygons (Byrne et al., 2009). The craters appear to have exposed ice on the inner slopes, coinciding with estimates from Gamma-Ray Subsystem, High-Energy Neutron Detector, and Neutron Spectrometer that the Martian surface's upper 1 m contains 75% (v/v) water ice (Byrne et al., 2009).

Geometric figures photographed by the Mars Reconnaissance Orbiter at the bottom of a 14-km diameter impact crater are similar but much larger than polygonal patterns observed in some of Earth's permafrost contractions (Baker, 2001). See Figure 10.3.

Note especially in Figure 10.3 that smaller polygonal features appear within the larger patterns at the bottom of the crater.

The most familiar polygonal patterns on Earth may appear in water/ice of freezing/melting sequences, or dehydration of soil, but 3D seismic data also details polygonal patterns in hydrate zones of deepwater seafloors.

Interestingly, polygonal patterns developed in the hydrate test cell of the Mississippi State University (MSU) Hydrate Laboratory during ultrasonic irradiation of a test solution at hydrate-forming conditions. (Cavitation effects of the ultrasonic waves possibly created crystal nucleation sites.) See Figure 10.4. In a stainless steel laboratory test cell, pressurized water solution with several hundred parts per million of emulsan biosurfactant was cooled slowly until temperature extended into the hydrate stability envelope, possibly in a supersaturated condition. Hydrates spontaneously covered the solution surface in polygonal patterns when ultrasonic pulses irradiated the pressurized/chilled solution.

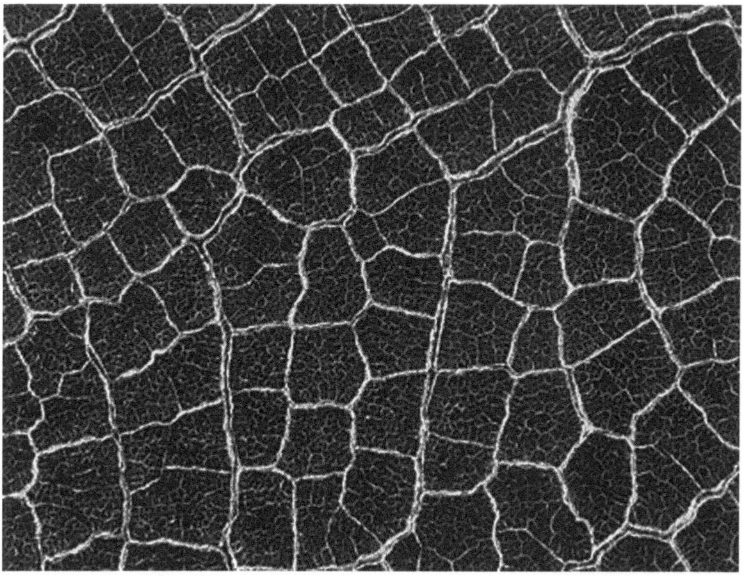

Figure 10.3 ***Polygonal patterns, bottom of Martian crater.*** *(Image credit: NASA/JPL-Caltech/University of Arizona; http://mars.jpl.nasa.gov/images/41_ESP_016641_2500-br2.jpg.)*

10.4.3 Brines

To explain the elemental compositions of regolith at Martian landing sites, as well as the discoveries that sulfate and chloride minerals exist over wide areas of Martian surface, water–electrolyte solutions having depressed freezing points possibly developed from basalt weathering during the Noachian period (Squyres et al., 2004; Hecht et al., 2009; Gendrin et al., 2005). Later, sustained Martian surface cooling may have cyclically lowered freezing points of the solutions, leading to successively higher salinities of the incremental remnant water after each ice formation (Fairen et al., 2009). With the same reasoning, if hydrates formed from these brines when a suitably low temperature was reached, then residual saline water excluded from the hydrate structure also became progressively more saline in successively lower temperature regimes.

Since brines might lower freezing points of water by as much as 70 K, liquid brine possibly remains in the current Martian subsurface, where the highly saline solutions may be expelled from the subsurface on hydrate decomposition. Because the hydrate unit cell structure contains only pure water, hydrate decomposition releases nonsaline water to mix heterogeneously with brine, giving solutions of nonuniform salinities. In such case, the brines

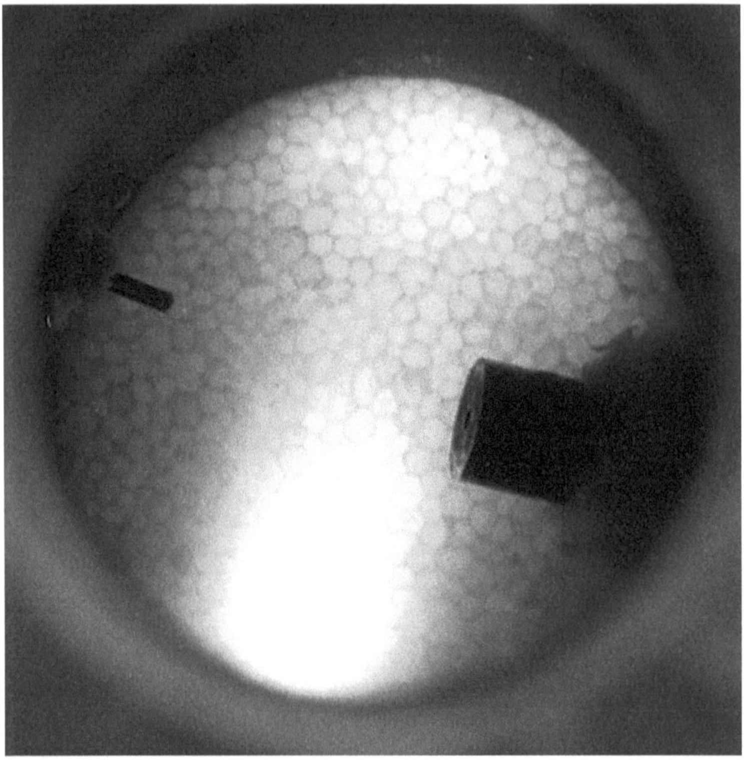

Figure 10.4 ***Polygonal patterns, bottom laboratory test cell.*** Pressurized with ethane; hydrates formed by irradiating test cell contents with six ultrasonic pulses; water supersaturated with hydrate nuclei in cell; photo taken through 5.1-cm-thick pure quartz transparent top; hydrates formed from 200 ppm emulsan in water. *(Courtesy MSU Gas Hydrate Laboratory.)*

would accompany mass ejections from hydrate decomposition overpressures and participate in surface erosions. On the regolith surface, brines are at least an order of magnitude less evaporative than water (Altheide et al., 2009) with eutectic temperatures estimated to range from 205 to 250 K (McEwen et al., 2011). Thus, Hansen et al. (2011) anticipate that brine flow depends on temperatures exceeding approximately 200 K.

To validate their model of brines concentrating progressively during the Noachian period, Fairen et al. (2009) used in laboratory tests water solutions containing materials, excluding Mn, P, and Ti, found from compositional analyses of soils at Mars' landing sites. By decreasing experimental temperatures in 40 steps from 270 to 215 K under pressures of 2 bar of carbon dioxide, they observed significant amounts of stable, liquid-phase brine remaining after the incremental changes. Quantities of remnant brine are presented in Table 10.5.

Table 10.5 Brine remaining after incremental temperature decreases (Fairen et al., 2009)

Temperature (K)	**Liquid phase brine (% of initial volume)**
263	78
250	22
245	14
223	6

Fairen et al. concluded that significant brine volumes could selectively exist spatially and seasonally at current temperatures on Mars. Because these brines at low temperatures retain free-flowing properties, they remain rheologically capable of performing as erosional liquids and sculpting Martian topographical features. Even though the brines may have formed in an earlier epoch, low evaporative properties and association with hydrates could extend their subsurface life.

Remnant brines from stepwise freezing at progressively lower temperatures were found by Fairen et al. to progressively increase salinity from 51.8 g/l at the initial test temperature of 275 K to the final 879.9 g/l salinity of remnant brine at 233 K.

The preceding results impact on Martian subsurface gas hydrates. Subsurface accumulations of highly saline brines, as a thermodynamic hydrate inhibitor (THI), surrounding hydrates have thermodynamic implications for periodic hydrate decompositions on hydrate–brine contact – especially as brine salinities increase. For example, consider the mobile brines to be in permeable, porous, and fractured strata. If brines originate at lower and warmer depths but migrate to contact shallower hydrates, then spontaneous hydrate decomposition would occur (Madden et al., 2007). That is, advective heat transfer of mobile brines as well as thermodynamic hydrate inhibition of saline waters decreases hydrate stability. As a precedent, advective heat transfer plus elevated salinities are observed to deteriorate hydrate stability in acoustic wipeout zones above salt domes of the Gulf of Mexico (GOM) and along flow paths of the Cascadia margin at South Hydrate Ridge (Liu and Flemings, 2006).

10.5 UNCONSOLIDATED MINERALS AND SOLIDS (REGOLITH)

10.5.1 Regolith Composition

At the Pathfinder landing site, soil origin is primarily weathered basalt. Elemental compositions at regolith top from two landing sites, Pathfinder

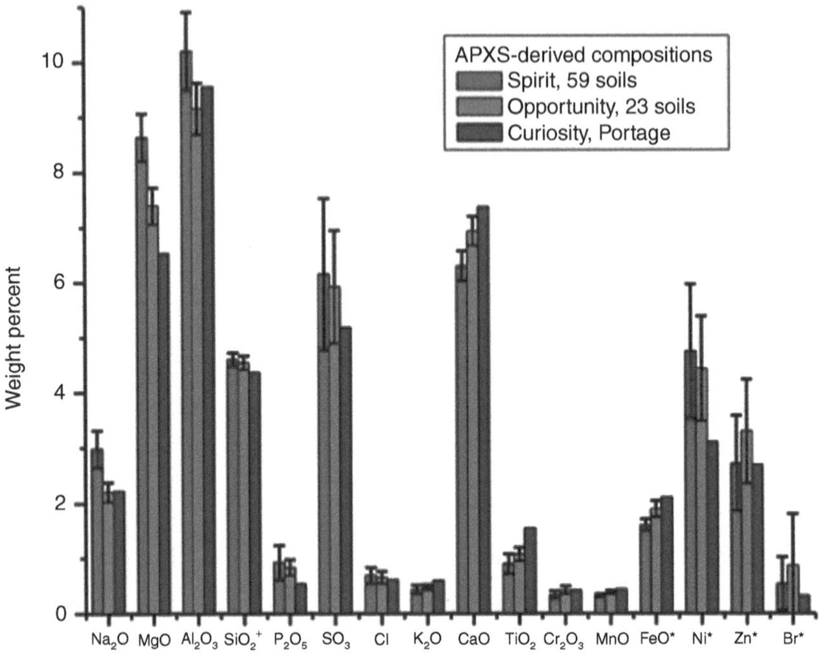

Figure 10.5 ***Mars' soil composition comparisons of three landing sites, where concentrations of SiO_2 and FeO are divided by 10, and concentrations of Ni, Zn, and Br are multiplied by 100.*** *(Image credit: NASA/JPL-Caltech/University of Guelph; http://photojournal.jpl.nasa.gov/catalog/PIA16572.)*

and Spirit, show similar elemental compositions where silicon and iron are prevalent on a mole percentage basis (Fairen et al., 2009).

The commonality of soil composition is further substantiated from a NASA analysis of three landing sites: Spirit, Opportunity, and Curiosity. Martian soil composition with attendant error bars defining accuracy of the analysis is given in Figure 10.5.

10.5.2 Hydrated Phyllosilicates, Smectites

The OMEGA instrument package on Mars Express Orbiter distinguishes phyllosilicates with an imaging spectrometer. During an orbital survey, the Mars Express identified phyllosilicate clay minerals, primarily the smectite clays nontronite and montmorillonite, to be widespread on the surface of Mars (Fairen et al., 2009; Bibring et al., 2006). Phyllosilicates are associated with freestanding water during the clay's formation. Therefore, they probably formed during the first part of the Noachian period (Poulet et al., 2005), suggesting a cold and wet environment in the planet's early history (Fairen et al., 2009).

Nontronite and montmorillonite require long-term water contacts to form. Smectites are favored over kaolinites during initial crystallization from flocculated electrolyte solutions under the following conditions: (1) silica and alumina together in flocculated state, (2) silica:alumina ratio of 2:1 in flocculated state, same ratios as in smectites, (3) high solution pH, (4) high electrolyte content, (5) more Mg^{2+} and Ca^{2+} than Na^{+} and K^{+} in the electrolyte, and (6) environmental promotion of evaporation over precipitation (Mitchell, 1993). The preceding conditions could have been met on Mars in the Noachian period.

The Si:Al ratio is 4.7:1, and the Mg^{2+} and Ca^{2+} are greater by a factor of 4 than Na^{+} and K^{+} at the Pathfinder site. Fairen et al. (2009) hypothesize that highly saline solutions developed during evaporation processes of the Noachian period, when both nontronite and montmorillonite formed early in the period.

The phyllosilicates have been identified in multiple locations on Mars – so abundant that they influence planning for future Mars missions (Kerr, 2010). In all cases, the appearances of phyllosilicates coincide with Noachian-aged topology. Dark areas in Martian topology indicate erosion having uncovered Fe-rich nontronite and other Fe/Mg smectites, as well as identifying localized deposits of Al-rich montmorillonite. Smectites are found on the periphery of cratered terrain among the secondary erosion features, but they are not found in the channels *per se* (Poulet et al., 2005).

Methane is released locally from central sources in Martian northern summers (Mumma et al., 2009). For example, Nili Fossae is a localized area of methane release, where Mars Express and Mars Reconnaissance Orbiter also distinguished rich deposits of phyllosilicates (Poulet et al., 2005; Bibring et al., 2006; Mustard et al., 2008).

These observations of smectites on Mars are significant because laboratory studies conclusively show that nontronite and montmorillonite catalyze gas hydrate formation, especially in the presence of anionic bioproducts (Dearman et al., 2009; Rogers et al., 2003).

The finding of Martian smectite clays helps define the geologic past of the planet, but the presence of these phyllosilicates may have a more far-reaching inference. The research of hydrates in Earth's ocean sediments reveals nucleation and kinetic enhancement of gas hydrate crystals by smectites, especially when associated with bioproducts. These facts combined with data of a Mars orbiter associating the same clays with surface erosional features encourage feasibility of an advanced role hydrates may have on Mars.

10.5.3 Hydrated Sulfates

Sulfates are prevalent near the Martian regolith surface, protruding in outcrops over large areas of the planet. Their widespread surface occurrences have been substantiated by the OMEGA spectrometer in the higher latitudes and in the equatorial midlatitudes (Arvidson et al., 2005). Further substantiations come from the data of Opportunity in Meridiani Planum.

Data from the Thermal Emission Spectrometer on the Opportunity Rover indicate that 15–35 wt% of the bedrocks analyzed in Meridiani were magnesium sulfates and calcium sulfates (Christensen et al., 2004). Data from the Alpha Particle X-Ray Spectrometer (APXS) on Opportunity determined that outcrops in Meridiani contain 7.7 wt% sulfates, and those percentages increased as samples were taken from deeper within the outcrops (Rieder et al., 2004). The observed spectra closely match those from laboratory samples of gypsum $CaSO_4{\cdot}2H_2O$ and kieserite $MgSO_4{\cdot}H_2O$ (Rieder et al., 2004; Arvidson et al., 2005). Interpretation of the spectra, along with previous data showing hydrogen-rich deposits (Feldman et al., 2004), points toward the hydrated sulfate form.

Similar to the phyllosilicates, the hydrated sulfates would have required large amounts of water for their formation process but probably formed in the latter part of the Noachian period after the phyllosilicates (Bibring et al., 2006). In fact, the most likely scenario of sulfate generation would have been during the long presence of electrolyte solutions described by Fairen et al. (2009), where acidic solutions contacted an atmosphere containing sulfur (Langevin et al., 2005). These conditions are believed to have occurred during active volcano eruptions (Neukum et al., 2004), where the ensuing HCl and H_2SO_4 generated the required acidic conditions (Rieder et al., 2004). A bedrock of hydrated sulfates resulted, which was subsequently covered with erosional dust during the Amazonian period.

The theory of hydrated sulfate formation on Mars has indirect implication for gas hydrate formation at the time because of simultaneous presence of abundant water, phyllosilicates, and SO_2/H_2S released by volcanic activity in a cold environment. Such sulfurous gases help form and stabilize hydrate structures because of their minimal *P–T* equilibrium requirements.

10.6 TOPOGRAPHICAL FEATURES

10.6.1 Images of Valley Networks

The orbiting Mariner 9 in 1972 returned the first Martian surface images of an extensive valley network created by fluid flow. Numerous space probes in following years verified the phenomena with high-clarity photographs showing similar large fluid conduits in other Martian locations.

Initial speculations focused on free-flowing water as the source of the erosional features, but current Martian surface conditions dictate any surface water to exist as concentrated brines (Fairen et al., 2009). Slurries containing 10–30% brine would have rheological and depositional capabilities to create such erosional features (Fink et al., 1981). Subsequent fluvial deposits collected by the Curiosity Rover from near the Gale Crater and analyzed by Williams et al. (2013) substantiate erosional features probably derived from brine-based slurries containing abrasive sand. They found rounded particles partitioned by size estimated to have been sculptured during short flows from catastrophic subsurface outbursts.

In our brief discourse, properties of subsurface gas hydrates are suggested as capable of creating these severe topographical features.

10.6.2 Alcoves, Channels, Gullies, Aprons

Examples of erosional valley systems in Figures 10.7 and 10.8 illustrate extensive gullies crafted from large volumetric slurry flows.

Several hundred meters below surface crust of an impact crater, parallel alcoves hover indented near the top of a steep slope. See Figure 10.6,

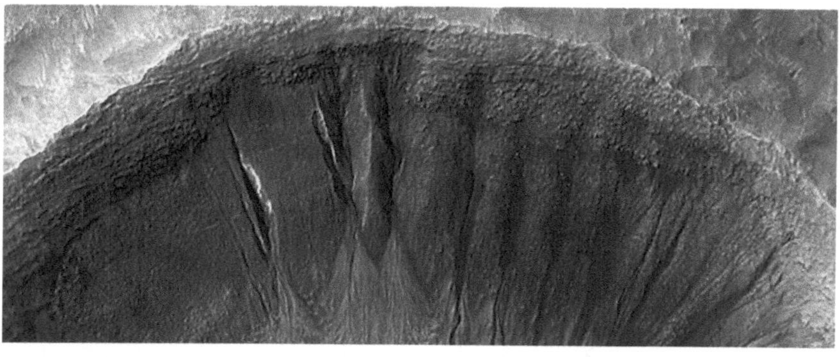

Figure 10.6 ***Alcoves, gullies, aprons from Martian crater wall eruptions.*** *(Image: NASA/JPL/University of Arizona; http://hirise.lpl.arizona.edu/ESP_023665_1410.)*

Figure 10.7 ***Wall erosions from Hale Crater eruptions.*** *(Image credit: NASA/JPL-Caltech/University of Arizona; http://mars.jpl.nasa.gov/spaceimages/details.php?id=PIA13536.)*

taken by HiRISE camera. The alcoves, gullies, and channels originate on the northern and eastern walls of the crater.

The elliptical, crater-like alcoves orient their long axes in a downslope direction, narrowing at their lower elevation. Slurries ejected from the alcoves flow by gravity to form a downslope gully system having major and secondary channels. Viscous slurries flow through the channels, forming triangular depositional aprons at the bottom. Interestingly, some aprons may even overlie older polygonal fracture patterns (Malin and Edgett, 2000).

Figure 10.7 presents a high-resolution photograph from the Mars Reconnaissance Orbiter, HiRISE experiment, of an erosional gulley network on the upper walls of the elliptical Hale Crater, a 125 × 150 km diameter crater in the southern latitudes. Multiple channels, some with large boulders in their rubble, emanate from an alcove below surface crust to mark paths of ejected materials. These major and secondary channels merge downslope into deep V-shaped canyons, finally dissipating as alluvial fans at their base.

In general, some secondary channels run less than or equal to 10 km wide and 5–1000 km long; major channels run less than or equal to 150 km wide and 2000 km long (Baker, 2001). These are conduit dimensions consistent with later estimates of Williams et al. (2013) to account for morphology and size distributions of particles analyzed from an equatorial Martian

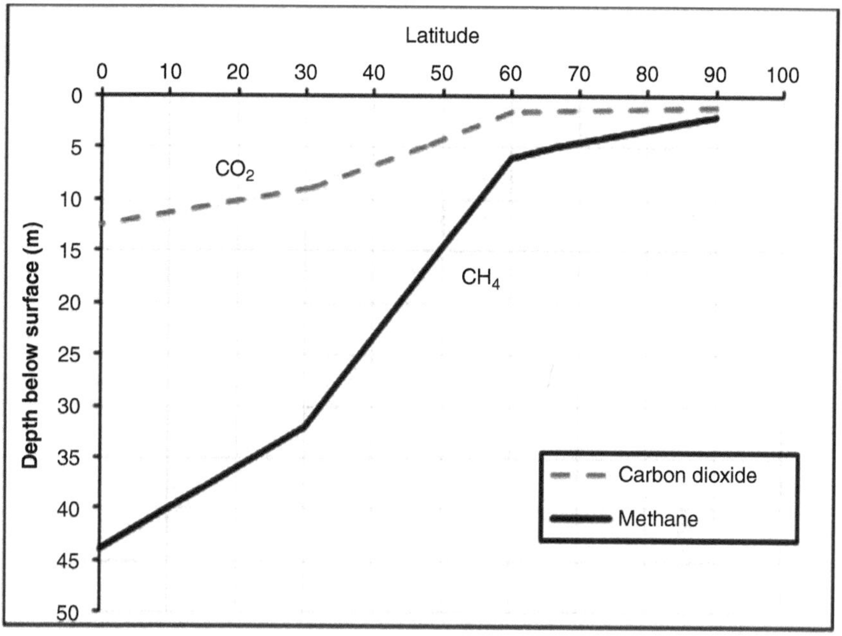

Figure 10.8 ***Subsurface depth to hydrate stability boundary depends on thermal gradient specific to planet location (Root and Elwood Madden, 2012).***

crater. Baker estimates flow rates that formed the largest Martian channels to have exceeded by a factor of 30,000 the normal capacity of the Mississippi River and possibly to have accommodated volumetric flows comparable to the Gulf Stream. In general, major conduits appear to have been formed from fluid slurries flowing at rates exceeding volumetric flows of river systems on Earth.

Malin and Edgett (2000) observed from Mars Orbiter Camera photographs that erosional networks predominantly orient on slopes facing the poles, perhaps suggesting slurries whose brine base is susceptible to delayed dissipation from solar insolation. They further observed that most erosional features occur south of the equator beyond 30° latitude.

10.6.3 McEwen Recurring Slope Lineae

Based on observations of the Mars Reconnaissance Orbiter, McEwen et al. (2011) define the erosional feature recurring sope lineae (RSL), so named because of its recurrence during summer seasons as flow conduits of narrow width and limited length. Much smaller scales of fluid flow are exhibited than in the creation of alcove–gully–channel systems.

RSLs occur exclusively in the summer in the southern hemisphere between 32°S and 48°S latitude on warmer slopes facing the equator where temperatures reach 250–300 K. The feature has been confirmed at seven different locations, all having steep inclines of 25–40°, which is near the angle of repose of solids. Additionally, RSLs exhibit the following characteristics (McEwen et al., 2011):

1. Width 0.5–5 m; length hundreds of meters
2. Extend downslope from bedrock outcrops, rocky areas
3. Form and grow in summer; disappear in winter
4. Associated with small channels but not larger gullies
5. Numerous (perhaps 1000) at single site
6. Flow terminated on steep slopes; material volumes restricted
7. Flows generating RSL diverted around boulders; flow lines split, rejoin

The mechanism causing the RSL feature remains speculative, but characteristics indicate brine involvement and fluid expulsion triggered by thermal change of fluid phase. These erosional occurrences will be explored further in the discussion of subsurface gas hydrates.

10.7 POTENTIAL HYDRATE STABILITY AT MARTIAN CONDITIONS

10.7.1 Surface

Considering gaseous composition of the atmosphere, low atmospheric pressures, and seasonal temperature fluctuations, Herri et al. (2011) predict gas hydrate instability on most Martian surfaces during the entire Martian year or during seasonal changes.

An exception is the southern polar region. Three gases compose greater than 99% of the Martian atmosphere – carbon dioxide, nitrogen, and argon. See Figure 10.1. At the consensus 6.1 mbar average surface pressure, the three major atmospheric gases form sI gas hydrates stable at 149.3 K (Herri and Chassefiére, 2012). According to Table 10.4, temperatures reach 143–148 K in the winter at high altitudes of the southern pole – sufficient to seasonally form and stabilize hydrates.

As a consequence of the southern polar region being colder than its northern counterpart, there is a net transfer of water vapor from north to south. That is, summer temperatures sublimate some ice in the north and the vapors migrate to the colder southern polar region where phase changes convert to solid ice. Herri and Chasefiére predict that portions of the ice transferred to the south react with atmospheric gases to form gas hydrates, giving cumulative hydrate increases. Interestingly, a planet obliquity change would be necessary to reverse the trend.

10.7.2 Subsurface

Thermal gradient effects on Martian subsurface hydrate stability at various latitudes are shown in Figure 10.8. Tops of the hydrate zones for CO_2 hydrates and CH_4 hydrates extend to the surface of higher southern latitudes (Root and Elwood Madden, 2012).

After an increase in planet obliquity, thermal gradients would not reach equilibrium directly thereafter but continue increasing with time, possibly for the obliquity period duration (Root and Elwood Madden, 2012). In the meantime until reaching equilibrium temperatures, marginal hydrates in the subsurface dissociate and release occluded gases. The dispensation of released gases then becomes highly dependent on gas diffusivity through the overburden, and if confining strata has low permeability, extraordinary pressures develop subsurface. Seafloor pockmarks on Earth's ocean floors above hydrate zones attest to such overpressure developments.

Martian hydrates are thought to occur under a protective coating of regolith, remaining marginally but at least seasonally stable in the near subsurface. Although near-surface, hydrate-forming conditions may prevail in the winter, Herri and Chassefiére (2012) predict near-surface hydrate dissociation in the summer except at the permanent south polar cap. Their calculations depend on the following: (1) atmospheric pressure 600 Pa at surface; (2) atmospheric gases for forming hydrates, primarily carbon dioxide and secondarily nitrogen as well as argon with traces of methane; (3) maximum temperatures 155–156 K to stabilize hydrates; (4) winter surface temperatures less than or equal to 148 K; (5) summer surface temperatures greater than 160 K for most areas of Mars.

If subsurface hydrates were formed during an earlier era under different atmospheric gas compositions, environment of volcanic activity, or microbial activity, then hydrate-occluded methane concentrations at depth could be substantially higher than anticipated.

A gas hydrate stability zone as a function of depth can be estimated knowing thermal and pressure gradients along with contemporary atmospheric or assumed gas compositions of another geologic period. Prieto-Ballesteros et al. (2006), as well as Max and Clifford (2000), superpose temperature versus Mar's subsurface depth on the methane hydrate phase boundary determined with the help of CMSHYD, a program limited to minimum temperatures near 170 K (Sloan, 1998). Their resultant graphs show depths of methane hydrate stability. A 170 K surface temperature with its consequential subsurface thermal gradient is correlated with methane hydrate equilibrium, while making the following assumptions (Prieto-Ballesteros et al., 2006): (1) methane is the only guest gas in the hydrates. (2) Heat flux through the subsurface is a uniform 30 mW/m^2. (*Note*: Clifford et al. (2010) also estimate the Martian geothermal heat flux to be 15–30 mW/m^2 and the crust thermal conductivity to be similar to the permafrost of 2–3 W/m/K at Mallik.) (3) Hydrates form beneath rocky permafrost. (4) Equivalent pressures can be substituted for subsurface depths.

Because warmer surface temperatures necessitate greater subsurface depths to attain hydrate stability, Prieto-Ballesteros et al. (2006) estimate that at a surface temperature of 200 K the shallowest methane hydrates could occur would be −15 m below permafrost instead of the −2.5 m for the cooler 170 K surface temperature. Also, overburden affects hydrate stability depths, primarily because of its density and thermal conductivity. Finally, atmospheric gas compositions when hydrates form influence stability zone depths. For example, 1% H_2S in methane hydrates halves stability depth below the Martian surface – the shorter distance becomes important if diffusion controls methane release into the atmosphere to relieve pressure from decomposed hydrates (Elwood Madden et al., 2011).

Hydrate calculations based solely on methane as the guest gas may seem unrealistic because Martian atmosphere is now mostly CO_2, but such calculations are useful to establish more stringent criteria for their formation. Carbon dioxide forms hydrates at lower pressures and higher temperatures. See Figure 10.8 for a consequence of the relative ease of formation.

Pellenbarg et al. (2003) estimate that stabilities of CH_4 and CO_2 gas hydrates may extend from 10 m to 1 km below the Martian surface. But Clifford et al. (2010) estimate hydrate stability depths reaching 12–24 and 5–10 km depths at the southern pole and at the equator, respectively.

10.7.3 High Methane Content Hydrates

To estimate hydrate content and stability more accurately, Thomas et al. (2009) used a program that incorporated experimental equilibrium curves rather than calculated equilibrium curves. With CO_2/N_2, CO_2/Ar, and N_2/Ar proportions based on their ratios in the Martian atmosphere, they predicted hydrate dissociation pressure–temperature when these gases were mixed with methane in concentrations of 0.01, 0.1, 1, 10, 50, or 90%. Hydrate stability potential was calculated to temperatures as low as 130 K, a temperature which Keiffer and Titus (2001) estimate to be attainable at the poles. Thomas et al. concluded that current methane atmospheric concentrations, as denoted by the lower end of the preceding series of experimental methane concentrations, could not form stable hydrates. However, methane hydrates could form from the series' upper end of methane concentrations.

Therefore, the case remains open-ended because high methane content hydrates could have formed in the subsurface and remained stable throughout geologic time (Prieto-Ballesteros et al., 2006; Chastain and Chevrier, 2007) from subsurface volcanic or microbial methane sources (Oze and Sharma, 2005; Krasnopolsky et al., 2004).

10.7.4 Ultrastability

Carbon dioxide gas hydrates in the Martian regolith may have anomalous stability similar to the ultrastability of methane, natural gas, or carbon dioxide hydrates observed in laboratories at low temperatures (Falenty and Kuhs, 2009; Kuhs et al., 2004). Feasibly, such phenomenon on Mars has significant consequences on hydrate preservation in the subsurface (Falenty et al., 2011).

In the ultrastability mechanism, ice formation generated by partial hydrate dissociation contributes to subsequent hydrate self-preservation. Cubic ice Ic preferentially forms rather than hexagonal ice Ih at temperatures of 150–160 K, but Ic transforms to Ih at progressively higher rates as temperature increases thereafter to 240 K. At temperatures warmer than 240 K, hexagonal ice preferentially forms. Previous works suggest anomalous methane hydrate stability at 240 K or warmer temperatures where Ih ice forms on hydrate dissociation (Stern et al., 2001).

Falenty et al. (2011) determined particle sizes of cubic ice Ic to be 2–3 μm compared with 6–10 μm for Ih. Whether these small cubic ice particles anneal at the low temperatures of CO_2 hydrates in the Martian subsurface to retard gas diffusion, stymie hydrate dissociation, and result in anomalous stability remains inconclusive.

10.7.5 Analogies with Earth's Seafloor Hydrates

A cascade of post-1964 data indicates that Martian hydrates exist in the polar regions and in the subsurface. Martian scientific probes reveal basic requirements for gas hydrates to form and stabilize on the planet.

It is possible that brines developed from the Noachian period now exist in the Martian subsurface (Fairen et al., 2009; Squyres et al., 2004; Hecht et al., 2009; Gendrin et al., 2005). It is feasible that hydrates formed from subsurface brines, further concentrating the electrolytes, yet leaving the remnant above their eutectic points (McEwen et al., 2011). Other calculations predict water to be secured in hydrate structures beneath the regolith, forming an important part of the Martian cryosphere (Max and Clifford, 2004; Prieto-Ballesteros et al., 2006), where near-surface hydrates may decompose because of planet obliquity change (Root and Elwood Madden, 2012) or seasonal temperature change or by subsurface migration of the THI brines (Madden et al., 2007).

Hydrate discoveries since the 1980s in Earth's extreme deep-ocean and Arctic environments led to large increases in hydrate scientific data that help interpret information being received on Martian extreme environments.

For example, scientific probes reveal smectite clays to be abundant and widespread in the Martian soil (Poulet et al., 2005) – a significant discovery substantiating that free water once existed on the planet. But also related to gas hydrates, smectite clays in Earth's deep ocean sediments interact with bioproducts to catalyze gas hydrate formation (Rogers et al., 2003).

Here, another important analog arises between the topographical features of some Martian landscapes and observed deep seafloors above hydrate zones on Earth. Pockmarks and large polygonal features help define Martian topographical features, appearing to originate from periodic catastrophic gas and material expulsions (Allen et al., 2012; Oehler and Allen, 2010). On Earth's seafloors above gas hydrate zones, craters and mud volcanoes pockmark the surface, and gas plumes mark active mud volcanoes. Periodic surges of methane believed to exist in the Martian atmosphere may come from similar gas releases.

Pertinent to mud volcanoes, craters, and polygonal faults on ocean seafloors is the fine-sediment theory by Cathles et al. (2010) explaining gas diffusion and small fracture formation that leads to catastrophic release of fluidized mass and pressurized gases from the bottom gas hydrate stability (BGHS). See Section 6.3. Polygonal faults on Earth develop in fine sediments of the hydrate zone, and it is interesting that the craters and surface-expressed polygonal faults on Mars occur in the fine sediments believed to have been deposited during the Hesperian age (Allen et al., 2012). Such fine sediments were defined by the Curiosity Rover in the fluvial deposits of Gale Crater (Grotzinger and MSL Team, 2014).

10.8 METHANE ON MARS

10.8.1 Detection

Microbes generate 90% of Earth's methane (Mumma et al., 2009). On Mars, methane is diligently sought as evidence of a microbial presence, but its quantitative detection and source verification is tantalizingly elusive.

Mars Express, while orbiting and landing on Mars in 2003, provided data from its Planetary Fourier Spectrometer that showed methane in the atmosphere at a mixing ratio of 10 ± 5 ppbv (Formisano et al., 2004). Independent studies using high-dispersion infrared spectrometers at three Hawaii telescopes scanned for 3 years the Martian atmosphere for methane, finally verifying the methane discovery (Mumma et al., 2004, 2009; Krasnopolsky et al., 2004). The data indicate methane releases as plumes, especially emanating from tropical regions during Martian summers. The plumes appear periodically from central sources. For example, a plume of about 19,000 metric ton of methane was released in northern latitudes during midsummer, observed at a rate greater than or equal to 0.6 kg/s. Overall, Geminale et al. (2011) estimate a total of 103,272 ton of methane to exist in the Martian atmosphere.

Although Krasnopolsky (2006) suggests that the methane variations might be unduly influenced by instrumental error, independent sources lend validation to the periodicity (Formisano et al., 2004; Mumma et al., 2004; Lefevre and Forget, 2009), especially the substantiations taken from instruments in orbiting spacecraft. Geminale et al. (2011) assimilated 2 Martian years of data from the orbiting Planetary Fourier Spectrometer that verified uneven CH_4 distributions over the planet's surface. With a Thermal Emission Spectrometer collecting 3 years of data from orbiting spacecraft, Fonti and Marzo (2010) found uneven spatial distributions of methane as well as emissions seasonally episodic.

Because of methane's long lifetime on Mars, the CH_4 mixing ratio should remain globally constant. That is, adequate time exists for atmospheric methane to mix and equalize its concentration on the planet. However, data indicate that the equalization does not happen, as methane concentrations vary significantly by place and season (Lefevre and Forget, 2009), maximizing at the end of the northern summer (Mumma et al., 2009). In fact, methane variation in the Martian atmosphere exceeds methane variation on Earth by approximately 20 orders of magnitude (Krasnopolsky, 2006).

Therefore, concentration variations raise questions of possible sources and sinks for methane. If methane concentrations vary seasonally and geographically on Mars, local sites must generate methane and other local sinks must dissipate methane.

10.8.2 Methane Sink Possibilities

Methane dissipating sites have not been confirmed, although Geminale et al. (2011) suggest the following sinks: (1) photochemical reaction from ultraviolet radiation at 70–90 km altitude, (2) oxidation of methane at a strongly oxidizing Martian surface, (3) dust storms where violently swirling mineral particles generate local and seasonal electrical discharges from strong electrical fields to dissociate methane, and (4) methane diffusion into regolith from atmosphere – adsorbs in winter and desorbs in summer.

A high flux of ultraviolet radiation strikes the Martian surface, and consequent photochemical reactions decompose methane (Krasnopolsky, 2006), shortening its atmospheric longevity. It is estimated that methane's mean life in the Martian atmosphere is 340 years, an approximation that could range over 250–430 years (Krasnopolsky et al., 2004; Chastain and Chevrier, 2007; Prieto-Ballesteros et al., 2006). An independent estimate by Thomas et al. (2009) suggests a mean life of 300–600 years. In contrast, methane on Earth has a mean life of 10 years.

10.8.3 Methane Generation Possibilities

The production of methane seems to come primarily during ice sublimation of the northern polar cap in spring and summer (Geminale et al., 2011), where it is estimated that 16,000–24,000 metric ton of methane per year is put into the atmosphere. The colder and permanent southern polar cap does not generate methane.

Other feasible sources of the methane and reasons for its periodicity in the Martian atmosphere have been suggested (Atreya et al., 2007; Oze and Sharma, 2005; Kress and McKay, 2004; Chastain and Chevrier, 2007). Possible sources include the following: (1) volcanic, (2) hydrothermal alteration of basalt, (3) comet impact or meteorite impact with interplanetary dust, and (4) methanogens (Gainey and Elwood Madden, 2012).

Evaluated in Table 10.6 is the probable methane generated from various sources, derived by the given references.

Cumulative methane generation must replenish large quantities of the gas that are lost by photochemical reactions. An estimated 126 ton/year of methane is needed to replace photochemical losses in the atmosphere (Formisano et al., 2004). An even higher estimate is given by Mumma et al.

Table 10.6 Possible methane sources to replenish atmospheric losses

Methane source	Comment	References
Volcanic, volcanic seepage, hydrothermal	No current hydrothermal activity No current hot spots Low seepage	Krasnopolsky (2006), Oze and Sharma (2005)
Comet impact	1 ton/year; 2% of total CH_4 needed	Formisano et al. (2004), Krasnopolsky et al. (2004)
Meteorite impact, interplanetary dust	4% of total CH_4 needed	Kress and McKay (2004), Krasnopolsky et al. (2004)

(2004), who calculate a 270 ton/year makeup requirement – indicating necessity of additional methane source(s).

There are two other plausible methane origins.

First, gas hydrates are possible sources. Large hydrate reservoirs in the Martian subsurface may contain methane preserved over geologic time – the methane originating from volcanic action in the Noachian epoch. Only circumstantial evidence supports this hypothesis (Tang et al., 2006), and it is dependent on a questionable prior methane-rich atmosphere. However, episodic decompositions of long-preserved hydrates would fit the observed periodicity of atmospheric methane (Prieto-Ballesteros et al., 2006). Episodic hydrate decompositions commonly occur within gas hydrate zones of Earth's ocean floors and are accompanied by emissions of large methane gas plumes.

Second, microbial methanogenesis is a feasible source (Krasnopolsky et al., 2004). Methane could possibly occur in the deep subsurface within a hydrate mass where occluded carbon dioxide occupies hydrate cavities and methanogens occupy the hydrate interstices – a situation where large extensive carbon sources are readily accessible to the microbes. Methanogens would convert occluded carbon dioxide to methane. In such an arrangement, hydrates surround the microbial habitat, sheltering the microbes from extreme external temperature variations while providing unrestricted access to the internal carbon source. Laboratory experiments with GOM hydrates support this mechanism (Radich et al., 2009).

10.9 FEASIBLE HYDRATE–MICROBE ASSOCIATIONS

A basic question regarding Mars has inspired the imaginations of people for many years: Does primary life in the form of microbes exist on the planet? Questions regarding simple life forms on Mars remain unanswered, but the information being gained by scientific missions makes answers less obscure.

No microbial life has been detected on the cold Martian surface, although microbes are known to survive temperatures on Earth as low as 230 K (Rhode and Price, 2007). But on the Martian surface, intense ultraviolet radiation and hydrogen peroxide presence are primary deterrents to microorganism survival (Formisano et al., 2004).

For microbial existence, essential water and the other supporting elements have been verified on the planet (Grotzinger and MSL Team, 2014).

If microbes do exist on Mars, hydrates in the subsurface would be their most probable survivable habitat, based on the study of microbes and hydrates in extreme deep-ocean environments on Earth. Prevailing conditions within hydrates of the Martian subsurface would not be too harsh to feasibly support microbial colonies, including the methanogens. A suitable environment within hydrate masses would provide the following accommodations: (1) form of water; (2) sources of energy, nutrients, and carbon; (3) compatible temperatures; (4) tolerable temperature fluctuations; and (5) a surrounding protective environment.

Most of the conditions for microbial existence have been positively addressed. Gale Crater's sedimentary nature defined by the Curiosity Rover indicates feasibility of a Noachian biosphere (Grotzinger and MSL Team, 2014). Since the 1964 Mariner 4 flyby, scientific probes show conclusively the presence of water or historical evidence of it (Boynton et al., 2002; Byrne et al., 2009; Chastain and Chevrier, 2007). Stable brines probably exist in the subsurface (Fairen et al., 2009) where microbes might survive in these aquifers (Max and Clifford, 2000; Weiss et al., 2000; Michalski et al., 2013). Stable temperatures might best be supplied by water solutions occupying hydrate interstitial spaces.

Krasnopolsky et al. (2004) further considered possible sources of methane:

$$4H_2 + CO_2 \rightarrow CH_4 + 2H_2O \quad (+39.5\,\text{kcal/mol}) \tag{10.2}$$

and

$$2H_2O + 4CO \rightarrow CH_4 + 3CO_2 \quad (+11.4\,\text{kcal/mol}) \tag{10.3}$$

If Equations 10.2 and 10.3 occur through nonmicrobial processes, prohibitively high temperatures on the order of 350°C are required. Therefore, Martian methane production in sufficient quantities by means of Equations 10.2 and 10.3 is not considered likely (Krasnopolsky et al., 2004).

In Earth's ocean sediments, methanogens convert acetate to methane via Equation 9.1 (Whiticar et al., 1986; Claypool and Kaplan, 1974) or methanogens convert carbon dioxide to methane via Equation 9.2 (Whiticar et al., 1986; Burdige, 2006; Borowski et al., 1999; Claypool and Kaplan, 1974). Similarly, a likely carbon source for methanogens in the subsurface of the Martian regolith is carbon dioxide occluded in hydrates, where methanogens could convert carbon dioxide to a methane product. Possibly, hydrates of interstitial brines solubilize the produced methane until saturation; any methane exceeding brine saturation then forms methane hydrates. It is feasible that brine diffusing through hydrate capillaries supplies nutrients to microbes sequestered in interstitial spaces.

Therefore, based on studies of gas hydrate/microbial/mineral synergies in Earth's ocean sediments, the probability of microbial life forms on Mars would increase greatly if gas hydrates exist extensively in the subsurface.

10.10 SUMMARY

Extraordinary progress has been made in understanding the extensive occurrences of gas hydrates in Earth's deep ocean sediments. There, a synergy has been found to exist between minerals, microbes, and gas hydrates. Meanwhile, a foundation of supporting data defining the Martian environment has developed. It is informative to compare findings from each of the two extreme environments.

In Earth's deep ocean sediments, gas hydrates primarily occur in the following forms: (1) massive, (2) dispersed within interstitial pores of sand media, (3) filling fractures and veins, and (4) stratified. Hydrates on Mars have not yet been verified, but it is believed that they do exist in the regolith subsurface in the most likely form of fracture and vein fillings (Max et al., 2011). At Gale Crater, the Curiosity Rover identified millimeter-sized void spaces in the sediment, as well as fractures and vugs filled with calcium sulfate precipitate (Grotzinger and MSL Team, 2014). Such vugs, fractures, and veins could house hydrates in the subsurface.

In Earth's ocean sediments, waterborne microbes proliferate around gas hydrates and seeps of carbon-containing fluids. Depending on carbon of the insoluble hydrocarbons for metabolism and survival, the microorganisms solubilize the carbon compounds by secretion of anionic surfactants, enhancing accessibility (Rogers et al., 2003). The biosurfactants perform another favor by acting as nucleating aids to form hydrate crystals. Particularly in conjunction with smectite clays, nontronite and montmorillonite, the association catalyzes hydrate formation. These smectite clays have been found in abundance on Mars (Fairen et al., 2009). From Curiosity Rover data, Grotzinger and MSL Team (2014) determined the Sheepbed formation of Gale Crater to be fine-particle mudstone containing about 20% smectite clays. Interestingly, they found magnetite (approximately 7%) and sulfide minerals to be in this Martian formation.

Microbes access the interior of hydrates and become active within the hydrate mass. In the GOM, microbial entry into the hydrate is afforded by associations of microbe, biosurfactant, and smectite clay (Radich, 2009). Smectite platelets adsorb edgewise and perpendicular onto microbial cell walls (Lahav and Keynan, 1962); then, biosurfactant absorbs onto the outer fringes of the smectite platelets and the total association becomes a hydrate crystal nucleation site, around which hydrates grow – but importantly as a consequence of the configuration, the microbes become enclosed within the hydrate mass. In Earth's oceans, microbes are sustained within the hydrate mass by nutrients diffusing from the exterior through capillaries and by the abundant carbon source occluded within the hydrate structure (Radich et al., 2009; Dearman et al., 2009). If hydrates exist in the Martian subsurface, it would be feasible for microorganisms to be sequestered within interstitial spaces of hydrate crystals – protected from the vagaries of extreme temperature changes and protected from incident radiation. With these accommodations, microbes might continue within the hydrate confinement *ad infinitum*.

Consider a possible scenario. Subsurface gas hydrates of carbon dioxide derive from a predominantly carbon dioxide atmosphere. Methanogens and waters of varying salinity occupy hydrate crystal interstices. The methanogens deplete dissolved CO_2 in the interstitial water solutions, replacing it with methane. The adjacent hydrate unit cells incrementally decompose, releasing occluded CO_2 to maintain equilibrium with interstitial water while cooling the immediate surroundings by the phase change. The cooler temperature facilitates re-formation of hydrate crystals but with higher occluded methane concentration because of the methanogen product. Thus, over long periods of time, massive methane hydrates develop below the regolith, having replaced CO_2 hydrates originally formed from atmospheric gases. A related mechanism of methanogenesis (Whiticar et al., 1986; Burdige, 2006; Borowski et al., 1999; Claypool and Kaplan, 1974) has been proposed for hydrates in Earth's ocean sediments (Radich et al., 2009).

Subsurface waters could play an important role in the presence of microbes on Mars (Michalski et al., 2013). It becomes noteworthy, therefore, that large volumes of concentrated brine may have developed throughout the Noachian era, being concentrated by successively colder periods (Fairen et al., 2009). Concentrations of the brines would have increased as hydrates formed or decreased on hydrate decomposition.

There are implications for Martian erosional features as a consequence of subsurface brines and gas hydrates. The solid hydrates cement Martian regolith, as happens in ocean sediments, but stresses develop fractures throughout the mass. Similar to the Earth system, the Martian BGHS may present barriers to upward fluid flow, both liquid and gaseous, that develop high pressures below the hydrate zones. As in the acoustic wipeout zones of ocean gas hydrates, weak points in the BGHS barrier develop to release the high pressures. The weak points may be faults, fractures, or fingering through a fine-particle matrix according to the Cathles Theory (Cathles et al., 2010). Note that the Curiosity Rover found a fine-grained (<50 μm grains) matrix of rock in the fluvial dispersions of Gale Crater (Grotzinger and MSL Team, 2014). Deep brines of varying salinity flowing upward decompose hydrates on contact by raising temperatures beyond the *P–T* stability envelope and by performing as thermodynamic inhibitors. Forced hydrate decomposition in confined spaces creates extraordinary pressures. When pressures periodically exceed confining strengths of overburden, blowouts of the Martian regolith occur. Gases released from decomposed hydrates, nonsaline structural waters, slurries of highly saline electrolyte water solutions, and regolith solids violently erupt. The result is erosional valleys and craters at the surface, similar to the large expulsion craters on Earth seafloors above gas chimneys of hydrate zones.

REFERENCES

Allen, C.C., Oehler, D.Z., Etiope, G., Van Rensbergen, P.V., Baciu, C., Feyzullayev, A., Martinelli, G., Tanaka, K., Van Rooij, D., 2012. Fluid expulsion in terrestrial sedimentary basins: a process providing potential analogs for giant polygons and mounds in the Martian lowlands. Icarus 224 (2), 424–432.

Altheide, T., Chevrier, V., Nicholson, C., Denson, J., 2009. Experimental investigation of the stability and evaporation of sulfate and chloride brines on Mars. Earth Planetary Sci. Lett. 282, 69–78.

Arvidson, R.E., Poulet, F., Bibring, J.-P., Wolff, M., Gendrin, A., Morris, R.V., Freeman, J.J., Langevin, Y., Mangold, N., Bellucci, G., 2005. Spectral reflectance and morphologic correlations in eastern Terra Meridiani, Mars. Science 307, 1591–1594.

Atreya, S.K., Mahaffy, P.R., Wong, A.-S., 2007. Methane and related trace species on Mars: origin, loss implications for life, and habitability. Planetary Space Sci. 55, 358–369.

Baker, V.R., 2001. Water and the Martian landscape. Nature 412, 228–236.

Bibring, J.-P., Langrevin, Y., Mustard, J.F., Poulet, F., Arvidson, R., Gendrin, A., Gondet, B., Mangold, N., Pinet, P., Forget, F., OMEGA Team, 2006. Global mineralogical and aqueous Mars history derived from OMEGA/Mars Express data. Science 312, 400–404.

Borowski, W.S., Paull, C.K., Ussler, III, W., 1999. Global and local variations of interstitial sulfate gradients in deep-water, continental margin sediments; sensitivity to underlying methane and gas hydrates. Mar. Geol. 159, 131–154.

Boynton, W.V., Feldman, W.C., Squyres, S.W., Prettyman, T.H., Bruckner, J., Evans, L.G., Reedy, R.C., Starr, R., Arnold, J.R., Drake, D.M., Englert, P.A.J., Metzger, A.E., Mitrofanov, I., Trombka, J.I., d'Uston, C., Wanke, H., Gasnault, O., Hamara, D.K., Janes, D.M., Marcialis, R.L., Maurice, S., Mikheeva, I., Taylor, G.J., Tokar, R., Shinohara, C., 2002. Distribution of hydrogen in the near surface of Mars: evidence for subsurface ice deposits. Science 297, 81–85.

Burdige, D.J., 2006. Geochemistry of Marine Sediments. Princeton University Press, Princeton, Oxford.

Byrne, S., Dundas, C.M., Kennedy, M.R., Mellon, M.T., McEwen, A.S., Cull, S.C., Daubar, I.J., Shean, D.E., Seelos, K.D., Murchie, S.L., Cantor, B.A., Arvidson, R.E., Edgett, K.S., Reufer, A., Thomas, N., Harrison, T.N., Posiolova, L.V., Seelos, F.P., 2009. Distribution of mid-latitude ground ice on Mars from new impact craters. Science 325, 1674–1676.

Cathles, L.M., Zheng, S., Chen, D., 2010. The physics of gas chimney and pockmark formation, with implications for assessment of seafloor hazards and gas sequestration. Marine Pet. Geol. 27 (1), 82–91.

Chastain, B.K., Chevrier, V., 2007. Methane clathrate hydrates as a potential source for Martian atmospheric methane. Planetary Space Sci. 55, 1246–1256.

Chinworth, H.E., Katz, D.L., 1947. Refrigerant hydrates. Refrig. Eng. 54, 359–362.

Christensen, P.R., Bandfield, J.L., Bell, III, J.F., Gorelick, N., Hamilton, V.E., Ivanov, A., Jakosky, B.M., Kieffer, H.H., Lane, M.D., Malin, M.C., McConnochie, T., McEwen, A.S., McSween, Jr., H.Y., Mehall, G.L., Moersch, J.E., Nealson, K.H., Rice, Jr., J.W., Titus, T.N., Wyatt, M.B., 2003. Morphology and composition of the surface of Mars: Mars Odyssey THEMIS results. Science 300, 2056.

Christensen, P.R., Wyatt, M.B., Glotch, T.D., Rogers, A.D., Anwar, S., Arvidson, R.E., Bandfield, J.L., Blaney, D.L., Budney, C., Calvin, W.M., Fallaacaro, A., Fergason, R.L., Gorelick, N., Graff, T.G., Hamilton, V.E., Hayes, A.G., Johnson, J.R., Knudson, A.T., McSween, Jr., H.Y., Mehall, G.L., Mehall, L.K., Moersch, J.E., Morris, R.V., Smith, M.D., Squyres, S.W., Ruff, S.W., Wolff, M.J., 2004. Mineralogy at Meridiani Planum from the Mini-TES Experiment on the Opportunity Rover. Science 306, 1733–1739.

Claypool, G.E., Kaplan, I.R., 1974. The origin and distribution of methane in sediments. In: Kaplan, I.R. (Ed.), Natural Gases in Marine Sediments. Plenum, New York, pp. 99–11139.

Clifford, S., Lasue, J., Heggy, E., Boisson, J., McGovern, P., Max, M.D., 2010. The depth of the Martian cryosphere: revised estimates and implications for the existence and detection of sub-permafrost groundwater. J. Geophys. Res. 115, 1–17.

Davidson, D.W., Desando, M.A., Gough, S.R., Handa, Y.P., Ratcliffe, C.I., Ripmeester, J.A., Tse, J.S., 1987. A clathrate hydrate of carbon monoxide. Nature 328, 418–419.

Dearman, J.L., Wilson, W.W., Rogers, R.E., Zhang, G., 2009. Gas-hydrate promotion by smectite–bioproduct interactions. Mar. Chem. 115 (1–2), 21–30.

Duxbury, N.S., Abyzov, S.S., Romanovskh, V.E., Yoshikawa, K., 2004. A combination of radar and thermal approaches to search for methane clathrate in the Martian subsurface. Planetary Space Sci. 52, 109–115.

Elwood Madden, M.E., Leeman, J.R., Root, M.J., Gainey, S., 2011. Reduced sulfur–carbon–water systems on Mars may yield shallow methane hydrate reservoirs. Planetary Space Sci. 59, 203–206.

Fairen, A.G., Davilla, A.F., Gago-Duport, L., Amils, R., McKay, C., 2009. Stability against freezing of aqueous solutions on early Mars. Nature 459, 401–404.

Falenty, A., Kuhs, W.F., 2009. Self-preservation of CO_2 gas hydrates – surface microstructure and ice perfection. J. Phys. Chem. B 113, 15975–15988.

Falenty, A., Kuhs, W.F., Kristallographie, A., Hansen, T.C., 2011. CO_2 hydrate dissociation at low temperatures – formation and annealing of ice Ic. In: Proceedings of the 7th International Conference on Gas Hydrates (ICGH 2011), Edinburgh, Scotland, UK, July 17–21. Paper 360.

Feldman, W.C., Prettyman, T.H., Maurice, S., et al.,2004. The global distribution of near-surface hydrogen on Mars. J. Geophys. Res. 109 (E9), 1991–2012.

Fink, J.H., Malin, M.C., D'Alliand, R.G., 1981. Rheological properties of mudflows associated with the spring 1980 eruptions of Mount St. Helens volcano, Washington. Geophys. Res. Lett. 8 (1), 43–46.

Fonti, S., Marzo, G.A., 2010. Mapping the methane on Mars. Astron. Astrophys. 512, A51.

Formisano, V., Atreya, S., Encrenaz, T., Ignatiev, N., Giuranna, M., 2004. Detection of methane in the atmosphere of Mars. Science 306, 1758–1761.

Gaillard, F., Seaillet, B., 2009. The sulfur content of volcanic gases on Mars. Earth Planetary Sci. Lett. 279, 34.

Gainey, S.R., Elwood Madden, M.E., 2012. Kinetics of methane clathrate formation and dissociation under Mars relevant conditions. Icarus 218, 513–524.

Geminale, A., Formisano, V., Sindoni, G., 2011. Mapping methane in Martian atmosphere with PFS-ME data. Planetary Space Sci. 59, 137–148.

Gendrin, A., Mangold, N., Bibring, J.-P., Langevin, Y., Gondet, B., Poulet, F., Bonello, G., Quantin, C., Mustard, J., Arvidson, R., LeMousélic, S., 2005. Sulfates in Martian layered terrains: the OMEGA/Mars Express view. Science 307, 1587–1591.

Grotzinger, J.P., MSL Team, 2014. A habitable fluvio-lacustrine environment at Yellowknife Bay, Gale Crater, Mars. Science 343, 1242777-1–11242777-14.

Haberle, R.M., Murphy, J.R., Schaeffer, J., 2003. Orbital change experiments with a Mars general circulation model. Icarus 161, 66–89.

Hansen, C.J., Bourke, M., Bridges, N.T., Byrne, S., Colon, C., Diniega, S., Dundas, C., Herkenhoff, K., McEwen, A., Mellon, M., Portyankina, G., Thomas, N., 2011. Seasonal erosion and restoration of Mars' northern polar dunes. Science 331, 575–578.

Hecht, M.H., Kounaves, S.P., Quinn, R.C., West, S.J., Young, S.M.M., Ming, D.W., Catling, D.C., Clark, B.C., Boynton, W.V., Hoffman, J., DeFlores, L.P., Gospodinova, K., Kapit, J., Smith, P.H., 2009. Detection of perchlorate and the soluble chemistry of Martian soil at the Phoenix Lander site. Science 325, 64–67.

Herri, J.-M., Chassefiére, E., 2012. Carbon dioxide, argon, nitrogen and methane clathrate hydrates: thermodynamic modeling, investigation of their stability in Martian atmospheric conditions and variability of methane trapping. Planetary Space Sci. 73, 376–386.

Herri, J.-M., Cournil, M., Chassefiere, E., 2011. Thermodynamic modeling of clathrate hydrates in the atmosphere of Mars. In: Proceedings of the 7th International Conference on Gas Hydrates (ICGH 2011), Edinburgh, Scotland, UK, July 17–21. Paper 436.

Keiffer, H.H., Titus, T.N., 2001. TES mapping of Mars' north seasonal cap. Icarus 154, 162–180.

Kerr, R.A., 2010. A roller-coaster plunge into Martian water – and life? Science 330, 1617.

King, P.L., McLennan, S.M., 2010. Sulfur on Mars. Elements 6, 107–112.

Krasnopolsky, V.A., 2005. A sensitive search for SO_2 in the Martian atmosphere: implications for seepage and origin of methane. Icarus 178, 487–492.

Krasnopolsky, V.A., 2006. Some problems related to the origin of methane on Mars. Icarus 180, 359–367.

Krasnopolsky, V.A., Feldman, P.D., 2001. Detection of molecular hydrogen in the atmosphere of Mars. Science 294, 1914–1917.

Krasnopolsky, V.A., Maillard, J.P., Owen, T.C., 2004. Detection of methane in the Martian atmosphere: evidence for life? Icarus 172, 537–547.

Kress, M.E., McKay, C.P., 2004. Formation of methane in comet impacts: implications for Earth, Mars, and Titan. Icarus 168, 475–483.

Kuhs, W.F., Genov, G., Staykova, D.K., Hansen, T.C., 2004. Ice perfection and onset of anomalous preservation of gas hydrates. Phys. Chem. Chem. Phys. 6, 4917–4920.
Lahav, N., Keynan, 1962. Adsorption of sodium bentonite particles and *Bacillus subtilis*. Plant Soil XVII (2), 191–208.
Langevin, Y., Poulet, F., Bibring, J.-P., Gondet, B., 2005. Sulfates in the north polar region of Mars detected by OMEGA/Mars Express. Science 307, 1584–1586.
Lefevre, F., Forget, F., 2009. Observed variations of methane on Mars unexplained by known atmospheric chemistry and physics. Nature 460, 720–723.
Leovy, C., 2001. Weather and climate on Mars. Nature 412, 245–249.
Liu, X., Flemings, P.B., 2006. Passing gas through the hydrate stability zone at southern Hydrate Ridge, offshore Oregon. Earth Planetary Sci. Lett. 241, 211–226.
Madden, M.E.E., Ulrich, S.M., Onstott, T.C., Phelps, T.J., 2007. Salinity-induced hydrate dissociation: a mechanism for recent CH_4 release on Mars. Geophys. Res. Lett. 34 (11), L11202/1–L11202/5.
Malin, M.C., Edgett, K.S., 2000. Evidence for recent groundwater seepage and surface runoff on Mars. Science 288, 2330–2335.
Max, M.D., Clifford, S.M., 2000. The state, potential distribution, and biological implications of methane in the Martian crust. J. Geophys. Res. Planets 105, 4165–4171.
Max, M.D., Clifford, S.M., 2004. The origin and distribution of methane hydrate in the Martian crust. In: Second Conference on Early Mars: Geologic, Hydrologic, and Climatic Evolution and the Implications for Life, Abstract #8083.
Max, M., Johnson, A., Clifford, S., 2011. Methane hydrate on Mars; a resource-rich stepping stone to the outer planets? In: Proceedings of the 7th International Conference on Gas Hydrates (ICGH 2011), Edinburgh, Scotland, UK, July 17–21. Paper 638.
McEwen, A.S., Ojha, L., Dundas, C.M., Mattson, S.S., Byrne, S., Wray, J.J., Cull, S.C., Murchie, S.L., Thomas, N., Gulick, V.C., 2011. Seasonal flows on warm Martian slopes. Science 333, 740–743.
Michalski, J.R., Cuadros, J., Niles, P.B., Parnell, J., Rogers, A.D., Wright, S.P., 2013. Groundwater activity on Mars and implications for a deep biosphere. Nat. Geosci. 6, 133–138.
Miller, S.L., 1961. The occurrence of gas hydrates in the solar system. Proc. Natl. Acad. Sci. 47 (11), 1798–1808.
Mitchell, J.K., 1993. Fundamentals of Soil Behavior, second ed John Wiley & Sons, Inc., New York, NY, pp. 46–47.
Mohammadi, A.H., Richon, D., 2011. Equilibrium data of sulfur dioxide and methyl mercaptan clathrate hydrates. J. Chem. Eng. Data 56, 1666–1668.
Mohammadi, A.H., Anderson, R., Tohidi, B., 2005. Carbon monoxide clathrate hydrates: equilibrium data and thermodynamic modeling. AIChE J. 51 (10), 2825–2833.
Moroz, V.I., 1998. Chemical composition of the atmosphere of Mars. Adv. Space Res. 22 (3), 449–457.
Mumma, M.J., Novak, R.E., DiSanti, M.A., Bonev, B.P., Dello Russo, N., 2004. Detection and mapping of methane and water on Mars. American Astronomical Society, DPS Meeting #36, #26.02. Bull. A Astron. Soc. 36, 1127.
Mumma, M.J., Villanueva, G.L., Novak, R.E., Hewagama, T., Bonev, B.P., DiSanti, M.A., Mandell, A.M., Smith, M.D., 2009. Strong release of methane on Mars in northern summer 2003. Science 323, 1041–1045.
Muromachi, S., Ohmura, R., Takeya, S., Mori, Y.H., 2010. Clathrate hydrates for ozone preservation. J. Phys. Chem. B 114 (35), 11430–11435.
Mustard, J.F., Murchie, S.L., Pelkey, S.M., Ehlmann, B.L., Milliken, R.E., Grant, J.A., Bibring, J.-P., Poulet, F., Bishop, J., Dobrea, E.N., Roach, L., Seelos, F., Arvidson, R.E., Wiseman, S., Green, R., Hash, C., Humm, D., Malaret, E., McGovern, J.A., Seelos, K., Clancy, T., Clark, R., Des Marais, D., Izenberg, N., Knudson, A., Langevin, Y., Martin, T., McGuire, P., Morris, R., Robinson, M., Roush, T., Smith, M., Swayze, G., Taylor, H., Titus, T., Wolff,

M., 2008. Hydrated silicate minerals on Mars observed by the Mars Reconnaissance Orbiter CRISM instrument. Nature 454, 305–309.
Nakagawa, H., Kasaba, Y., Maezawa, H., Hashimoto, A., Sagawa, H., Murata, I., Okano, S., Aoki, S., Moribe, N., Mizuno, A., Momose, M., Ohnishi, T., Mizuno, N., Nagahama, T., 2009. Search of SO_2 in the Martian atmosphere by ground-based submillimeter observation. Planetary Space Sci. 57 (14–15), 2123–2127.
Neukum, G., Jaumann, R., Hoffmann, H., Hauber, E., Head, J.W., Basilevsky, A.T., Ivanov, B.A., Werner, S.C., van Gasselt, S., Murray, J.B., McCord, T., The HRSC Co-Investigator Team, 2004. Recent and episodic volcanic and glacial activity on Mars revealed by the High Resolution Stereo Camera. Nature 432, 971.
Newman, C., Lewis, S., Read, P., 2005. The atmospheric circulation and dust activity in different orbital epochs on Mars. Icarus 174, 135.
Oehler, D.Z., Allen, C.C., 2010. Evidence for pervasive mud volcanism in Acidalia Planitia, Mars. Icarus 208 (2), 636–657.
Oze, C., Sharma, M., 2005. Have olivine, will gas: serpentinization and the abiogenic production of methane on Mars. Geophys. Res. Lett. 32, 10203.
Pellenbarg, R.E., Max, M.D., Clifford, S.M., 2003. Methane and carbon dioxide hydrates on Mars: potential origins, distribution, detection, and implications for future in situ resource utilization. J. Geophys. Res. 108, 8042–8047.
Perron, J.T., Huybers, P., 2009. Is there an orbital signal in the polar layered deposits on Mars? Geology 37 (2), 155–158.
Phillips, R.J., Davis, B.J., Tanaka, K.L., Byrne, S., Mellon, M.T., Putzig, N.E., Haberle, R.M., Kahre, M.A., Campbell, B.A., Carter, L.M., Smith, I.B., Holt, J.W., Smrekar, S.E., Nunes, D.S., Plaut, J.J., Egan, A.F., Titus, T.N., Seu, R., 2011. Massive CO_2 ice deposits sequestered in the south polar layered deposits of Mars. Science 332, 838–841.
Poulet, F., Bibring, J.-P., Mustard, J.F., Gendrin, A., Mangold, N., Langevin, Y., Arvidson, R.E., Gondet, B., Gomez, C., OMEGA Team, 2005. Phyllosilicates on Mars and implications for early Martian climate. Nature 438, 623–627.
Prieto-Ballesteros, O., Kargel, J.S., Fairen, A.G., Fernandez-Remolar, D.C., Dohm, J.M., Amils, R., 2006. Interglacial clathrate destabilization on Mars: possible contributing source of its atmospheric methane. Geology 34 (3), 149–152.
Radich, J.G., 2009. Laboratory and theoretical investigations of direct and indirect microbial influences on seafloor gas hydrates. A thesis in partial fulfillment of the requirements for the degree of Master of Science in Chemical Engineering in the Dave C. Swalm School of Chemical Engineering, Mississippi State University, Mississippi State, MS.
Radich, J.G., Rogers, R.E., French, W.T., Zhang, G., 2009. Biochemical reaction and diffusion in seafloor gas hydrate capillaries: implications for gas hydrate stability. Chem. Eng. Sci. 64, 4278–4285.
Rhode, R.A., Price, P.B., 2007. Diffusion-controlled metabolism for long-term survival of single isolated microorganisms trapped within ice crystals. Proc. Natl. Acad. Sci. U. S. A. 104, 16592–16597.
Richardson, M.I., Wilson, R.J., 2002. Investigation of the nature and stability of the Martian seasonal water cycle with a general circulation model. J. Geophys. Res. 107 (E5), 5031.
Rieder, R., Gellert, R., Anderson, R.C., Bruckner, J., Clark, B.C., Dreibus, G., Economou, T., Klingelhofer, G., Lugmair, G.W., Ming, D.W., Squyres, S.W., d'Uston, C., Wanke, H., Yen, A., Zipfel, J., 2004. Chemistry of rocks and soils at Meridiani Planum from the Alpha Particle X-Ray Spectrometer. Science 306, 1746–1749.
Ripmeester, J.A., Ding, L., Klug, D.D., 1996. A clathrate hydrate of formaldehyde. J. Phys. Chem. 100 (32), 13330–13332.
Rogers, R.E., Kothapalli, C., Lee, M.S., Woolsey, J.R., 2003. Catalysis of gas hydrates by biosurfactants in seawater-saturated sand/clay. Can. J. Chem. Eng. 81, 1–8.

Root, M.J., Elwood Madden, M.E., 2012. Potential effects of obliquity change on gas hydrate stability zones on Mars. Icarus 218, 534–544.

Sloan, E.D., 1998. Clathrate Hydrates of Natural Gases, second ed Marcel Dekker, Inc., New York, 628 pp.

Squyres, S.W., Grotzinger, J.P., Arvidson, R.E., Bell, III, J.F., Calvin, W., Christensen, P.R., Clark, B.C., Crisp, J.A., Farrand, W.H., Herkenhoff, K.E., Johnson, J.R., Klingelhöfer, G., Knoll, A.H., McLennan, S.M., McSween, Jr., H.Y., Morris, R.V., Rice, Jr., J.W., Rieder, R., Soderblom, L.A., 2004. In situ evidence for an ancient aqueous environment at Meridiani Planum, Mars. Science 306, 1709–1714.

Stern, L.A., Circone, S., Kirby, S.H., 2001. Anomalous preservation of pure methane hydrate at 1 atm. J. Phys. Chem. B 105, 1756–1762.

Tamman, G., Krige, G.J.R., 1925. Die gleichgewichtsdrucke von gas hydraten. Z. Anorg. Allg. Chem. 146, 179–195.

Tang, Y., Chen, Q., Huang, Y., 2006. Early Mars may have had a methanol ocean. Icarus 180, 88–92.

Thomas, P.C., 2011. Cold-trapping Mars' atmosphere. Science 332, 797–798.

Thomas, C., Mousis, O., Picaud, S., Ballenegger, V., 2009. Variability of the methane trapping in Martian subsurface clathrate hydrates. Planetary Space Sci. 57, 42–47.

Weiss, B.P., Yung, Y.L., Nealson, K.H., 2000. Atmospheric energy for subsurface life on Mars? Proc. Natl. Acad. Sci. U. S. A. 97, 1395–1399.

Whiticar, M.J., Faber, E., Schoell, M., 1986. Biogenic methane formation in marine and freshwater environments: CO_2 reduction vs acetate fermentation – isotope evidence. Geochim. Cosmochim. Acta 50, 693–709.

Williams, R.M.E., MSL Science Team, et al., 2013. Martian fluvial conglomerates at Gale Crater. Science 340, 1068–1072.

SUBJECT INDEX

D

E

F

G

H

Q

R

S

T

FSC
www.fsc.org
MIX
Papier | Fördert gute Waldnutzung
FSC® C083411

Kolibri 360 GmbH
Ferdinand-Jühlke-Straße 7
99095 Erfurt, Deutschland
produktsicherheit@kolibri360.de